NEUROMETHODS

Series Editor
Wolfgang Walz
University of Saskatchewan
Saskatoon, SK, Canada

For further volumes:
http://www.springer.com/series/7657

Neuromethods publishes cutting-edge methods and protocols in all areas of neuroscience as well as translational neurological and mental research. Each volume in the series offers tested laboratory protocols, step-by-step methods for reproducible lab experiments and addresses methodological controversies and pitfalls in order to aid neuroscientists in experimentation. *Neuromethods* focuses on traditional and emerging topics with wide-ranging implications to brain function, such as electrophysiology, neuroimaging, behavioral analysis, genomics, neurodegeneration, translational research and clinical trials. *Neuromethods* provides investigators and trainees with highly useful compendiums of key strategies and approaches for successful research in animal and human brain function including translational "bench to bedside" approaches to mental and neurological diseases.

Metabolomics

Edited by

Paul L. Wood

Metabolomics Unit, College of Veterinary Medicine, Lincoln Memorial University, Harrogate, TN, USA

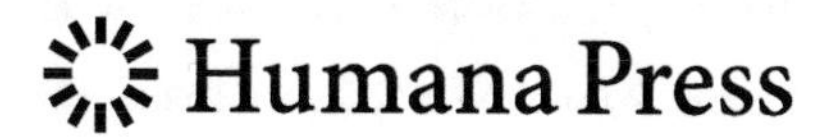

Editor
Paul L. Wood
Metabolomics Unit
College of Veterinary Medicine
Lincoln Memorial University
Harrogate, TN, USA

ISSN 0893-2336 ISSN 1940-6045 (electronic)
Neuromethods
ISBN 978-1-0716-0866-1 ISBN 978-1-0716-0864-7 (eBook)
https://doi.org/10.1007/978-1-0716-0864-7

This Humana imprint is published by the registered company Springer Science+Business Media, LLC, part of Springer Nature.
The registered company address is: 1 New York Plaza, New York, NY 10004, U.S.A.

Preface to the Series

Experimental life sciences have two basic foundations: concepts and tools. The *Neuromethods* series focuses on the tools and techniques unique to the investigation of the nervous system and excitable cells. It will not, however, shortchange the concept side of things as care has been taken to integrate these tools within the context of the concepts and questions under investigation. In this way, the series is unique in that it not only collects protocols but also includes theoretical background information and critiques which led to the methods and their development. Thus it gives the reader a better understanding of the origin of the techniques and their potential future development. The *Neuromethods* publishing program strikes a balance between recent and exciting developments like those concerning new animal models of disease, imaging, *in vivo* methods, and more established techniques, including, for example, immunocytochemistry and electrophysiological technologies. New trainees in neurosciences still need a sound footing in these older methods in order to apply a critical approach to their results.

Under the guidance of its founders, Alan Boulton and Glen Baker, the *Neuromethods* series has been a success since its first volume published through Humana Press in 1985. The series continues to flourish through many changes over the years. It is now published under the umbrella of Springer Protocols. While methods involving brain research have changed a lot since the series started, the publishing environment and technology have changed even more radically. *Neuromethods* has the distinct layout and style of the Springer Protocols program, designed specifically for readability and ease of reference in a laboratory setting.

The careful application of methods is potentially the most important step in the process of scientific inquiry. In the past, new methodologies led the way in developing new disciplines in the biological and medical sciences. For example, Physiology emerged out of Anatomy in the nineteenth century by harnessing new methods based on the newly discovered phenomenon of electricity. Nowadays, the relationships between disciplines and methods are more complex. Methods are now widely shared between disciplines and research areas. New developments in electronic publishing make it possible for scientists that encounter new methods to quickly find sources of information electronically. The design of individual volumes and chapters in this series takes this new access technology into account. Springer Protocols makes it possible to download single protocols separately. In addition, Springer makes its print-on-demand technology available globally. A print copy can therefore be acquired quickly and for a competitive price anywhere in the world.

Saskatoon, SK, Canada ***Wolfgang Walz***

Preface

Metabolomics was initially envisioned as an "omics" platform that would provide a snapshot of the entire metabolome. This vision has not been achieved since there is no single analytical platform that can monitor the wide diversity of chemical structures that compose the metabolome. As a result, metabolomics has evolved into a number of subfields that include lipidomics (see Volume 125 in this series), small molecule metabolomics, oxidative metabolomics, fluxomics, metallomics, and mass spectrometric imaging of small molecular weight metabolites and lipids.

In this volume, we will address the current status of analytical approaches utilized to sample defined molecular populations of metabolites via functional group derivatization, specialized chromatographic methods, and ionization techniques. All of these approaches require detailed experimental design to utilize the optimal methodology to address specific scientific questions. This is essential since many metabolites yield incredibly variable observations as a result of their rapid ongoing dynamics. Examples include rapid postmortem degradation (e.g., acetylcholine levels) and postmortem accumulation (e.g., GABA). Therefore, designing analytical platforms to monitor stable end products that are destined for excretion (e.g., dipeptides as biomarkers of protein metabolism and uric acid as an end product of purine metabolism), stable oxidized metabolites (e.g., isoprostanes), non-rate-limiting precursor pools, non-reutilized modified nucleosides as cancer biomarkers (e.g., 7-methylguanine), metabolites indicative of abnormal intermediary metabolism (e.g., keto-glutaramic acid as a biomarker of hyperammonemia), biomarkers of immune activation (e.g., kyneurenine and cyclic phosphatidic acid 16:0), biomarkers of ECM collagen degradation (e.g., deoxypyridinoline), inflammatory biomarkers of nitrosative damage (e.g., 8-nitroguanine), biomarkers of polyamine metabolism and diamine transporter function (e.g., putrescine), sulfur amino acid status in redox reactions (e.g., cysteine), biomarkers of uremia (e.g., guanidino acids), biomarkers of abnormal peroxisomal activity (e.g., dicarboxylic acid 16:0 and plasmalogens), and biomarkers of gut microbial metabolism (e.g., p-cresol sulfate).

To monitor this diversity of complex metabolites, a number of analytical approaches are required. Mass spectrometry is the most sensitive and specific methodology for metabolite identification and quantitation. In this volume, we review the key methods for sample introduction to the ion source, including direct flow, gas chromatography, liquid chromatography, and capillary electrophoresis. In addition, derivatization methods required for these sample introduction techniques are presented in detail, including complex enantioselective assays. Both nontargeted and targeted analyses are included in our volume as well as the emerging field of metallomics.

Finally, it is important to note that reporting metabolite levels from metabolomics studies is always reflective of relative and not absolute levels. Absolute levels cannot be reported until an isotopic variant of each metabolite of interest has been included in the assay to correct for extraction efficiency, ionization potential, and in the case of derivatization, reaction completion. With regard to stable isotope internal standards, it should be stressed that ^{13}C and/or ^{15}N are the optimal isotopes since deuterated compounds have different solubility properties to endogenous metabolites (i.e., altered hydrogen bonding activity).

Many investigators report their metabolomics data as absolute levels utilizing a single internal standard or as a mol%. These data are deceptive in that they are not corrected for the different ionization capacities between metabolites. Development of absolute quantitation assays for critical metabolites is essential in translational research for the advancement of biomarkers to improve the delivery of healthcare in clinical and veterinary medicine. These will include antecedent, screening, diagnostic, and prognostic assays as well as improved analytical approaches to improve the clinical stratification of patient subpopulations.

In summary, our *Metabolomics* volume is of great value to students, researchers, practicing physicians and veterinarians, as well as administrators involved in the funding of research.

Harrogate, TN, USA ***Paul L. Wood***

Contents

Preface to the Series .... *v*
Preface .... *vii*
Contributors .... *xi*

1 Flow Infusion ESI High Resolution Mass Spectrometry Metabolomics Analytical Platforms 1
Paul L. Wood

2 High-Throughput Metabolomics Using Flow Injection Analysis and Fourier Transform Ion Cyclotron Resonance Mass Spectrometry.. 9
Estelle Rathahao-Paris, Sandra Alves, and Alain Paris

3 Two-Dimensional Liquid Chromatography in Metabolomics and Lipidomics 25
Miriam Pérez-Cova, Romà Tauler, and Joaquim Jaumot

4 Metabolomics of Small Numbers of Cells Using Chemical Isotope Labeling Combined with Nanoflow LC-MS 49
Xian Luo and Liang Li

5 Untargeted Metabolomics to Interpret the Effects of Social Defeat: Integrating Chemistry and Behavioral Approaches in Neuroscience 61
Allen K. Bourdon and Brooke N. Dulka

6 Flow Infusion Electrospray Ionization High-Resolution Mass Spectrometry of the Chloride Adducts of Sphingolipids, Glycerophospholipids, Glycerols, Hydroxy Fatty Acids, Fatty Acid Esters of Hydroxy Fatty Acids (FAHFA), and Modified Nucleosides 69
Paul L. Wood and Randall L. Woltjer

7 High Accurate Mass Gas Chromatography–Mass Spectrometry for Performing Isotopic Ratio Outlier Analysis: Applications for Nonannotated Metabolite Detection 77
Yunping Qiu and Irwin J. Kurland

8 Enantioseparation and Detection of (R)-2-Hydroxyglutarate and (S)-2-Hydroxyglutarate by Chiral Gas Chromatography–Triple Quadrupole Mass Spectrometry 89
Shinji K. Strain, Morris D. Groves, and Mark R. Emmett

9 Enantioselective Supercritical Fluid Chromatography (SFC) for Chiral Metabolomics 101
Robert Hofstetter, Andreas Link, and Georg M. Fassauer

10 Capillary Electrophoresis–Mass Spectrometry of Hydrophilic Metabolomics 113
Masahiro Sugimoto

11 Lipidomic Analysis of Oxygenated Polyunsaturated Fatty Acid–Derived Inflammatory Mediators in Neurodegenerative Diseases 121
Mauricio Mastrogiovanni, Estefanía Ifrán, Andrés Trostchansky, and Homero Rubbo

12 Characterization and Quantification of the Fatty Acid Amidome 143
Kristen A. Jeffries, Emma K. Farrell, Ryan L. Anderson, Gabriela Suarez, Amanda J. Goyette Osborne, Mc Kenzi K. Heide, and David J. Merkler

13 Metabolomics Analysis of Complex Biological Specimens Using Nuclear Magnetic Resonance Spectroscopy 155
Khushboo Gulati, Sharanya Sarkar, and Krishna Mohan Poluri

14 Untargeted Metabolomics Methods to Analyze Blood-Derived Samples 173
Danuta Dudzik and Antonia García

15 Multicompartmental High-Throughput Metabolomics Based on Mass Spectrometry 189
Raúl González-Domínguez, Álvaro González-Domínguez, Ana Sayago, and Ángeles Fernández-Recamales

16 GC-MS Nontargeted Metabolomics of Neural Tissue 199
Carolina Gonzalez-Riano, Mª. Fernanda Rey-Stolle, Coral Barbas, and Antonia García

17 GC-MS of Pentafluourobenzyl Derivatives of Phenols, Amines, and Carboxylic Acids 221
Paul L. Wood

18 GC-MS of *tert*-Butyldimethylsilyl (tBDMS) Derivatives of Neurochemicals 229
Paul L. Wood

19 UHPLC-MS/MS Method for Determination of Biologically Important Thiols in Plasma Using New Derivatizing Maleimide Reagent 235
Dominika Olesova, Andrej Kovac, and Jaroslav Galba

20 Untargeted Metabolomics Determination of Postmortem Changes in Brain Tissue Samples by UHPLC-ESI-QTOF-MS and GC-EI-Q-MS 245
Carolina Gonzalez-Riano, Antonia García, and Coral Barbas

21 Metallomics Imaging 267
Valderi Luiz Dressler, Graciela Marini Hiedrich, Vinicius Machado Neves, Eson Irineu Müller, and Dirce Pozebon

Index *305*

Contributors

SANDRA ALVES • *Sorbonne Université, Faculté des Sciences et de l'Ingénierie, Institut Parisien de Chimie Moléculaire (IPCM), Paris, France*
RYAN L. ANDERSON • *Department of Chemistry, University of South Florida, Tampa, FL, USA; Department of Biochemistry and Molecular Genetics, University of Colorado—Denver Anschutz Medical Campus, Aurora, CO, USA*
CORAL BARBAS • *Center for Metabolomics and Bioanalysis (CEMBIO), Facultad de Farmacia, Universidad San Pablo CEU. CEU Universities, Madrid, Spain; Centre for Metabolomics and Bioanalysis (CEMBIO), University San Pablo CEU, Madrid, Spain*
ALLEN K. BOURDON • *Stepan Company, Northfield, IL, USA*
VALDERI LUIZ DRESSLER • *Department of Chemistry, Federal University of Santa Maria, Santa Maria, RS, Brazil*
DANUTA DUDZIK • *Department of Biopharmaceutics and Pharmacodynamics, Faculty of Pharmacy, Medical University of Gdańsk, Gdańsk, Poland*
BROOKE N. DULKA • *University of Wisconsin-Milwaukee, Milwaukee, WI, USA*
MARK R. EMMETT • *Department of Biochemistry and Molecular Biology, University of Texas Medical Branch, Galveston, TX, USA*
EMMA K. FARRELL • *Department of Chemistry, University of South Florida, Tampa, FL, USA; The McCrone Group, Westmont, IL, USA*
GEORG M. FASSAUER • *Pharmaceutical and Medicinal Chemistry, Institute of Pharmacy, University Greifswald—Germany, Greifswald, Germany*
ÁNGELES FERNÁNDEZ-RECAMALES • *Department of Chemistry, Faculty of Experimental Sciences, University of Huelva, Huelva, Spain; International Campus of Excellence CeiA3, University of Huelva, Huelva, Spain*
JAROSLAV GALBA • *Institute of Neuroimmunology, Slovak Academy of Sciences, Bratislava, Slovak Republic*
ANTONIA GARCÍA • *Centre for Metabolomics and Bioanalysis (CEMBIO), Department of Chemistry and Biochemistry, Facultad de Farmacia, Universidad San Pablo-CEU, CEU Universities, Madrid, Spain; Center for Metabolomics and Bioanalysis (CEMBIO), Facultad de Farmacia, Universidad San Pablo CEU. CEU Universities, Madrid, Spain; Centre for Metabolomics and Bioanalysis (CEMBIO), University San Pablo CEU, Madrid, Spain*
ÁLVARO GONZÁLEZ-DOMÍNGUEZ • *Department of Pediatrics, Hospital Universitario Puerta del Mar., Cádiz, Spain; Institute of Research and Innovation in Biomedical Sciences of the Province of Cádiz (INiBICA), Cádiz, Spain*
RAÚL GONZÁLEZ-DOMÍNGUEZ • *Department of Chemistry, Faculty of Experimental Sciences, University of Huelva, Huelva, Spain; International Campus of Excellence CeiA3, University of Huelva, Huelva, Spain*
CAROLINA GONZALEZ-RIANO • *Center for Metabolomics and Bioanalysis (CEMBIO), Facultad de Farmacia, Universidad San Pablo CEU. CEU Universities, Madrid, Spain; Centre for Metabolomics and Bioanalysis (CEMBIO), University San Pablo CEU, Madrid, Spain*
AMANDA J. GOYETTE OSBORNE • *Department of Chemistry, University of South Florida, Tampa, FL, USA*

MORRIS D. GROVES • *Austin Brain Tumor Center, Texas Oncology/US Oncology Research, Austin, TX, USA*
KHUSHBOO GULATI • *Department of Biotechnology, Indian Institute of Technology Roorkee, Roorkee, Uttarakhand, India*
MC KENZI K. HEIDE • *Department of Chemistry, University of South Florida, Tampa, FL, USA*
GRACIELA MARINI HIEDRICH • *Department of Chemistry, Federal University of Santa Maria, Santa Maria, RS, Brazil*
ROBERT HOFSTETTER • *Pharmaceutical and Medicinal Chemistry, Institute of Pharmacy, University Greifswald—Germany, Greifswald, Germany*
ESTEFANÍA IFRÁN • *Departamento de Bioquímica, Facultad de Medicina and Centro de Investigaciones Biomédicas (CEINBIO), Universidad de la República, Montevideo, Uruguay*
JOAQUIM JAUMOT • *Department of Environmental Chemistry, IDAEA-CSIC, Barcelona, Spain*
KRISTEN A. JEFFRIES • *Department of Chemistry, University of South Florida, Tampa, FL, USA; Agricultural Marketing Service, USDA, Gastonia, NC, USA*
ANDREJ KOVAC • *Institute of Neuroimmunology, Slovak Academy of Sciences, Bratislava, Slovak Republic*
IRWIN J. KURLAND • *Stable Isotope and Metabolomics Core Facility, Diabetes Center, Department of Medicine, Albert Einstein College of Medicine, Bronx, NY, USA*
LIANG LI • *Department of Chemistry, University of Alberta, Edmonton, AB, Canada*
ANDREAS LINK • *Pharmaceutical and Medicinal Chemistry, Institute of Pharmacy, University Greifswald—Germany, Greifswald, Germany*
XIAN LUO • *Department of Chemistry, University of Alberta, Edmonton, AB, Canada*
MAURICIO MASTROGIOVANNI • *Departamento de Bioquímica, Facultad de Medicina and Centro de Investigaciones Biomédicas (CEINBIO), Universidad de la República, Montevideo, Uruguay*
DAVID J. MERKLER • *Department of Chemistry, University of South Florida, Tampa, FL, USA*
ESON IRINEU MÜLLER • *Department of Chemistry, Federal University of Santa Maria, Santa Maria, RS, Brazil*
VINICIUS MACHADO NEVES • *Department of Chemistry, Federal University of Santa Maria, Santa Maria, RS, Brazil*
DOMINIKA OLESOVA • *Institute of Neuroimmunology, Slovak Academy of Sciences, Bratislava, Slovak Republic*
ALAIN PARIS • *Muséum national d'Histoire naturelle, CEA, Unité Molécules de Communication et Adaptation des Microorganismes (MCAM), Paris, France*
MIRIAM PÉREZ-COVA • *Department of Environmental Chemistry, IDAEA-CSIC, Barcelona, Spain*
KRISHNA MOHAN POLURI • *Department of Biotechnology, Indian Institute of Technology Roorkee, Roorkee, Uttarakhand, India; Centre for Nanotechnology, Indian Institute of Technology Roorkee, Roorkee, Uttarakhand, India*
DIRCE POZEBON • *Department of Chemistry, Federal University of Santa Maria, Santa Maria, RS, Brazil; Chemistry Institute, Federal University of Rio Grande do Sul, Porto Alegre, RS, Brazil*
YUNPING QIU • *Stable Isotope and Metabolomics Core Facility, Diabetes Center, Department of Medicine, Albert Einstein College of Medicine, Bronx, NY, USA*

ESTELLE RATHAHAO-PARIS • *Université Paris Saclay, CEA, INRAE, Département Médicaments et Technologies pour la Santé (DMTS), SPI, Gif-sur-Yvette, France; Sorbonne Université, Faculté des Sciences et de l'Ingénierie, Institut Parisien de Chimie Moléculaire (IPCM), Paris, France*

Mª. FERNANDA REY-STOLLE • *Center for Metabolomics and Bioanalysis (CEMBIO), Facultad de Farmacia, Universidad San Pablo CEU. CEU Universities, Madrid, Spain*

HOMERO RUBBO • *Departamento de Bioquímica, Facultad de Medicina and Centro de Investigaciones Biomédicas (CEINBIO), Universidad de la República, Montevideo, Uruguay*

SHARANYA SARKAR • *Department of Biotechnology, Indian Institute of Technology Roorkee, Roorkee, Uttarakhand, India*

ANA SAYAGO • *Department of Chemistry, Faculty of Experimental Sciences, University of Huelva, Huelva, Spain; International Campus of Excellence CeiA3, University of Huelva, Huelva, Spain*

SHINJI K. STRAIN • *Department of Neuroscience, Cell Biology and Anatomy, University of Texas Medical Branch, Galveston, TX, USA*

GABRIELA SUAREZ • *Department of Chemistry, University of South Florida, Tampa, FL, USA*

MASAHIRO SUGIMOTO • *Research and Development Center for Minimally Invasive Therapies Health Promotion and Preemptive Medicine, Tokyo Medical University, Shinjuku, Tokyo, Japan; Institute for Advanced Biosciences, Keio University, Tsuruoka, Yamagata, Japan*

ROMÀ TAULER • *Department of Environmental Chemistry, IDAEA-CSIC, Barcelona, Spain*

ANDRÉS TROSTCHANSKY • *Departamento de Bioquímica, Facultad de Medicina and Centro de Investigaciones Biomédicas (CEINBIO), Universidad de la República, Montevideo, Uruguay*

RANDALL L. WOLTJER • *Department of Neurology, Oregon Health Science University and Portland VA Medical Center, Portland, OR, USA*

PAUL L. WOOD • *Metabolomics Unit, College of Veterinary Medicine, Lincoln Memorial University, Harrogate, TN, USA*

Chapter 1

Flow Infusion ESI High Resolution Mass Spectrometry Metabolomics Analytical Platforms

Paul L. Wood

Abstract

Flow infusion ESI high resolution (0.4–3 ppm mass error) mass spectrometric metabolomics allows for the analysis of a broad array of metabolites of interest. The issue of isobars must always be assessed carefully and can be managed by tandem mass spectrometry, by derivatization prior to direct infusion analyses, and/or via the use of chromatographic systems.

Key words High resolution mass spectrometry, Metabolomics, Flow injection analysis, Tandem mass spectrometry

1 Introduction

Metabolomics platforms are incredibly valuable in that they allow investigators to interrogate precious clinical samples in addition to samples from animal models of disease [1]. This approach is utilized by investigators to examine biofluids and tissues from different disease populations and thereby determine which animal models are optimal for evaluating experimental treatment paradigms prior to clinical trials. This approach also is invaluable in establishing biomarkers of disease, of disease progression (temporal axis), of the therapeutic response in patients (therapeutic axis), for defining patient subpopulations (patient stratification at the molecular level), and for defining new molecular therapeutic targets (drug discovery).

In our Metabolomics Unit, we utilize nontargeted flow infusion electrospray ionization coupled with high-resolution mass spectrometry (FI-ESI-HR-MS) for pilot metabolomics studies. For validation studies, a target set of molecules were identified using our nontargeted strategy. Next we further refine the number of metabolites to monitor and establish absolute quantitation assays

Paul L. Wood (ed.), *Metabolomics*, Neuromethods, vol. 159, https://doi.org/10.1007/978-1-0716-0864-7_1,

for each. In many cases this requires the custom synthesis of an analytical standard and/or a stable isotope internal standard.

We have built a database in Excel (Microsoft) of key metabolites that focus on stable end products of metabolic pathways that are exported from a compartment and/or excreted, and large precursor pools. Intermediate metabolites that turn over rapidly and are most susceptible to sample collection artifacts are not included in the database. The categories monitored in positive electrospray ionization (PESI) are choline metabolites, carnitines, basic collagen metabolites, biomarkers of GI flora, lysine metabolites/modifications, reactive aldehyde metabolites, and polyamine metabolites. Metabolites monitored in Negative ion ESI (NESI), are acylglycines, acidic collagen metabolites, histamine acidic metabolites, nucleotides, organic acids, proline metabolites, taurine metabolites, uremic toxins, modified amino acids, and dipeptides.

2 Materials

2.1 Buffers and Standards

- Extraction Buffer (*see* **Notes 1** and **2**): acetonitrile–methanol–formic acid (800:200:2.5).
- Direct Infusion Buffer: acetonitrile–methanol (1:1).
- Syringe and Infusion Line Wash Buffer: Acetonitrile–methanol–water (240:60:100).
- Standards and Internal Standards are dissolved in ACN–MeOH–6 N-HCl (1:1:0.2).
- Stable isotope internal standards (Listed in Table 1).
- Pierce ESI Negative Ion Calibration Solution (Thermo Fisher; *see* **Note 3**).
- Pierce ESI Positive Ion Calibration Solution (Thermo Fisher).

2.2 Analytical Equipment and Supplies

- Mass spectrometer: Q-Exactive (Thermo Fisher). The Q-Exactive is a benchtop quadrupole orbitrap mass spectrometer with high mass resolution, high mass accuracy (0.4–3 ppm), high sensitivity, and excellent stability, all with a maintenance-free analyzer (*see* **Note 4**).
- Centrifugal dryer (Eppendorf Vacufuge Plus).
- Refrigerated microfuge capable of 30,000 × *g* (Eppendorf 5430R).
- Sonicator (Thermo Fisher FB50).
- 1.5 mL screw-top microfuge tubes (Thermo Fisher).

Table 1
Stable isotope internal standards utilized in our metabolomics studies

Int. Std.	nmoles	Exact mass	$[M + H]^+$	MS/MS	$[MS/MS]^+$
$[^2H_9]$Choline	10	112.1562	113.1635	$[C_3ND_9]$	69.13726
$[^2H_9]$Propionylcholine	0.05	168.1824	169.1897	$[M-[^2H_9]TMA]$	101.0603
$[^2H_3]$N-Acetyl-Methionine	0.2	194.0844	195.0917	[Met]	150.0583
$[^2H_5]$Phenylalanine	40	170.1103	171.1176	$[M-CO_2H_2]$	125.0021
$[^2H_3]$Hydroxyproline	10	134.0770	135.0843	$[C_4H_4NOD_3]$	87.0789

Int. Std.	nmoles	Exact mass	$[M-H]^-$	MS/MS	$[MS/MS]^-$
$[^2H_3]$N-Acetyl-Methionine	0.2	194.0844	193.0772	$[^2H_3]$Met	151.0626
$[^2H_5]$Phenylalanine	40	170.1103	169.1030	$[M-NH_3]$	152.0765
$[^{13}C_5,^{15}N_1]$Gly-Pro	6	178.0986	177.0913	$[^{13}C_5,^{15}N_1]$Pro	120.0699
$[^2H_5]$N-Acetyl-Alanine	2	135.0353	134.0280	$[^2H_4]$Ala	92.0654

All isotopes were purchased from CND Isotopes except for $[^{13}C_5,^{15}N_1]$Gly-Pro which was purchased from Creative Peptides. *TMA* trimethylamine

3 Methods

3.1 Sample Preparation for HR-MS

- Plasma or serum (100 μL), in 1.5 mL microfuge tubes, is vortexed with 1 mL of cold Extraction Buffer and 50 μL of the Internal Standard solution.
- Tissues (500–800 mg), in 1.5 mL microfuge tubes, are sonicated in 1 mL of cold Extraction Buffer 1 mL of cold Extraction Buffer.
- Next samples are centrifuged at 30,000 × *g* and 4 °C for 30 min.
- 750 μL of the supernatant are transferred to a clean 1.5 mL microfuge tube.
- The samples are next dried by vacuum centrifugation.
- 150 μL of Infusion Buffer are added to the dried samples with vortexing.
- The tubes are centrifuged at 30,000 × *g* for 10 min to precipitate any potential suspended particulates.

3.2 ESI-HR-MS

- The prepared samples are directly infused at 12 μL per minute (*see* **Note 5**).
- Scans in PESI and NESI are for 30 s each with a mass range of 60–900 amu.
- The syringe and all tubing are flushed between samples first with 1 mL of methanol and then 1 mL of Syringe and Infusion Line Wash Buffer.

Table 2
Examples of MS/MS discrimination of isobars in PESI

Exact mass	[M + H]	Isobars	Product	Product anion
102.11569	103.12297	• Cadaverine • N-Methyl-Putrescine	• [C_5H_9] • [C_4H_8N]	• 69.07043 • 70.06567
131.05824	132.06552	• Hydroxyproline • N-Propionyl-Gly	• [M-CO_2H_2] • [Gly]	• 86.0600 • 76.0393
145.07389	146.08117	• N-Propionyl-Ala • N-Formyl-Val • N-Butyryl-Gly	• [Ala] • [M- CO_2H_2] • [Gly]	• 90.0549 • 100.0757 • 76.0393
147.08954	148.09682	• 6-Hydroxy-Norleucine	• [NLeu-CH_2O_2]	• 84.0818

3.3 MS/MS Experiments

- For MS/MS experiments, parent ions are selected with a 0.4 amu window (*see* **Note 6**) and product ions are monitored with HR-MS (140,000 resolution).
- In the lower mass range of 60 to 400 many metabolites have multiple isobars and MS/MS is essential for accurate quantitation. Examples of MS/MS discrimination of a number of critical isobars is presented in Tables 2 and 3.

3.4 Data Reduction

- We have built a database in Excel (Microsoft) of key metabolites that focus on stable end products of metabolic pathways and large precursor pools. Intermediate metabolites that turn over rapidly and are most susceptible to sample collection artifacts in plasma and serum are not included in the database.
- Exact masses for the metabolites are from the online databases HMDB (http://www.hmdb.ca) and ChemSpider (http://www.chemspider.com).
- For pilot studies only relative quantitation is utilized such that the ratio of the ion intensity of the biomolecule to that of an appropriate internal standard is recorded, only if the mass error is less than 3 ppm.
- For absolute quantitation, a stable isotopic analog of the biomolecule of interest is utilized as the internal standard and a seven-point standard curve generated.
- In the case of a metabolite of interest that has a number of isobars we utilize MS/MS to achieve accurate quantitation. LC-HR-MS is an alternative if isobars demonstrate the same fragmentation pattern.
- Potential isobars are identified utilizing the exact mass search function in the HMDB.

Table 3
Examples of MS/MS discrimination of isobars in NESI

Exact mass	[M-H]	Isobars	Product	Product anion
117.0425	116.0352	• N-Acetyl-Gly • Asp-semialdehyde	• [Gly] • [Asp-CO_2]	• 74.0247 • 88.0404
131.0582	130.0509	• N-Acetyl-Ala • Propionyl-Gly • Glu-semialdehyde	• [Ala] • [Gly] • [Glu-CO_2]	• 88.0404 • 74.0247 • 102.0560
132.05349	131.04622	• N-Carbamoyl-Src • Ureidopropionate • Gly-Gly	• [Sarcosine] • [Propionate] • [Gly]	• 88.0404 • 73.0295 • 74.02475
159.08954	158.08227	• N-Formyl-Leu • N-Acetyl-Val • N-Valeryl-Gly	• [Leu] • [Val] • [Gly]	• 130.08735 • 116.0717 • 74.02475
174.10044	173.09317	• Val-Gly • Gly-Val • N-Formyl-Lys	• [Gly] • [Val] • [M- CO_2]	• 74.02475 • 116.0717 • 129.10334
175.0480	174.0407	• N-Acetyl-Asp	• [Asp-CO_2]	• 88.0404
176.0797	175.07242	• N-Acetyl-Lys • Ala-Val • Val-Ala • Leu-Gly • Gly-Leu	• [Lys] • [Val] • [Ala] • [Gly] • [Leu]	• 145.09825 • 116.0717 • 88.0404 • 74.02475 • 130.08735
189.0637	188.0564	• N-Acetyl-Glu	• [Glu-CO_2]	• 102.0560
191.0616	190.0542	• N-Acetyl-Met	• [Met]	• 148.0437
221.0721	220.0648	• HPMA • N-Acetyl-galactosamine	• [$C_5H_7NO_3$] • [$C_4H_8NO_2$]	• 128.0353 • 102.0555
304.0906	303.0833	• N-Acetyl-Asp-Glu	• [Glu-H_2O]	• 128.0353.

3.5 Bioanalytical Example

We utilized this analytical platform for a postmortem analysis of N-acetyl amino acids in the frontal cortex (Brodmann's area 10) of control subjects and schizophrenia patients (Fig. 1). Flow infusion analysis generated data with RSD values of 100% as a result of the potential isobars (Table 2). However, with flow infusion analysis and tandem mass spectrometric analyses with acquisition of high-resolution data for the product anions, the RSD values were reduced to 25–35%. As presented in Fig. 1, only N-acetylaspartylglutamate (NAAG) levels were altered. The decrements in this acetylated dipeptide were directly comparable to those reported previously by Zhang et al. [3].

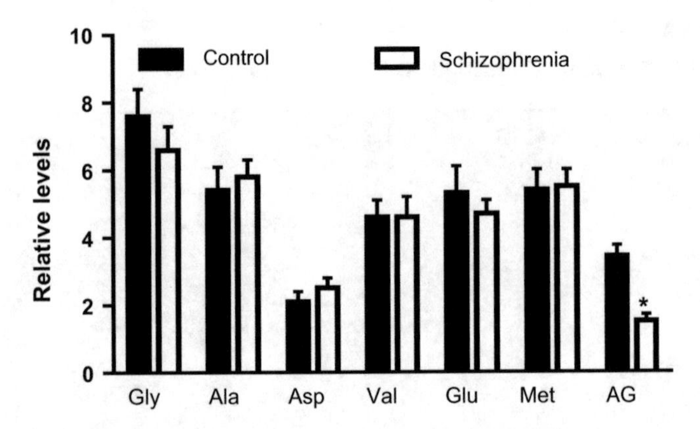

Fig. 1 Relative levels of N-acetyl amino acids in the frontal cortex of control and schizophrenia subjects ($N = 10$). Gly, N-acetylglycine; Ala, N-acetylalanine; Asp, N-acetylaspartate; Val, N-acetylvaline; Glu, N-acetylglutamate; Met, N-acetylmethionine; AG, N-acetylaspartylglutamate. *, $p = 0.018$

4 Notes

1. All organic solvents should be of LC-MS grade.
2. The extraction buffer must be cold to stabilize metabolites from further metabolism.
3. Pierce ESI Negative Ion Calibration Solution does not have a low mass calibrant. We dilute the NESI standard 1:1 with our lipidomics infusion buffer [2] (80 mL 2 propanol +40 mL methanol +20 mL chloroform +0.5 mL water containing 164 mg ammonium acetate) and calibrate the low end of the mass axis using the buffer impurity $C_3H_6O_3$ (anion = 89.02442) which is a robust anion.
4. The MS/MS fragmentation pattern of metabolites can demonstrate significant differences, with regard to intensities, between mass spectrometer types. The data presented in this chapter are specific to the Q-Exactive orbitrap.
5. For the infusion lines for the ESI, we have switched from PEEK tubing to PEEKsil tubing to further decrease the risk of "memory effects" in the tubing.
6. The 0.4 amu window utilized in this analytical platform, significantly decreases the number of interfering biomolecules with a similar mass from contaminating the MS/MS analysis of target molecules. A large number of investigators utilize windows of 1.5–3 amu reducing the specificity of the MS/MS analysis, particularly with triple quadrupole instrumentation.

Acknowledgments

I wish to thank my clinical collaborators and the many students whose rotations in the Metabolomics Unit have aided in building and validating the metabolomics platform described in this chapter. In addition this work would have not been possible without the financial support of Lincoln Memorial University and the DOJ.

References

1. Wood PL (2014) Mass spectrometry strategies for targeted clinical metabolomics and lipidomics in psychiatry, neurology, and neuro-oncology. Neuropsychopharmacology 39:24–33
2. Wood PL (2017) Non-targeted lipidomics utilizing constant infusion high resolution ESI mass spectrometry. In: Wood PL (ed) Springer protocols, Neuromethods: lipidomics, vol 125, pp 13–19
3. Zhang R, Zhang T, Ali AM, Al Washih M, Pickard B, Watson DG (2016) Metabolomic profiling of post-mortem brain reveals changes in amino acid and glucose metabolism in mental illness compared with controls. Comput Struct Biotechnol J 14:106–116

Chapter 2

High-Throughput Metabolomics Using Flow Injection Analysis and Fourier Transform Ion Cyclotron Resonance Mass Spectrometry

Estelle Rathahao-Paris, Sandra Alves, and Alain Paris

Abstract

The hyphenation of flow injection analysis (FIA) with Fourier transform ion cyclotron resonance mass spectrometry (FTICR-MS) is an efficient approach usable to perform high throughput and very high resolution metabolomic data acquisition. Instrumental and analytical conditions for performing FIA-MS are provided. The procedure to optimize dilution factor of biological samples as well as to evaluate quality of acquisition data are also described. In this protocol, urine is chosen as a matrix example to illustrate the application of procedures. Last, some indications on an adapted data processing are given.

Key words Metabolomics, High throughput, Fourier transform ion cyclotron resonance mass spectrometry (FTICR), Flow injection analysis

1 Introduction

Flow injection analysis (FIA) hyphenated with mass spectrometry (MS) offers a high sampling rate with a minimal sample consumption [1]. Its ability to perform high throughput analyses represents a main advantage in metabolomics, especially for large scale analyses such as in cohort studies [2]. The hyphenation of FIA with very high performance mass analyzer such as Fourier transform ion cyclotron resonance mass spectrometry (FTICR-MS) [3] enabling distinction of isobaric ions (very close mass-to-charge ratios (*m/z*)) should deeply improve the generated data quality in terms of number of detectable peaks and accurate mass measurements.

Metabolomics aims to detect at the metabolome level the organism response to perturbations mediated by external or internal factors [4, 5]. It proceeds through the study of metabolomic profiles, but it is not straightforward because the variations in the metabolite level are often very small. Hence, the reliability

Paul L. Wood (ed.), *Metabolomics*, Neuromethods, vol. 159, https://doi.org/10.1007/978-1-0716-0864-7_2,

and reproducibility of the data produced are essential in metabolomics, especially to reveal the small differences between groups.

In this protocol, procedures are proposed to perform high-throughput metabolomics using FIA-FTICR-MS, including determination of the optimized dilution factor and evaluation of the data quality. Indeed, a major drawback of such a direct approach when using an electrospray ionization (ESI) source is the matrix effects which can hide relevant information [6]. This can be overcome by determining an appropriate dilution factor which minimizes matrix effects and therefore to obtain reliable and informative data from biological matrix analysis.

2 Chemical and Biological Materials

Contaminations from solvent, containers (e.g., polyethylene glycol, phthalates) or other sources can produce artifacts in the results of the subsequent analysis. Handling must be done with care and unavoidable chemical noise should be evaluated before analysis.

- General recommendations before starting experiments.
  - Check that the used containers and solvents do not release high contaminant levels and use the same batch of consumables (purchase a sufficient amount of consumables).
  - Use ultrapure water (typically with a resistivity of 18 Meg-ohm.cm) and mass spectrometry or HPLC grade organic solvents to prepare MS calibration and reference compound solutions, all samples, as well as solvent mixture for FIA experiments (*see* **Note 1**).
  - Prepare in advance your materials: labeled vials, solvents, ice, and so on.
  - Work under cold conditions during sample preparation: kept biological samples as cold as possible in a fridge or stored on ice to prevent sample degradation.

2.1 MS Calibration Solution

Several MS calibration solutions can be used to calibrate a mass spectrometer. There are commercially available calibration solutions, for example, ESI Tuning Mix (G1969-85000, Agilent Technologies, Santa Clara, CA, USA). The solution used for calibration depends on the instrument and the ionization mode, it should cover the *m/z* range of interest and display weak carryover. Low mass range calibration (below *m/z* 1000) is recommended for performing metabolomic fingerprints on an ion trapping system such as FTICR instrument. To do this, a calibration based on sodium ion clusters could be a good solution. Manufacturer recommends to fleshly prepare the calibration solution according to the following procedure:

- Prepare a solution of sodium formate and acetate clusters.
  - Stock solution: Mix 500 μL of 1 M NaOH, 25 μL of formic acid, 75 μL of glacial acetic acid, and 50 mL of isopropanol–water (1:1, v/v).
  - Due to the high sensitivity of the FTICR instrument, a solution at 1:50 dilution is used: Dilute 100 μL of the stock solution in 4.900 mL of methanol–water (1:1, v/v) (*see* **Note 2**).

2.2 Reference Compound Mixture for Spiking Biological Samples

Reference compound mixture is used to control the MS data quality in terms of the mass measurement accuracy and the signal stability. The three used compounds [7–9] are (a) isoniazid (Sigma-Aldrich Chimie, Saint-Quentin Fallavier, France), CAS number: 54-85-3, (b) oxytetracycline (Sigma-Aldrich Chimie, Saint-Quentin Fallavier, France), CAS number: 79-57-2, and (c) 2-amino-l-methyl-6-phenylimidazo[4,5-b]pyridine (PhIP, Toronto Research Chemicals, Ontario, Canada), CAS number: 105650-23-5 (*see* Table 1). Those compounds are selected since they are not naturally present in biological samples and they have different molecular weights, thus covering a wide mass range. In addition, these molecules are detectable in both positive and negative ionization modes, with the exception of PhIP which is not detected in negative mode. Of course, other reference compounds could be chosen but they should provide sensitive MS response at very low concentrations

Table 1
List of reference compounds

Name	Structure	Elemental compound	$[M + H]^+$	$[M\text{-}H]^-$
Isoniazid	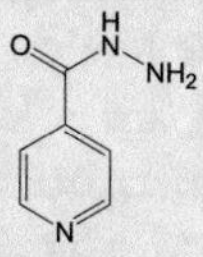	$C_6H_7N_3O$	138.0662	136.0516
2-amino-l-methyl-6-phenylimidazo[4,5-b] pyridine	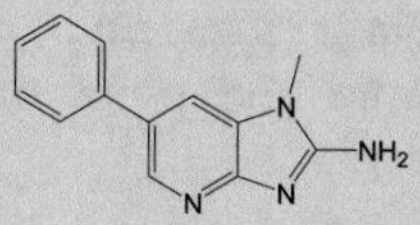	$C_{13}H_{12}N_4$	225.1135	[a]n.d.
Oxytetracycline	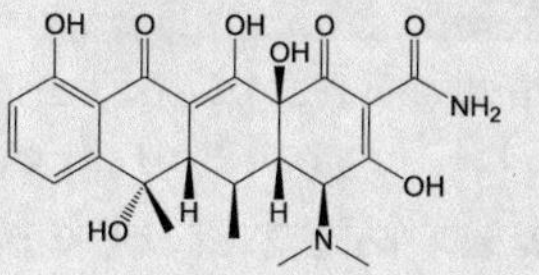	$C_{22}H_{24}N_2O_9$	461.1555	459.1409

[a]*n.d.* not detected

since the matrix effects could be induced by their addition. The final concentrations of the reference compounds should be adjusted depending on the MS response.

- Prepare the *stock solutions*.
  - Weigh 1 mg of isoniazid, add 1 mL of methanol and vortex mix until total dissolution to obtain solution *I-1* at 1 μg/μL.
  - Weigh 1 mg of 2-amino-l-methyl-6-phenylimidazo[4,5-b] pyridine, add 10 mL of methanol and vortex mix until total dissolution to obtain solution *PhIP-1* at 0.1 μg/μL.
  - Weigh 1 mg of oxytetracycline, add 2 mL of water–methanol (1:1, v/v) and vortex mix until total dissolution to obtain solution *O-1* at 0.5 μg/μL.
- Prepare the *working solutions*.
  - Mix 100 μL of the solution *I-1* and 900 μL of water–methanol (1:1, v/v) to obtain solution *I-2* at 100 ng/μL.
  - Mix 100 μL of the solution *PhIP-1* and 900 μL of water–methanol (1:1, v/v) to obtain solution *PhIP-2* at 10 ng/μL.
  - Mix 200 μL of the solution *O-1* and 800 μL of water–methanol (1:1, v/v) to obtain solution *O-2* at 100 ng/μL.
- Prepare the solution *Mix* of the mixture of the three reference compounds: Mix 4 μL of the solution *I-2*, 12 μL of the solution *PhIP-2*, 4 μL of the solution *O-2*, and 1.980 μL of water–methanol (1:1, v/v). The final solution *Mix* contains isoniazid, PhIP, and oxytetracycline at 0.20 ng/μL, 0.06 ng/μL, and 0.20 ng/μL respectively.

2.3 Serially Diluted Biological Samples Spiked with Reference Compounds

Before performing series analyses, a preliminary step consists in the determination of the best dilution factor for the studied biological samples. To do this, two or three biological samples spiked with the reference compounds should be analyzed at different dilution factors. These biological samples are selected randomly from each group of studied subjects or in cohort samples. Minimal sample preparation should be performed whatever the studied matrix.

- For urine, the simple preparation is: centrifuge 1 mL of urine at 6000 × *g* for 5 min and pipet only the supernatant (e.g., 900 mL).
- Prepare samples spiked with the same amount of reference compounds at different dilution factors:
  - *1:50 dilution*: Mix 2 μL of supernatant, 48 μL of water–methanol (1:1, v/v) and 50 μL of solution *Mix*;
  - *1:100 dilution*: Mix 2 μL of supernatant, 98 μL of water–methanol (1:1, v/v) and 100 μL of solution *Mix*;

- *1:200 dilution*: Mix 2 μL of supernatant, 198 μL of water–methanol (1:1, v/v) and 200 μL of solution *Mix*;
- *1:500 dilution*: Mix 2 μL of supernatant and 498 μL of water–methanol (1:1, v/v); then pipet 100 μL of the previous solution and add 100 μL of solution *Mix*;
- *1:1000 dilution*: Mix 2 μL of supernatant and 998 μL of water–methanol (1:1, v/v); then pipet 100 μL of the previous solution and add 100 μL of solution *Mix*;
- *1:3000 dilution*: Mix 2 μL of supernatant and 2998 μL of water–methanol (1:1, v/v); then pipet 100 μL of the previous solution and add 100 μL of solution *Mix*;
- *1:5000 dilution*: Mix 2 μL of supernatant and 4998 μL of water–methanol (1:1, v/v); then pipet 100 μL the previous solution and add 100 μL of solution *Mix*.

Diluted urine samples spiked with the same amount of reference compounds are subsequently analyzed using FIA-MS to determine the best dilution factor (see below: *Optimization of sample dilution factor*).

2.4 Biological and Quality Control (QC) Samples

All biological and QC samples will be prepared after the "*Optimization of sample dilution factor*" step using the chosen dilution factor. Ensure to have a sufficient volume of water–methanol (1:1, v/v) mixture for the dilution of all samples as well as for the analytical blank sample. The QC sample is required to evaluate the instrument stability and also to correct eventual analytical drifts using post acquisition data mining tools.

For example, if a dilution factor of 1000 is chosen then prepare all samples as follows:

- Prepare biological samples by pipetting 2 μL of each biological sample (supernatant or extract) and add 1.998 mL of water–methanol (1:1, v/v).
- Prepare QC sample by pooling equal volumes of all biological samples for small-to-medium size study, (or a subset group for large scale study) followed by dilution to an appropriate factor. For example, dilute this pool sample at 500 dilution factor, then pipet 500 μL of this diluted pool sample and add 500 μL of solution *Mix* to have a 1:1000 dilution. Ensure to have a sufficient volume of QC sample for all repeated injections during the single or multiple experimental batches.

3 Analytical System

The procedures and settings described in this protocol are for a micro-HPLC UltiMate 3000 RSLCnano system (Thermo Fisher Scientific) coupled to a FTICR 7 T instrument (Solarix-XR, Bruker) equipped with a dynamically harmonized cell [10] and an electrospray ionization source. The chromatography (e.g., flow rate and sample injection volume) as well as Fourier transform mass spectrometry (FTMS) instrument parameters can be adjusted, depending on the studied sample and the used analytical platform (*see* **Note 3**).

3.1 Sample Introduction System

Flow injection analysis (FIA) is performed using an autosampler for sample injection and a liquid chromatography (LC) system for delivering a dedicated flow rate. Note that the FIA mode does require neither the chromatographic column nor the pressure measurement.

- Prepare the eluent for FIA-MS experiments.
  - For positive ionization mode: Mix 500 mL of methanol and 500 mL of water; and add 100 μL of formic acid (0.1% in volume).
  - For negative ionization mode: Mix 500 mL of methanol and 500 mL of water. No formic acid is added to the eluent.
- Prepare the needle and syringe washing solution.
  - Mix 500 mL of methanol and 500 mL of water.
- Conditioning the FIA system.
  - Degas the eluent and the needle washing solution.
  - Prime and purge the LC system.
  - Wash the injection needle and syringe with the washing solution.
  - Set the flow rate at 15 $\mu L.min^{-1}$ and the autosampler temperature at 4 °C.
- Place samples to be analyzed in the trays inside the autosampler.
- Program the injection system to load the 5-μL loop with 4 μL of sample for each analysis and to wash the injection needle and syringe with the washing solution between each injection.

3.2 Mass Spectrometry

Before starting each series of analyses, the ion source should be cleaned including the sweep cone, and the MS system should be calibrated. The calibration of the mass spectrometer is indispensable to guarantee the measurement accuracy. Optionally, an internal calibration procedure could be performed during data acquisition thank to the "lock mass" option based on the accurate *m/z* value of

Table 2
Parameters usually applied for full scan acquisition using FIA-FTICR-MS

Source conditions	Ion transmission parameters	Acquisition mode
Dry gas temperature, 200 °C; Dry gas flow, 4 L/min; Nebulizer gas, 3 bar; Endplate offset, −500 V; Capillary voltage, −4.5 kV for ESI(+) Or + 3.5 kV for ESI(−)	Time of flight, 0.6 s; Capillary exit, 150 V for ESI(+) Or − 150 V for ESI(−); Deflector plate, 180 V for ESI(+) Or − 180 V for ESI(−); Funnel, 1150 V for ESI(+) Or − 150 V for ESI(−); Skimmer, 15 V for ESI(+) Or − 15 V for ESI(−)	Broadband and serial modes; Acquisition time, 3 min; *m/z* range, 100 to 1000; Average scan, 1; Ion accumulation time, 0.2 s; Data size, 8 M; Save as "profile" or "line" spectra; Data reduction factor, 90%

selected ions (e.g., known contaminant ions). The mass error must be less than 1 ppm for an acceptable calibration. Typically, mass spectra are recorded in the profile mode with Gaussian shape peak acquisition. The centroid mode can also be used, providing a smaller data file size, but it generates poorer data quality by reducing each signal distribution to a single intensity, that is, a line shape. Eventually, the signal threshold removal option and no FID (free induction decay) acquisition could be applied to significantly reduce the data file size although a possible loss of information may occur. The MS parameters proposed in Table 2 are suitable for low *m/z* ion transmission (e.g., low funnel RF voltages), and to limit the ion trap charge space effects (e.g., low accumulation time) [11].

- Set the instrumental parameters before starting MS experiments.
  - Load an appropriate MS method and set the appropriate parameter values for the used ionization mode (*see* Table 2 for operating conditions in both positive and negative ESI modes).
  - Wait for the read-backs to reach the setting values.

The instrumental parameters are given here as an example; of course their values vary between FTMS instruments. Electrospray ionization is proposed here since it is well-adapted for polar molecules as metabolites. Other ionization sources can be also used instead or in complementary way. Typically, samples are analyzed using both positive and negative ionization modes to enhance the metabolome coverage.

- Calibrate the *m/z* range of the mass spectrometer

  The step is essential to obtain reliable mass spectra and must be performed for a given MS method, for example, at the used ionization mode.

- Use the syringe pump to infuse the calibration solution into the ESI source at 5 μL.min^{-1} rate.
- Acquire data in full-scan mode with *m/z* range from 50 to 1000. A series of sodium ion clusters (not necessary all the ions listed in **Note 4**) should be detected.
- Do the calibration using quadratic mode (i.e., at least three calibration points are requested) and save the method.

3.3 Analytical System Suitability

- Prepare the analytical system before starting analyses.
  - Connect the eluent tubing from the LC system to the ion source.
  - Start with several blank injections (at less 5 injections): use full loop injection mode for the first injection, and perform the following injections by setting the injection volume at 4 μL using partial loop injection mode. Note that the "blank" sample correspond the eluent composition.
- Check the calibration using the FIA mode.
  - Inject 4 μL of the solution Mix diluted twice in water–methanol (1:1, v/v).
  - Acquire data for 3 min under the MS conditions described in Table 2.
  - Ensure that the mass errors of the three reference compounds are less than 1 ppm; otherwise recalibrate the mass spectrometer.

4 Optimization of Sample Dilution Factor

- After checking the analytical system suitability (*see* Subheading 3.3), start the analytical sequence.
  - Start with two or three blank injections to check that there is no contamination of the analytical system.
  - Analyze samples diluted at different dilution factors and spiked with the same amount of reference compounds, at less two blank injections should be performed between sample analyses.
  - Repeat the sequence twice to compensate an eventual analytical drift or sample injection errors.
- Choose the best dilution factor (an example of urine analyses is given in the Fig. 1).
  - For low dilution (i.e., 1:50 dilution), the Total ion current (TIC) does not return to the baseline after 3 min, reflecting strong carryover and matrix effects confirmed by the

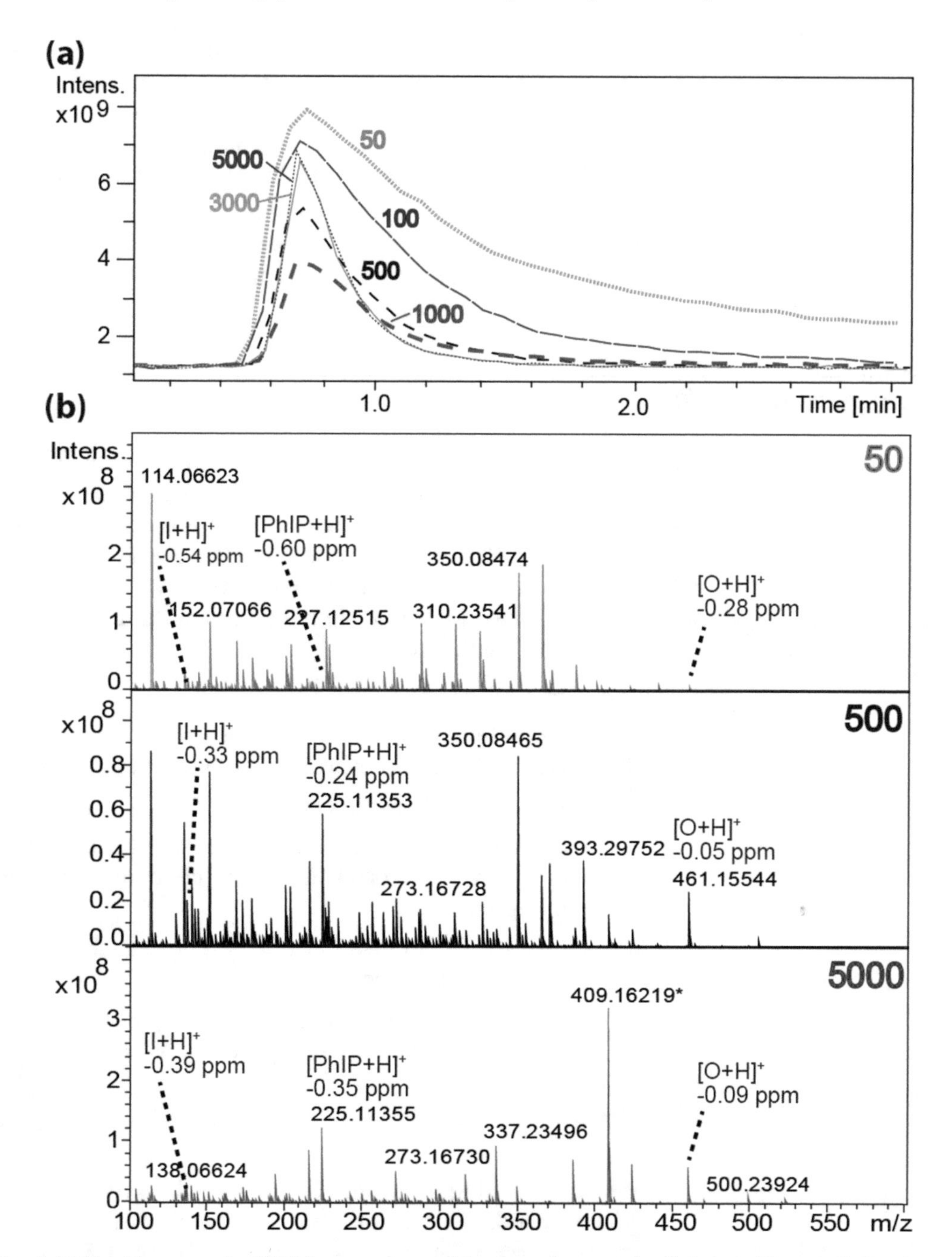

Fig. 1 (**a**) Total ion currents (TICs) from analyses of human urine sample diluted at different dilution factors (i.e., 50, 100, 500, 1000, 3000, and 5000) and spiked with the same reference compound quantity, and (**b**) the mass spectra for 50, 1000 and 5000 dilution factors using FIA hyphenated to ESI (+)-FTICR instrument (Solarix-XR 7T, Bruker). Note that the *m/z* 409.16219 ions are solvent contaminants

corresponding mass spectrum where the reference compound ions are detected with unexpected low abundances compared to the mass spectra of other diluted samples.

- For high dilution (i.e., 1:3000 or 1:5000 dilution), the shape of the TICs is relatively thin and the mass spectra are similar to that of the Mix solution diluted two folds, with an increased intensity of contaminant ions (e.g., 409.16219 *m/z* ions), probably due to high dilution effects.
- The optimal dilution factor should be accurately determined based on the signals of the reference compound ions and the number of detectable peaks; here, the 1:500 dilution appears to give the best results.

5 Fingerprint Acquisition

The QC sample is periodically analyzed during the sequence for quality control of the produced data. A sufficient number of QC injections can be used to correct both between-batch and within-batch effects [12]. Repeated blank injections are also performed to check possible system contamination or memory effects. More importantly, repeated injections should give identical mass spectra for both QC and blank analyses.

- Execute the batch queue.
  - Start repeated injections of QC sample (at least three injections) to check the reproducibility in the spectral fingerprints.
  - Continue with two blank injections.
  - Randomly analyze ten samples followed by a QC sample injection and then a blank injection; continue with this order during the batch.
- Evaluate the reliability of data based on the repeatability in the QC profiles during the batch.
  - Check the intensity variations for some ions (i.e., at least ten *m/z* values with the three reference standards among them) detected in the QC profiles.
  - Calculate the *m/z* value drifts for some ions during the sequence.
  - For example, Fig. 2 (from [8]) displays 15% of the mean relative standard deviation (RSD) value from the absolute intensity repeatability and a mass drift below 1.5 mu by considering approximately 500 ions commonly detected in 54 repeated QC injections during a batch of over 74 h (for a total number of about 650 injections). This RSD value is acceptable according the FDA (Food and Drug Administration) guidelines for biomarker studies (RSD of <20%) and is similar to the value obtained by Kirwan et al. by performing eight batches during 7 days [13].

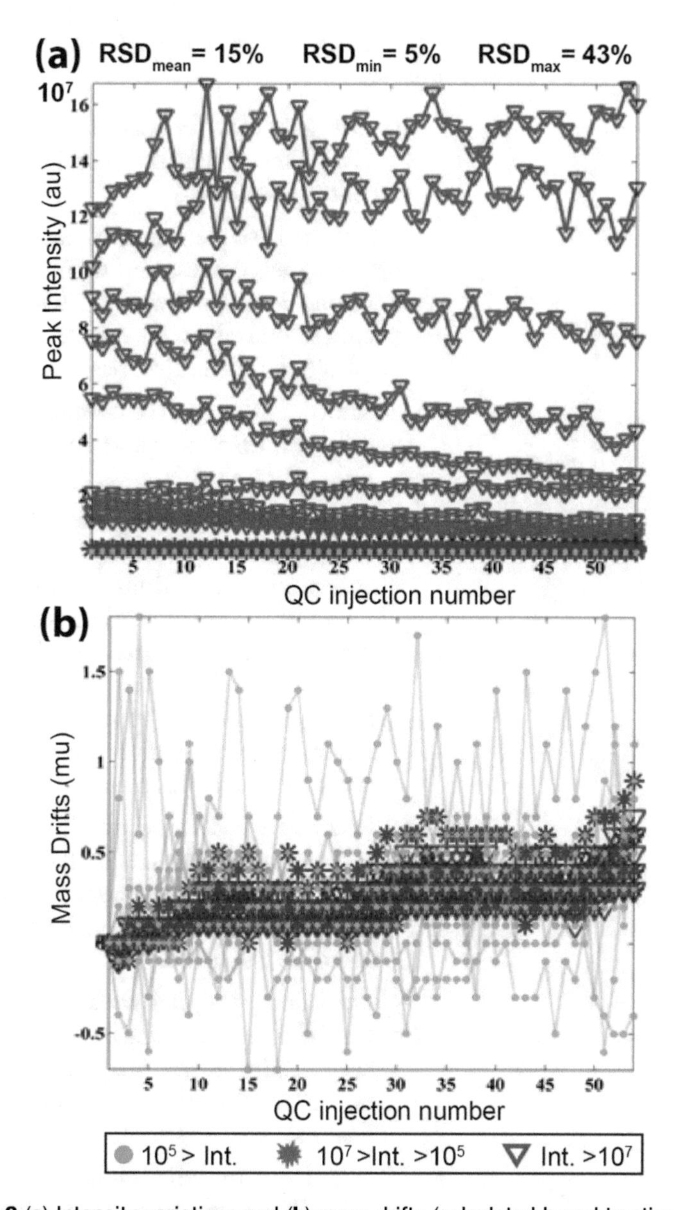

Fig. 2 (**a**) Intensity variations and (**b**) mass drifts (calculated by subtracting *m/z* values between the first and every QC injection) for approximately 500 ions detected in 54 QC injections during a single batch of 650 injections over 74 h (adapted from [8])

6 Data Processing

The generated data are processed using either in-house data mining tools [7] or software packages provided by instrument manufacturer. For example, MetaboScape 4.0 software can be used to process FIA-MS raw data generated from Bruker Daltonics FTICR instruments.

The general guidelines to process FIA-FTICR-MS data are as follows:

- Export all data obtained from "profile" or "line" mode to netCDF format files that can be easily imported in R or other statistical platforms.
- Adjust the baseline for each mass spectrum to discard noninformative signals corresponding to many *m/z* variables using the R functions, *loess()* or *lowess()*. This could be done locally (i.e., in a specific region of a mass spectrum) or globally given a constant threshold for the entire mass spectrum.
- Generate a data matrix containing information about sample identity, ion *m/z* value and abundance for each detected peak.
  - Extract the data features, that is, *m/z* variables by searching for the beginning and the end of each peak shape in every profile spectrum. Such a procedure is based on the detection of variations in consecutive values of the sign of slopes of the "profile" data set. All *m/z* values extracted from the same baseline signal correspond to a single *m/z* peak. The name of a so-calculated integrated variable is simply given by the *m/z* value of the maximum of the peak. In the "line" mode, this integration step is not necessary, the values of every *m/z* variables being already calculated.
  - Round all *m/z* values to three digits after the decimal point and sum all intensities of the variables having the same *m/z* value for every experiment. Put the value NA (Not Available) in the final data matrix when no intensity is detected for a given *m/z* variable.
  - Retain only the *m/z* variables encountered in the data matrix with significant intensity in the majority of the samples to keep a reduced-size final dataset usable for multivariate statistics. Generally, the total amount of empty unitary cells in the final data matrix should be lower than 20%.
- Use multivariate statistical analysis such as principal component analysis (PCA) to have an overview of the considered data set to detect global trends and eventually some outliers that would be discard in complementary statistical analyses. Note that univariate method (such as Student's t-test) can be used to detect level changes of individual metabolites or to select variables prior to multivariate statistical analysis, but it should be controlled for the redundancy-related alpha-risk which is modified according to the multiplicity of unitary test achieved systematically [14]. In R, the function *p.adjust()* can be used. With the resulting selected variables, the use of the "mixOmics" [15] or "opls" [16] R libraries is recommended to perform principal component analysis.

- Process the data matrix with a partial least squares-discriminant analysis (PLS-DA) to get the group separation and to detect biomarkers. Alternatively, apply a regularized canonical correlation analysis to detect global correlation between the first and possibly the second latent variables explaining the metabolomic dataset on the one hand and a second multivariate matrix built from exactly the same statistical individuals, on the other hand. Note that most of the time the PLS2 regression gives a comparable result of correlative links between latent variables coming from the two data sets.

7 Notes

1. Caution: Methanol is highly flammable and toxic. Formic acid is corrosive and volatile. 2-amino-1-methyl-6-phenylimidazo (4,5-b)pyridine (PhIP) is carcinogenic. All chemicals should be handled with gloves and under a fume hood.
2. The final concentration of the calibration and reference compound solutions depends on the MS instrument performances.
3. The acquisition time could be adjusted, depending on the TIC peak shape.
4. Accurate *m/z* values of positive and negative sodium formate (Fo) and acetate (Ac) cluster ions (list provided by the manufacturer, Table 3).

Table 3
M/z list of main calibrant ions of formate and acetate cluster ions (Fo = HCOO and Ac = CH_3COO)

Positive cluster ions					
n	$Na(FoNa)_n$	$Na(AcNa)_n$	$Na(NaAc)_1(NaFo)_n$	$Na(NaAc)_2(NaFo)_n$	...
1	90.976645	104.992295	172.979719	254.982793	
2	158.964069	186.995369	240.967143	322.970217	
3	226.951493	268.998443	308.954567	390.957641	
4	294.938917	351.001517	376.941991	458.945065	
5	362.926341	433.004591	444.929415	526.932489	
6	430.913765	515.007665	512.916839	594.919913	
7	498.901189	597.010739	580.904263	662.907337	
8	566.888613	679.013813		730.894761	
9				798.882185	
10				866.869609	
11				934.857033	

(continued)

Table 3 (continued)

Negative cluster ions					
n	Fo(NaFo)$_n$	Ac(NaAc)$_n$	Ac(NaFo)$_n$	Ac(NaAc)$_1$(NaFo)$_n$	...
1	112.985627	141.016927	127.001277	209.004351	
2	180.973051	223.020001	194.988701	276.991775	
3	248.960475	305.023075	262.976125	344.979199	
4	316.947899	387.026149	330.963549	412.966623	
5	384.935323	469.029223	398.950973	480.954047	
6	452.922747	551.032297	466.938397	548.941471	
7	520.91017	633.035371	534.925821	616.928895	
8	588.897594	715.038445	602.913244	684.916319	
9	656.885018	797.041519	670.900668	752.903742	
10	724.872442	879.044593	738.888092	820.891166	
11	792.859866		806.875516	888.878590	
12	860.84729		874.862940	956.866014	
13	928.834714		942.850364		
14	996.822138		1010.837788		

Acknowledgments

The authors thank Dr. Baninia Habchi for her collaboration, Dim ASTREA and the National FTICR network (FR 3624 CNRS) for financial support for research works conducted on our FTICR instrument.

References

1. Nanita SC, Kaldon LG (2016) Emerging flow injection mass spectrometry methods for high-throughput quantitative analysis. Anal Bioanal Chem 408:23–33
2. Habchi B, Alves S, Paris A, Rutledge DN, Rathahao-Paris E (2016) How to really perform high throughput metabolomic analyses efficiently? Trends Anal Chem 85:128–139
3. Marshall AG, Hendrickson CL, Jackson GS (1998) Fourier transform ion cyclotron resonance mass spectrometry: a primer. Mass Spectrom Rev 17:1–35
4. Oliver S, Winson MK, Kell DB, Baganz F (1998) Systematic functional analysis of the yeast genome. Trends Biotechnol 16:373–378
5. Goodacre R, Vaidyanathan S, Dunn WB, Harrigan GG, Kell DB (2004) Metabolomics by numbers: acquiring and understanding global metabolite data. Trends Biotechnol 22:245–252
6. Taylor PJ (2005) Matrix effects: the Achilles heel of quantitative high-performance liquid chromatography-electrospray-tandem mass spectrometry. Clin Biochem 38:328–334
7. Habchi B, Alves S, Jouan-Rimbaud Bouveresse D, Moslah B, Paris A, Lécluse Y et al (2017) An innovative chemometric method for processing direct introduction high resolution mass spectrometry metabolomic data: independent component-

discriminant analysis (IC–DA). Metabolomics 13:45

8. Habchi B, Alves S, Jouan-Rimbaud Bouveresse D, Appenzeller B, Paris A, Rutledge DN et al (2017) Potential of dynamically harmonized Fourier transform ion cyclotron resonance cell for high-throughput metabolomics fingerprinting: control of data quality. Anal Bioanal Chem 410:483–490
9. Rathahao-Paris E, Alves S, Boussaid N, Toutain P, Picard-Hagen N, Tabet JC et al (2018) Evaluation and validation of an analytical approach for high-throughput metabolomic fingerprinting using direct introduction-high-resolution mass spectrometry: applicability to classification of urine of scrapie-infected ewes. Eur J Mass Spectrom 25(2):251–258
10. Nikolaev EN, Jertz R, Grigoryev A, Baykut G (2012) Fine structure in isotopic peak distributions measured using a dynamically harmonized fourier transform ion cyclotron resonance cell at 7 T. Anal Chem 84:2275–2283
11. Ledford EB, Rempel DL, Gross ML (1984) Space charge effects in Fourier transform mass spectrometry. II. Mass calibration. Anal Chem 56:2744–2748
12. Wehrens R, Hageman JA, van Eeuwijk F, Kooke R, Flood PJ, Wijnker E et al (2016) Improved batch correction in untargeted MS-based metabolomics. Metabolomics 12:88
13. Kirwan JA, Broadhurst DI, Davidson RL, Viant MR (2013) Characterising and correcting batch variation in an automated direct infusion mass spectrometry (DIMS) metabolomics workflow. Anal Bioanal Chem 405:5147–5157
14. Chen SY, Feng Z, Yi X (2017) A general introduction to adjustment for multiple comparisons. J Thorac Dis 9:1725–1729
15. Rohart F, Gautier B, Singh A, Lê Cao KA (2017) mixOmics: an R package for 'omics feature selection and multiple data integration. PLoS Comput Biol 13:e1005752
16. Thevenot EA, Roux A, Xu Y, Ezan E, Junot C (2015) Analysis of the human adult urinary metabolome variations with age, body mass index and gender by implementing a comprehensive workflow for univariate and OPLS statistical analyses. J Proteome Res 14:3322–3335

Chapter 3

Two-Dimensional Liquid Chromatography in Metabolomics and Lipidomics

Miriam Pérez-Cova, Romà Tauler, and Joaquim Jaumot

Abstract

Multidimensional separation systems have arisen in the last years to overcome certain limitations of the classical one-dimensional separations. Multidimensional analytical approaches achieve a greater separation power, a crucial aspect when dealing with highly complex samples, as in metabolomics and lipidomics. Online comprehensive two-dimensional chromatography is a particularly interesting mode when pursuing untargeted analysis, which allows for separating, and consequently identifying a more extensive number of analytes. This chapter aims to summarize current applications of 2D-LC to the fields of metabolomics and lipidomics, and setups of different separation mechanism that are being employed, showing the most suitable combinations of chromatographic modes, depending on the target compounds.

Key words Two-dimensional liquid chromatography, LC×LC-MS, Comprehensive, Metabolomics, Lipidomics

1 Introduction

One-dimensional liquid chromatography (1D-LC) combined with mass spectrometry (MS) is the leading analytical technique employed in metabolomics and lipidomics studies. It is selective and versatile for nonvolatile analytes, due to the wide variety of different separation modes and adjustable parameters (e.g., mobile and stationary phases, pH, temperature, additives). However, the complexity of metabolomics (and lipidomics) samples with a large number of compounds and the presence of isobaric molecules difficult its identification and quantification. From a chromatographic viewpoint, a new hypothesis can be tested: "Extremely complex matrices require higher peak capacities to achieve a complete resolution." In the last years, different analytical techniques have been proposed to add another dimension to the separation, and, consequently, achieve a greater separation power (e.g., ion-mobility spectrometry, IMS [1]; supercritical fluid

Paul L. Wood (ed.), *Metabolomics*, Neuromethods, vol. 159, https://doi.org/10.1007/978-1-0716-0864-7_3,

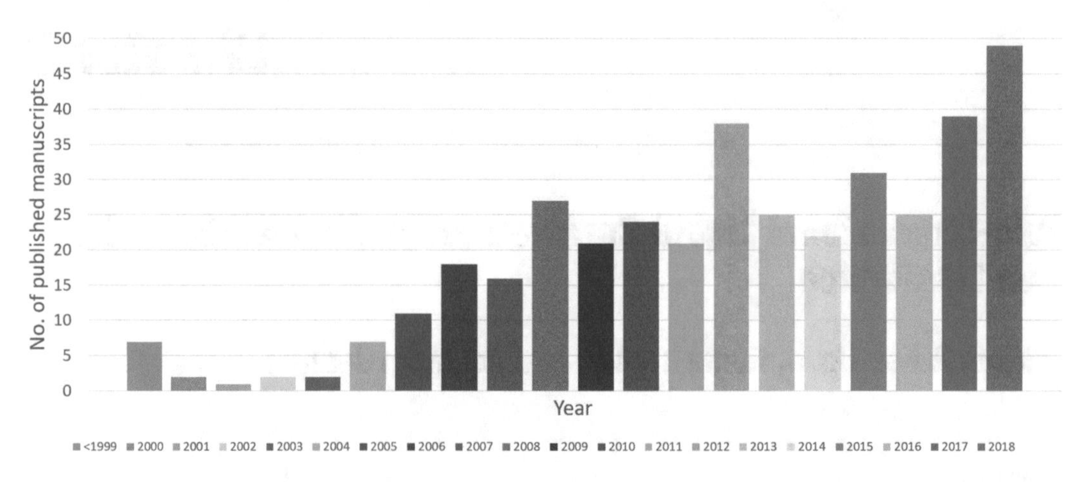

Fig. 1 Evolution of the number of published manuscripts related to 2D-LC (source Scopus, March 2019)

chromatography, SFC [2]; two-dimensional liquid chromatography, 2D-LC [3]).

Recently, 2D-LC has gained popularity for the analysis of metabolomics and lipidomics samples [4]. Two-dimensional chromatography is an attractive alternative for overcoming these limitations when characterizing these highly complex samples. The second column provides additional selectivity assuming that, ideally, both separations are orthogonal [5], which means that non-correlated retention time mechanisms are employed in the two dimensions. Consequently, the elution times from the two dimensions can be treated as statistically independent [6]. 2D-LC is commonly coupled to high-resolution mass spectrometry (HRMS) and/or tandem mass spectrometry (MS/MS), but other suitable detectors and analytical instruments include IMS-MS, ultraviolet–visible spectroscopy (UV-Vis), and fluorescence [7].

Figure 1 shows the continuous growth in the use of 2D-LC from the early 2000s. In the first years of the decade, only few manuscripts describing new instrumental developments allowing the evolution from offline to online analysis and from the UV-Vis to the MS detection (widely used nowadays) can be found. Also, pioneering works describing data analysis approaches for the processing of these data sets were published. During this period, most of the application of 2D-LC were related to proteomics and polymers analysis. In the second decade, the instrumental developments have allowed a major diversity of applications, in which metabolomics and lipidomics application stand out. However, these new examples cannot be explained without the recent technical progress that will be described below.

There are several ways of implementing 2D-LC that define the mainly used classification criteria.

The first classification approach dealing with the sample collection refers to offline, stop-flow, or online configurations.

Offline implies that fractions from the first-dimension (^{1}D) are collected, and afterward are transferred to the second-dimension (^{2}D). This setup is experimentally less complicated, and, as the two columns from the two dimensions are independent, greater peak capacities usually are achieved. The offline configuration also has the advantage of allowing for preconcentration, derivatization, and reconstitution into an appropriate mobile phase between the two separations [8].

In the stop-flow configuration, fractions are transferred directly to ^{2}D, but the flow is stopped in the ^{1}D when performing the analysis in the ^{2}D. This procedure is repeated for the whole ^{1}D separation [9]. In this approach, it is possible to increase the ^{1}D flow rate, as it will be continuously stopped before entering in the ^{2}D column. One of the most relevant drawbacks that this setup presents is the additional first dimension band broadening during stop-flow periods, especially when small molecules are analyzed. This is the main reason why stop-flow is less common than offline and online configurations [9]. However, it has been successfully employed for bigger molecules, such as peptides [10] or polymers [11].

In the online mode, the connection is made directly through an interface, that is, a high-pressure switching valve. ^{2}D separation needs to be very fast to analyze the whole fraction before the subsequent fraction is transferred. Therefore, short ^{2}D columns with large internal diameters and fast ^{2}D gradients are highly desirable, whereas long columns and low flow rates are commonly used in ^{1}D. Consequently, injection volume in ^{2}D is also reduced [12]. Chromatographic resolution is maintained from ^{1}D to ^{2}D because each peak is sampled 3–4 times, appearing in consecutive modulations (i.e., two successive injections into the second column [3]). This online arrangement allows for decreasing the total analysis time (an essential drawback in 2D-LC) thanks to full automatization, and also diminishes sample loss, cross-contamination, and degradation [8].

Another classification criterion is based on how fractions are collected and injected from first-dimension (^{1}D) to the second-dimension (^{2}D). When a single fraction from the ^{1}D effluent is reinjected in the ^{2}D column, it is a single heartcutting approach (LC—LC). Multiple heartcutting (mLC—LC) refers to the cases where several regions of the first separation are selected and introduced one at a time in the second separation. When all fractions are transferred at regular intervals and, therefore, are subjected to both separations is called comprehensive (LC $\times$ LC). Additionally, selective comprehensive (sLC $\times$ LC) implies that only specific and successive regions of the first separation are targeted for further separation, while the separation already obtained in the ^{1}D is maintained [3, 5].

This chapter is focused on online comprehensive 2D-LC. LC × LC is a mode particularly interesting when pursuing untargeted analysis, in which the main aim is to obtain the maximum information possible from the sample in one single run. This strategy allows separating and, consequently, identifying a more extensive number of analytes. Therefore, it is the main approach employed in screening and profiling, as it enables the separation of different classes of metabolites and their constituents [8]. The hyphenation of mass spectrometry analyzers and hybrid systems that permit fragmentation (e.g., Q-TOF, Orbitrap, Ion Trap, Triple Quadrupole) are also handy tools for achieving the identification of unknown compounds.

A generic scheme of a passive modulation interface is shown in Fig. 2. The concept of passive modulation implies that no modification takes place between the two dimensions. On the contrary, if some alterations have been made to the modulation process compared to the passive mode (i.e., presence of a bypass capillary, trapping columns, vacuum evaporation systems), it is called active modulation [4].

In the scheme, a setup including two six-port valves is detailed, as well as the two positions that valves acquire during subsequent modulations. Position 1 implies that ^{1}D effluent passes through

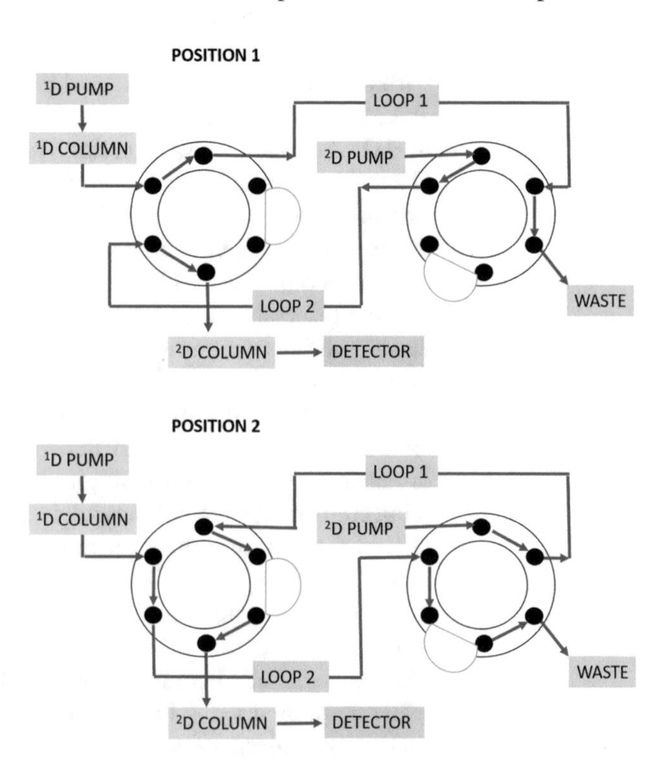

Fig. 2 Generic scheme of a passive-modulation interface in 2D-LC. Top—Position 1: Sampling loop 1— Injecting loop 2 in ^{2}D; Bottom—Position 2: Sampling loop 2—Injection ^{2}D

the ^{1}D column, and it is sampled afterward by loop 1, while ^{2}D effluent allows that content of loop 2 enters in the ^{2}D column; whereas in position 2, ^{1}D effluent passes through the ^{1}D column, but, in this case, it is sampled by loop 2, at the same time that ^{2}D effluent introduces the content of loop 2 in the ^{2}D column. Other common arrangements include the use of 8 or 10-port valves instead [4].

Although the predominant use of comprehensive 2D-LC is led by biopharmaceuticals, peptides, synthetic polymers, and traditional Chinese medicines (TCMs), metabolomics and lipidomics applications are currently widespread and are starting to gain importance. There is an increasing tendency in its use in both fields in the last few years [4]. Until 2010, the application of 2D-LC to metabolomics or lipidomics was negligible. A continuous increase in the number of published manuscripts within these omics fields can be observed (following a similar trend to the observed in Fig. 1 for the overall of 2D-LC references).

Setting up a 2D-LC is not straightforward. Many considerations are required to select and optimize different parameters (e.g., modulation interface, column dimensions, compatibility of mobile phases, isocratic or gradient elution in both dimensions, flow rates) [5].

Special attention is required in the choice of stationary phases and mobile phase composition. As there are two columns, the number of possible combinations is large, and uncorrelated separations (i.e., orthogonal mechanisms) are highly desirable. Depending on the application, some separation mode combinations could suit better than others, although there are some still unexplored. The preferred separation modes for 2D-LC metabolomics and lipidomics include reverse phase (RP), hydrophilic interaction liquid chromatography (HILIC), normal phase (NP), ion-exchange (IEC) and chiral chromatography [4]. Examples of different setups already employed will be described later in this chapter.

Optimization of comprehensive 2D-LC methods can be done through two different approaches: sample-independent (e.g., Pareto optimization [13]), or sample-dependent (e.g., the PIOTR program [14]). The first one aims to define two or more objective parameters, based on the impact that multiple method settings (e.g., gradient slope, column dimensions, flow rates) have through theoretical relations [13]. In contrast, the sample-dependent approach optimizes all chemical parameters that affect retention and selectivity, such as the mobile phase composition, temperature, pH, and buffer strength, for a specific sample. It is based on a very small number of experiments, modeling of the retention, and generalizing band-broadening behavior of individual sample components [14].

When MS detectors are coupled to LC × LC, some widespread experimental issues such as ion suppression effects (typical in LC-MS) are reduced. This hyphenation implies that quantitation is also more reliable, as matrix effects are minimized. However, in LC × LC-MS, the splitting ^{2}D flow before entering the ion source is frequent, which may cause a decrease in sensitivity and a significant peak broadening [12]. Other main drawbacks associated with comprehensive 2D-LC are related to conceptual and instrumental complexity, long analysis time (normally up to one hour, instead of a few minutes analysis offered by 1D-LC), solvent compatibility, lower detection sensitivity and increased difficulty in data analysis [4]. Different strategies have started to arise to solve these downsides.

A common problem is solvent strength mismatch, which causes peak distortion. In the cases that the ^{1}D effluent is a relatively strong injection solvent when compared to ^{2}D eluent, analytes are not strongly retained by the stationary phase and, consequently, the retention mechanism employed may be affected. If the unretained peak elutes mostly in the dead volume instead of in its normal location, this phenomenon is known as breakthrough [4].

For instance, active modulation has emerged as an effective manner of handling solvent strength mismatch, reducing breakthrough and possibly enhancing detection sensitivity. Several active modulation approaches have been recently proposed (e.g., active solvent modulation (ASM) [15], stationary-phase-assisted modulation (SPAM) [16], vacuum-evaporation modulation (VEM) [17]). More details about these approaches can be found in a review article by Pirok et al. [4].

The need for data analysis software is another bottleneck in 2D-LC, which is also currently being faced. Examples of commercially available software packages are Chromsquare from Shimadzu and GC Image LC × LC Edition Software from GC Image™ (including LC Image, LC Project, and Image Investigator). Other used approaches employ more sophisticated chemometric tools for the deconvolution and resolution of overlapping peaks. Examples of these methods include curve resolution algorithms such as the Multivariate Curve-Resolution by Alternating Least Squares (MCR-ALS) or Parallel Factor Analysis (PARAFAC) [18]. Both methods had been successfully used in the analysis of comprehensive gas chromatography data (GC × GC). MCR-ALS decomposes the experimental data following a bilinear model in the spectral and chromatographic contributions. In contrast, PARAFAC employs a trilinear model which needs higher data quality (i.e., high reproducibility in the observed retention times and peak shapes). Consequently, the use of these methods for the analysis of LC × LC data has also been proposed. However, the inherent properties of LC × LC data prevent from this totally trilinear behavior (i.e., minor changes in the retention time between modulations and different peak shapes) and, therefore, it seems that the use of MCR-ALS is more advisable (in particular, when dealing with LC × LC-MS, as in the case of metabolomics and lipidomics studies).

However, due to the huge size of the generated datasets, it is also highly advisable a preliminary step of data reduction. This compression can be applied either in the spectral mode, to reduce the total number of *m/z* values leaving only these that can be considered as more interesting (i.e., using, for instance, the Regions of Interest strategy), or in the time direction by compacting all the measured retention times (e.g., applying one-dimensional wavelet analysis strategy) [19].

Another challenge that is being faced is trying to reduce analysis time as much as possible, achieving in some cases, chromatographic runs of less than one hour [20–22].

All these recent instrumental and data analysis developments have allowed the application of this two-dimensional chromatography in new research fields. Examples in the fields of metabolomics and lipidomics are described more in detail throughout the chapter, with special emphasis in stationary phases chemistry and most employed ways to combine them, to obtain acceptable selectivities. It is important to notice that mobile phase composition is also a decisive parameter that needs to be taken into account when developing an LC × LC method. Although lipidomics is a branch of metabolomics, lipids are analyzed separately from the more water-soluble metabolites due to their generally hydrophobic character. The distinction between lipidomics and metabolomics 2D-LC methods is required though, pursuing a more specific approach for each group of compounds.

Thus, the following sections will be focused on comprehensive 2D-LC in lipidomics and metabolomics, respectively. There will be a special emphasis on strategies to set up methods for each field separately, according to most suitable combinations that have been applied.

2 Lipidomics

The main LC chromatographic modes employed in lipid analysis include RP, HILIC, NP, and silver-ion (Ag), which can be considered a particular type of NP [23]. Consequently, it is not surprising that the most common comprehensive 2D-LC combinations in lipidomic studies have been RP × HILIC, HILIC × RP, NP × RP, and Ag × RP. In addition, some studies using strong anion exchange (SAX) — RP, although not fully comprehensive, are also found in the literature [24]. All these combinations are considered orthogonal separations, which basically means that their selectivities are complementary enough and allows achieving enhanced peak capacities. A more detailed explanation of the main characteristics of each separation mode and their combinations is presented below.

RP is the most selected chromatographic mode for lipid analysis, and in general, in most of LC × LC applications. RP is a flexible and versatile retention mechanism, and therefore, it is suitable for any of the two chromatographic dimensions. In the particular lipidomics case, RP separates lipids according to the degree of hydrophobicity of fatty acids alkyl chains and the number and position of the double bonds present in these chains [23]. RP presents a series of experimental advantages as can be considered robust, fast, uses MS-compatible solvents, needs low reequilibration times, handles high temperatures and pressures, and provides high-resolution separations [5]. Hence, RP is widely employed as the ^{2}D.

NP and HILIC chromatographic mechanisms can be employed to separate lipids according to the different polarity of their head groups and, as a consequence, allows distinguishing between the different lipid classes [5]. Thus, the combination of the previously described RP separation with either of these two modes presents several benefits, due to the amphiphilic nature of lipids. Therefore, in the NP × RP and HILIC × RP, separation of lipids takes place firstly depending on their classes and polarities, and subsequently, according to the hydrophobicity of fatty acid alkyl chains and the number and positions of the double bonds.

NP × RP coupling is much more complicated, as the miscibility of the two mobile phases from the two chromatographic dimensions presents a high degree of incompatibility (i.e., the eluting strength of the ^{1}D eluent in NP mode is stronger than that of the mobile phase on the head of the ^{2}D RP column [17]). In an attempt to overcome these difficulties, some approaches with increased experimental drawbacks can be applied, that is, thermal and vacuum-evaporation modulations [25]. This NP × RP coupling also presents significant solvent-immiscibility problems that require complex interfaces [26]. For these reasons, RP × NP is usually discarded due to its slow separations (extremely long analysis times), slow column reequilibration times of the NP column and minor solvent-MS compatibility.

HILIC mode presents several improvements when compared to NP for 2D-LC. For instance, HILIC shows better repeatability, shorter equilibration times, and longer column lifetimes [26, 27]). Thereby, HILIC is becoming a popular alternative to NP in the lipidomics (and, also metabolomics) field, as shown in recent publications which are often based in combinations between RP and HILIC [18, 20–23, 26]. Nevertheless, in the case of HILIC × RP, some experimental difficulties should also be considered, that is, solvent strength mismatch. However, recent work has been focused on solving this drawback and the application of active modulation strategies (see above) has already been proposed.

Another less common setup is RP × HILIC. The main advantage of this coupling is that it does not require active modulation strategies due to good solvent compatibility and lower risk of solvent strength mismatch. However, the use of HILIC as the ^{2}D mechanism is less frequent because it may suffer from injection effects [28]. An alternative to overcome this issue is using a narrow gradient span [26]. Particularly in lipidomics, RP × HILIC could be a good option despite these limitations. HILIC in ^{2}D allows for partial separation of lipid species within individual classes [23], because more variations in the nonpolar side of the structure are found than due to the polar head-groups. Consequently, RP as ^{1}D would perform a longer separation and therefore, a higher separation power can be obtained [23].

Differences between these HILIC × RP and RP × HILIC couplings are more easily understood when comparing the obtained 2D-LC chromatograms. An example of each type is presented below (Fig. 3).

In HILIC × RP mode (Fig. 3a), the ^{1}D (HILIC) separates several lipid classes according to their polarity. For instance, it is possible to differentiate lysophosphatidylethanolamines (LPE) and lysophosphatidylcholines (LPC) according to their different head groups. Then, the ^{2}D (RP) separates lipids on the basis of the length of the fatty acyl chains and the number and position of the double bonds. This is the reason why carotenoids, cholesteryl esters (CEs), and triacylglycerides (TGs) are separated in the ^{2}D, despite they elute in the same fraction of the ^{1}D.

In RP × HILIC (Fig. 3b), an opposite separation is observed. For example, it becomes clearer how shorter fatty acids alkyl chains of smaller phosphatidylcholines (PC) elute before longer ones, as

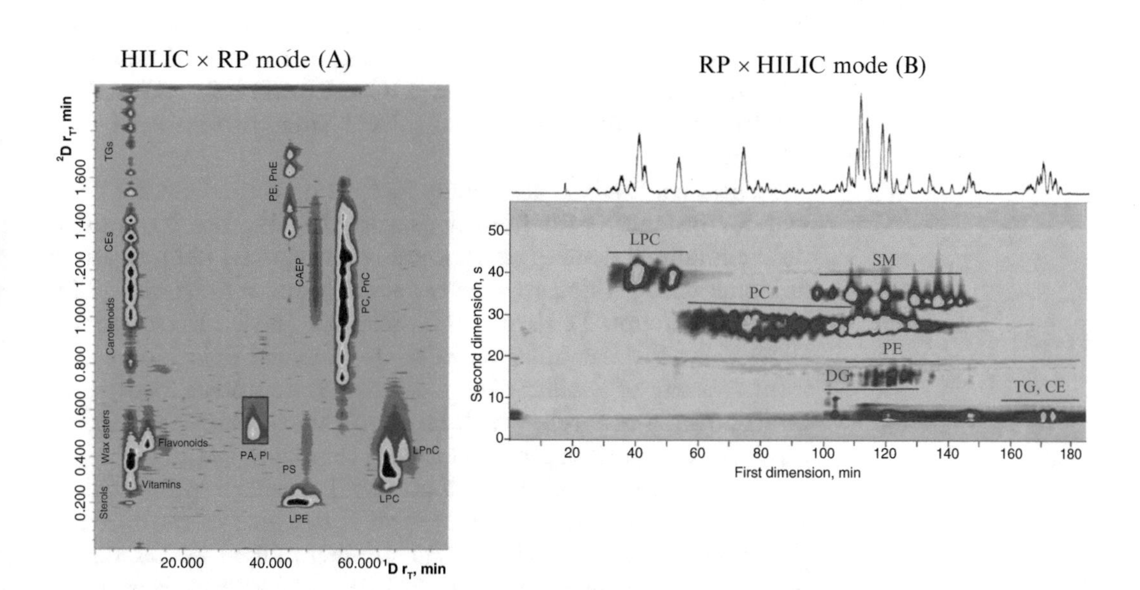

Fig. 3 Comparison of the application of HILIC × RP (**a**) and RP × HILIC (**b**) modes to lipidomic studies, and how lipids are though separated depending on their polarity and afterward by the length of the fatty acid alkyl chains, or vice versa. Reprinted with permission from [20] (**a**) and [26] (**b**)

they are less hydrophobic, and consequently, less retained by the RP stationary phase. On the contrary, the ^{2}D (HILIC) allows differentiation between phosphatidylcholines (PC) and sphingomyelins (SM), that otherwise will be overlapped, if only the ^{1}D (RP) was employed.

Finally, two chromatographic modes are less common in the literature but present some specific advantages for lipidomics studies. For instance, the use of Ag × RP is especially interesting when triacylglycerides (TGs) are the target analytes [29], as the ones with a different number of double bonds can be separated by the ^{1}D column. In constrast, those with different partition numbers are well resolved by the ^{2}D column. SAX resin chromatography allows for separating anionic lipids from neutral lipids. SAX × RP combination benefits from easier modulation strategies (i.e., relatively lack of difficulty in developing active modulation methods), although this coupling is slightly less orthogonal than other combinations [24].

Table 1 outlines the benefits and downsides of the chromatographic mode combinations already described, emphasizing its suitability in lipidomics analysis.

In contrast, the main aim of Table 2 is to illustrate examples of applications of 2D-LC to lipidomics, with special emphasis in comprehensive setups, including some details of the chromatographic conditions employed and some insights on the type of samples and target analytes.

Table 2 shows that the main separation mechanisms that have dominated lipidomic in 2D-LC include combinations of HILIC, RP, and NP. The recent improvements in solvent compatibility issues due to active modulation strategies have been especially relevant in the development of HILIC × RP and NP × RP, reducing solvent immiscibility issues, solvent strength mismatch, and breakthrough.

These applications are mainly focused on clinical studies dealing, for instance, with human plasma [23, 24, 26, 30] and in different animal tissues (e.g., mouse brain regions) [21], to study lipidome changes due to illnesses such as anxiety disorder [21], lacunar infarction [25], or atherosclerosis [32]). These studies allowed for the identification of some potential lipid biomarkers from a variety of families. There are also some studies focused on environmental research, which targeted some bioindicator organisms [18] or model organisms exposed to pollutants [20].

Comprehensive 2D-LC is a powerful approach in untargeted studies, such as lipid profiling, which is the goal of most of the publications shown in Table 2. The number of lipid species identified in these studies varies between 100 and 500, and from up to 17 lipid classes. Some examples where a specific group of compounds are targeted are also found (i.e., phospholipid and sphingomyelin [21], TGs [29]).

Table 1
Advantages and disadvantages of different combinations of separations in lipidomics

Advantages	Drawbacks
HILIC × RP	
• Separation based on polar head-groups of lipids in ^{1}D, and on hydrophobic interactions with nonpolar parts of lipids in ^{2}D • Good solvent compatibility • Analytes are preconcentrated in the interface when using traps (gain in sensitivity) • Possibility to do lipid profiling	• Requires active modulation • RP as ^{2}D cannot provide a highly efficient separation in a short period of time in the case of lipids
RP × HILIC	
• Separation of lipids based on hydrophobic interactions with the length of the fatty acyl chains and the number and positions of double bonds in ^{1}D, and on polar head-groups in ^{2}D, providing partial separation of lipid species within individual classes • Excellent solvent compatibility • Possibility to do lipid profiling	• HILIC as ^{2}D may suffer from injection effects
NP × RP	
• Separation of lipids based on the polar head groups (^{1}D) and on the different fatty-acyl chains (^{2}D) • Possibility to do lipid profiling • Analytes are enriched in the interface	• Requires thermal and/or vacuum-evaporation modulations • Competes with HILIC mode in ^{1}D. NP presents longer equilibration times, lower run-to-run repeatability, rapid column deterioration • Lower solvent compatibility
SAX × RP	
• Separation according to electrical propensities SAX resin, and molecular separation in RP • Capability of detecting low abundant species • Possibility to do lipid profiling	• Lower peak capacity than UPLC due to the use of nanoLC • Lower orthogonality
Ag × RP	
• Good separation of TGs, in ^{1}D according to different double bonds (critical pairs) and in ^{2}D with different partition number	• Applied in target analysis of a specific family (e.g., TGs)

3 Metabolomics

In 1D-LC metabolomics, RP is often used for the analysis of hydrophobic compounds, and the profiling of medium to nonpolar metabolites. However, for hydrophilic and neutral compounds, HILIC or NP modes are more suitable. Although NP presents

Table 2
Summary of applications of online 2D-LC in lipidomics

Year	Separation	Columns	Mobile phase solvents	Gradient	Detection	Sample	Ref.
2018	HILIC × RP	^{1}D: Ascentis Express HILIC (150 × 2.1 mm i.d., 2.7 μm d.p) ^{2}D: Titan C18 (50 × 4.6 mm i.d., 1.9 μm d.p)	^{1}D: (A) ACN:10 mM $HCOONH_4$ (98:2, v/v); (B) ACN: MeOH:10 mM $HCOONH_4$ (55:35:10, v/v/v) ^{2}D: (A) ACN:MeOH:10 mM $HCOONH_4$ (55:35:10, v/v/v); (B) IPA + 0.1% HCOOH	Both	HRMS; MS/MS	Mediterranean mussels; lipid profiling	[20]
2018	HILIC × RP	^{1}D: Kinetex HILIC (150 × 2.1 mm i.d., 2.6 μm, d.p) ^{2}D: Acquity UPLC BEH C18 (50 × 2.1 mm i.d., 1.7 μm d.p)	^{1}D: (A) 10 mM $HCOONH_4$; (B) Acetone ^{2}D: (A) H_2O:ACN:5 mM $HCOONH_4$ (50:50 v/v); (B) H_2O:ACN:5 mM $HCOONH_4$ (5:95 v/v)	Both	HRMS	Mouse brain; phospholipid and sphingomyelin profiling	[21]
2018	RP × HILIC	^{1}D: RP ZORBAX Eclipse XDB-C18 (150 × 2.1 mm i.d..; 5 μm d.p) ^{2}D: Kinetex HILIC (30 × 3 mm i.d.; 2.6 μm d.p)	^{1}D: (A) ACN:IPA (1:2, v/v) + 0.1% HCOOH; (B) H_2O + 0.1% HCOOH ^{2}D: (A) H_2O:5 mM CH_3COONH_4 pH = 5.5 (CH_3COOH); (B) ACN	Only ^{1}D	MS/MS	Rice exposed to As; lipid profiling	[18]
2017	NP – RP	^{1}D: Rx-SIL silica (150 × 2.1 mm i.d., 5 μm d.p) ^{2}D: Poroshell 120 EC C8 (50 × 2.1 mm i.d.., 2.7 μm d.p)	^{1}D: (A) Hexane; (B) IPA (2% H_2O + 5 mM $HCOONH_4$); (C) MeOH (2% H_2O + 5 mM $HCOONH_4$) ^{2}D: (A) MeOH:5 mM $HCOONH_4$; (B) H_2O:5 mM $HCOONH_4$	Both	HRMS; MS/MS	Human plasma (patients with lacunar infarction); lipid profiling	[25][a]

2017	RP × HILIC	^{1}D: Acquity UPLC (BEH) C18 (150 × 2.1 mm i.d., 1.7 μm d.p) ^{2}D: Acquity UPLC (BEH) HILIC (50 × 2.1 mm i.d., 1.7 μm d.p)	^{1}D: (A) ACN:H_2O:10 mM $HCOONH_4$ (60:40, v/v); (B) IPA:ACN:10 mM $HCOONH_4$ (90:10, v/v) ^{2}D: (A) H_2O:10 mM $HCOONH_4$; (B) ACN	Both	HRMS	lipid standards; pooled human plasma purchased	[26]
2015	HILIC – RP	^{1}D: Ascentis Express HILIC (150 × 2.1 mm i.d., 2.7 μm d.p) ^{2}D: Ascentis Express C18 (75 × 2.1 mm i.d., 2.7 μm d.p)	^{1}D: (A) ACN; (B) H_2O:10 mM $HCOONH_4$; pH = 3 (HCOOH) ^{2}D: (A) THF:ACN:IPA: H_2O:10 mM $HCOONH_4$; pH = 3 (HCOOH); 45 °C	Only ^{1}D	ozonolysis; HRMS	Egg yolk phospholipids extract	[22][b]
2015	NP – RP	^{1}D: Rx-SIL silica (150 × 2.1 mm i.d., 5 μm d.p) ^{2}D: Poroshell 120 EC C8 (50 × 2.1 mm i.d.., 2.7 μm d.p)	^{1}D: (A) Hexane; (B) IPA (2% H_2O + 5 mM $HCOONH_4$); (C) MeOH (2% H_2O + 5 mM $HCOONH_4$) ^{2}D: (A) MeOH:5 mM $HCOONH_4$; (B) H_2O:5 mM $HCOONH_4$	Both	HRMS; MS/MS	Human plasma (patients with benign breast tumor and breast cancer); lipid profiling	[30][a]
2015	RP × HILIC	^{1}D: Acquity UPLC (BEH) C18 (150 × 1 mm i.d., 1.7 μm d.p) ^{2}D: Core–shell silica Cortecs HILIC (50 × 3 mm i.d., 2.7 μm d.p)	^{1}D: (A) H_2O:5 mM $HCOONH_4$; (B) ACN:IPA (1:2, v/v) + 0.5% $HCOONH_4$ ^{2}D: (A) H_2O:5 mM $HCOONH_4$ pH = 5.5; (B) ACN	Both	HRMS; MS/MS	Human plasma; porcine brain; lipid profiling	[23]
2014	HILIC – RP	^{1}D: Ascentis Express HILIC (150 × 2.1 mm i.d., 2.7 μm d.p) ^{2}D: Ascentis Express C18 (75 × 2.1 mm i.d., 2.7 μm d.p)	^{1}D: (A) ACN; (B) H_2O:10 mM $HCOONH_4$; pH = 3 (HCOOH) ^{2}D: (A) THF:ACN:IPA: H_2O:10 mM $HCOONH_4$; pH = 3 (HCOOH); 45 °C	Only ^{1}D	ozonolysis; HRMS	Rat liver; elucidation of phosphatidylcholine isomers	[31][b]

(continued)

Table 2 (continued)

Year	Separation	Columns	Mobile phase solvents	Gradient	Detection	Sample	Ref.
2014	NP – RP	^{1}D: Rx-SIL silica (150 × 2.1 mm i.d., 5 μm d.p) ^{2}D: Poroshell 120 EC C8 (50 × 2.1 mm i.d.., 2.7 μm d.p)	^{1}D: (A) Hexane; (B) IPA (2% H_2O + 5 mM $HCOONH_4$); (C) MeOH (2% H_2O + 5 mM $HCOONH_4$) ^{2}D: (A) MeOH:5 mM $HCOONH_4$; (B) H_2O:5 mM $HCOONH_4$	Both	HRMS; MS/MS	Human plasma from atherosclerosis patients; lipid profiling	[32][a]
2013	NP – RP	^{1}D: Rx-SIL silica (150 × 2.1 mm i.d., 5 μm d.p) ^{2}D: Poroshell 120 EC C8 (50 × 2.1 mm i.d.., 2.7 μm d.p)	^{1}D: (A) Hexane:IPA:H_2O (30:70:2, v/v/v) + 5 mM $HCOONH_4$; (B) MeOH:H_2O (100:2, v/v) + 5 mM $HCOONH_4$ ^{2}D: (A) MeOH:H_2O:5 mM $HCOONH_4$ (50:50, v/v); (B) MeOH:5 mM $HCOONH_4$	Both	HRMS; MS/MS	Peritoneal dialysis patients; lipid profiling	[33][a]
2013	SAX – RP	^{1}D: lab made ^{2}D: lab made	^{1}D: From 10 mM to 1 M CH_3COONH_4 ^{2}D: (A) 0.05% NH_4OH; (B) MeOH/CH_3CN/IPA + 0.05% NH_4OH; and 5 mM $HCOONH_4$	Both	MS/MS	Healthy human plasma; lipid profiling	[24][b]
2012	Ag × RP	^{1}D: Microbore TSK gel SP-2SW column (150 × 1 mm i.d., 5 μm d.p) ^{2}D: Platinum™ EPS C18 (33 × 7 mm, i.d., 1.5 μm)	^{1}D: (A) Hexane +0.3% ACN; (B) Hexane +1.3% ACN ^{2}D: (C) ACN; (D) IPA–Hexane (2:1, v/v)	Only ^{1}D	HRMS; MS/MS	Peanut oil and mouse tissue; TGs analysis	[29]

[a]Non–stop-flow NP/RP 2D-LC system with vacuum evaporation interface
[b]Not fully comprehensive setups

some solvent incompatibilities issues when coupled to MS detectors and, consequently, other options such as aqueous normal phase chromatography (ANP) are preferred [8, 34]. Besides, for separating hydrophilic and charged small molecules, capillary electrophoresis seems a good option to be considered [34]. Chiral columns are also very useful in specific applications, for instance, in the pharmaceutical industry [35].

2D-LC combinations of separation modes more used in metabolomics include HILIC × RP, RP × RP, and Chiral × Chiral. The main advantages of HILIC × RP, as already described in the lipidomics section, are good orthogonality, good MS and solvent compatibility, and high applicability (i.e., for the simultaneous separation of polar and nonpolar metabolites). However, active modulation strategies are highly recommended when employing this combination. Below, a more detailed explanation of RP × RP and Chiral × Chiral is included, as they are not described in the previous lipidomics section.

In the case of highly polar metabolites, one of the most common combinations is HILIC × RP [19, 36–39], although NP × RP combinations have also been employed in other fields, such as in food analysis [40]. However, when dealing with hydrophobic compounds such as anabolic steroids, HILIC may not be the best option, as it is unlikely to provide sufficient separation [41].

RP × RP seems a good alternative in these cases. To overcome orthogonality problems that arise from the fact that both dimensions employed the same retention mechanism, some differences (i.e., different pH values, temperature, organic modifiers, or varying gradient composition between subsequent ^{2}D runs) are highly desirable from one chromatographic dimension to the other one [12, 28]. Thus, a better chromatographic resolution is achieved. Other advantages of RP × RP are an excellent robustness, repeatability, well-known separation mechanisms and adequate selectivity [12]. However, as well as HILIC × RP, the RP × RP coupling presents some solvent-compatibility issues, that can also be solved using active modulation strategies already described.

Chiral × Chiral has been used for enantioselective amino acid analysis of peptide and protein hydrolysates. Two chiral stationary phases are combined with similar chemoselectivity but orthogonal stereochemistry (i.e., opposite chiral recognition due to inverted configurations in C8 and C9 of the recognition site [35]). Therefore, this methodology is useful for stereoconfiguration profiling as, for instance, the analysis of chiral amino acids in peptide therapeutics [35].

Another potential strategy that has not been widely employed yet in metabolomics is HILIC × HILIC. This fact is mainly due to the shallow degree of orthogonality between both dimensions depending on the target analytes and the used stationary phases, For instance, D'Attoma et al. found low orthogonality in peptide analysis using a variety of chromatographic columns, organic modifiers, and temperatures [28]. However, this drawback was not found by Wang et al. separating saponins using a TSK gel Amide-80 and a polyhydroxyethyl A columns as the ^{1}D y ^{2}D separations respectively, employing acetonitrile and aqueous mobile phases, at acid pH [42]. Nevertheless, further improvements are needed in terms of chromatography efficiency [5].

When considering the metabolomics applications found in the literature, the most frequent combinations are HILIC × RP and RP × RP, depending on the type of metabolites targeted.

Table 3 includes a series of metabolomics examples of applications of online comprehensive 2D-LC, specifying chromatographic conditions employed, the type of samples and target analytes.

The main difference with the lipidomics field is that applications in metabolomics are usually more specific, as they target some groups of compounds such as polyphenols [12, 38], flavones [36, 48], or steroids [41, 43]. Matrices, where these compounds are found, are quite diverse, as it includes, for instance, green cocoa beans [36], bovine urine [41], apples [38], or wine [12]. Food analysis is the principal field of current applications, with some exceptions in environmental studies [19], microbial metabolites [45] or biological fluids [41, 43].

Although there are less frequent than in lipidomic analysis, there are also some examples of metabolic profiling in licorice [39, 44] or rice [19]. In these untargeted studies, the total number of metabolites detected ranges between 80 and 150, while it was possible to identify between 40 and 140.

4 Conclusions

The inherent complexity of metabolomics and lipidomics samples in which hundreds (or thousands) of compounds can coexist, has pushed to the limits the traditional one-dimensional separation techniques. The quest for a better description of these samples has led to the development of new technologies merging different separation strategies. Therefore, the combination of different chromatographic modes (i.e., GC × GC or LC × LC) or its combination with another separation technique such as capillary electrophoresis or ion mobility, has allowed for improved sample characterization.

Among all these new technologies, comprehensive 2D-LC seems a promising technique for the analysis of complex samples, with especial interest in untargeted lipidomics and metabolomics. Recent technical developments have increased their applicability overcoming experimental issues such as the solvent immiscibility and enhancing detection sensitivity. Also, the development of a new generation of chromatographic stationary phases increases the

Table 3
Summary of applications of online comprehensive 2D-LC in metabolomics

Year	Separation	Columns	Mobile phase solvents	Gradient	Detection	Sample and analyte	Ref.
2018	HILIC × RP	^{1}D: LiChrospher DIOL 5 (150 × 1.0 mm i.d., 5 μm d.p) ^{2}D: Ascentis Express C18 (30 × 4.6 mm, 2.7 μm d.p)	^{1}D: (A) ACN (2% CH_3COOH); (B) MeOH:H_2O:CH_3COOH (95:3:2 v/v) ^{2}D: (A) H_2O (0.05% TFA); (B) ACN (0.05% TFA)	Both	UV-Vis; MS/MS	Green cocoa beans; flavan-3-ol monomers and procyanidin oligomers	[36]
2018	HILIC × RP	^{1}D: Develosil 100 Diol column (250 × 1.0 mm i.d., 5 μm d.p) ^{2}D: Kinetex C18 (50 × 4.6 mm i.d., 2.6 μm d.p)	^{1}D: (A) ACN (99:1, v/v); (B) MeOH:H_2O:CH_3COOH (94.05:4.95:1, v/v/v) ^{2}D: (A) H_2O (0.1% HCOOH); (B) ACN (0.1% HCOOH)	Both	UV-Vis	Cocoa extract; procyanidins isomers	[37]
2018	RP × RP	^{1}D: Zorbax Eclipse Plus C18 (50 × 2.1 mm i.d., 1.8 μm d.p) ^{2}D: Chromolith Performance NH_2 (10 × 4.6 mm)	^{1}D: (A) H_2O (0.1% HCOOH); (B) MeOH ^{2}D: (A) 20 mM $HCOONH_4$ (pH = 4.3); (B) ACN	Both	UV-Vis	standards; phenolic and flavone antioxidants	[12]
2018	RP × RP	^{1}D: Waters HSS Cyano (150 × 1 mm i.d., 1.8 μm d.p) ^{2}D: Waters Phenyl column (50 × 1 mm i.d., 1.7 μm d.p)	^{1}D: (A) H_2O:ACN (90:10 v/v) (0.1% HCOOH); (B) H_2O:ACN (10:90 v/v) (0.1% HCOOH) ^{2}D: (A) H_2O:ACN (90:10 v/v) (0.1% HCOOH); (B) H_2O:ACN (10:90 v/v) (0.1% HCOOH)	Both	MS	Bovine urine; residues of sulfonamides, beta-agonists, and steroids	[43]
2018	RP × RP	^{1}D: Ascentis Cyano column (150 × 1.0 mm i.d., 2.7 μm d.p) ^{2}D: Ascentis Express C18 column (50 × 2.1 mm i.d., 2.7 μm d.p)	^{1}D: (A) H_2O (0.1% CH_3COOH); (B) ACN (0.1% CH_3COOH) ^{2}D: (A) H_2O (0.1% CH_3COOH); (B) ACN (0.1% CH_3COOH)	Both	UV-Vis; MS/MS	Liquorice; metabolic profiling	[44]
2018	RP × RP	^{1}D: Ascentis Express C8 column (150 × 2.1 mm i.d., 2.7 μm d.p) ^{2}D: Cyano column Zorbax SB-CN (30 × 4.6 mm i.d., 1.8 μm d.p)	^{1}D: (A) H_2O (0.1% HCOOH); (B) MeOH (0.1% HCOOH) ^{2}D: (A) H_2O (0.1% HCOOH); (B) ACN (0.1% HCOOH) 1-butanol	Both	HRMS	Bovine urine; hydrophobic compounds; anabolic-steroid residues	[41]

(continued)

Table 3
(continued)

Year	Separation	Columns	Mobile phase solvents	Gradient	Detection	Sample and analyte	Ref.
2018	Chiral × Chiral	lab made	^{1}D: MeOH:CH_3COOH:$HCOONH_4$ (98:2:0.5, v/v/w) ^{1}D: MeOH:CH_3COOH:$HCOONH_4$ (98:2:0.5, v/v/w)	None	UV-Vis	Peptide antibiotic drugs gramicidin and bacitracin; amino acids	[35]
2017	HILIC × RP	^{1}D: HILIC TSK gel amide-80 column (250 mm × 2.0 mm i.d., 5 μm d.p) ^{2}D: Kinetex C18 (50 mm × 2.1 mm i.d., 1.7 μm d.p)	^{1}D: (A) ACN; (B) 5 mM CH_3COONH_4 pH = 5.5 (CH_3COOH) ^{2}D: (A) H_2O (0.1% HCOOH); (B) ACN (0.1% HCOOH)	Both	HRMS	Rice; metabolite profiling	[19]
2017	HILIC × RP	^{1}D: Luna HILIC (150 mm × 2.0 mm i. d., 3.0 μm d.p) ^{2}D: Titan™C18 (50 mm × 3.0 mm i.d., 1.9 μm d.p)	^{1}D: (A) H_2O:ACN (80:20 v/v) (0.1% CH_3COOH); (B) ACN (0.1% CH_3COOH) ^{2}D: (A) H_2O (0.1% CH_3COOH); (B) ACN (0.1% CH_3COOH)	Both	MS/MS	Typical Italian apple variety; polyphenols	[38]
2016	HILIC × RP	^{1}D: SeQuant ZIC HILIC (150 × 1.0 mm i.d., 3.5 μm d.p) ^{2}D: Ascentis Express C18 (50 × 4.6 mm i.d., 2.7 μm d.p)	^{1}D: (A) ACN; (B) H_2O:10 mM $HCOONH_4$ (pH = 5.0) ^{2}D: (A) H_2O (0.1% HCOOH); (B) ACN	Both	MS/MS	Roots licorice; metabolic profiling	[39]

2016	RP × RP	^{1}D: COSMOSIL silica-based reversed phase (250 mm × 4.6 mm i.d., 5 μm d.p) ^{2}D: COSMOSIL C18-MS-II UPLC (50 mm × 30 mm i.d., 2.6 μm d.p)	^{1}D: (A) H_2O; (B) MeOH ^{2}D: (A) H_2O; (B) ACN	Both	HRMS	Rice medium of *C. globosum* SNSHI-5; microbial metabolites	[45]
2016	RP × RP	^{1}D: monolithic zwitterionic polymethacrylate 0.53 mm i.d. BIGDMA–MEDSA microcolumn ^{2}D: Five short commercial core–shell or silica-based monolithic columns/three commercial silica-based monolithic columns with octadecyl stationary phases	^{1}D: (A) H_2O:10 mM $HCOONH_4$ (pH = 3.1 HCOOH); (B) ACN:10 mM $HCOONH_4$ (pH = 3.1 HCOOH) ^{2}D: (A) H_2O:10 mM $HCOONH_4$ (pH = 3.1 HCOOH); (B) ACN:10 mM $HCOONH_4$ (pH = 3.1 HCOOH)	Both	UV-Vis	Standards; flavones and related polyphenolic compounds	[46]
2016	RP × RP	^{1}D: Brownlee Choice C18 (50 mm × 2.1 mm i.d., 5 μm d.p) ^{2}D: Xbridge Shield C18 (50 mm × 2.1 mm i.d., 5 μm d.p)	^{1}D: H_2O:MeOH (87:13, v/v) (0.1% CH_3COOH) ^{2}D: (A) H_2O:MeOH (90:10, v/v) (0.1% HCOOH)	None	UV-Vis	Standards; formic acid, benzyl alcohol, catechol, vanillin, and guaiacol	[47]

number of potential setup combinations, allowing for an expanded number of applications either in a targeted or an untargeted manner. Finally, it is worth to mention the complementarity of the information provided by the same stationary phases depending on the order used in the analysis. A clear example in lipidomics is the possibility of obtaining the primary separation on the lipid class or the chain length/hydrogen bonds depending on the first chromatographic mode (HILIC or RP, respectively) which can be a significant help for the compound identification.

However, there are still some weaknesses that should be improved to achieve a major popularization in omics communities. On the one hand, a simplification in the required instrumentation will ease its use for new research groups, which could be promoted by vendor companies. New developments in data mining tools could help the automation of the entire pipeline, which could also favor the use of these techniques in routine analysis. On the other hand, more work has to be done in order to improve experimental factors such as the analysis time, because it can limit the use of 2D-LC in high-throughput studies. If these limitations can be overcome, a successful future can be expected for the 2D-LC.

Acknowledgments

The research leading to these results has received funding from the Spanish Ministry of Science and Innovation (MCI, Grant CTQ2017-82598-P). MPC acknowledges a predoctoral FPU 16/02640 scholarship from Spanish Ministry of Education and Vocational Training (MEFP).

References

1. Zhang H, Jiang JM, Zheng D et al (2019) A multidimensional analytical approach based on time-decoupled online comprehensive two-dimensional liquid chromatography coupled with ion mobility quadrupole time-of-flight mass spectrometry for the analysis of ginsenosides from white and red ginsengs. J Pharm Biomed Anal 163:24–33. https://doi.org/10.1016/j.jpba.2018.09.036
2. Sarrut M, Corgier A, Crétier G et al (2015) Potential and limitations of on-line comprehensive reversed phase liquid chromatography×supercritical fluid chromatography for the separation of neutral compounds: An approach to separate an aqueous extract of bio-oil. J Chromatogr A 1402:124–133. https://doi.org/10.1016/j.chroma.2015.05.005
3. Stoll DR, Carr PW (2017) Two-dimensional liquid chromatography: a state of the art tutorial. Anal Chem 89(1):519–531. https://doi.org/10.1021/acs.analchem.6b03506
4. Pirok BWJ, Stoll DR, Schoenmakers PJ (2019) Recent developments in two-dimensional liquid chromatography – fundamental improvements for practical applications. Anal Chem 91(1):240–263. https://doi.org/10.1021/acs.analchem.8b04841
5. Pirok BWJ, Gargano AFG, Schoenmakers PJ (2018) Optimizing separations in online comprehensive two-dimensional liquid chromatography. J Sep Sci 41:68–98. https://doi.org/10.1002/jssc.201700863
6. Marriott PJ, Wu Z-Y, Schoenmakers P (2012) Nomenclature and conventions in

comprehensive multidimensional chromatography – an update. Chromatogr Online 25 (5):266–275

7. Cheng C, Liao CF (2018) Novel dual two-dimensional liquid chromatography online coupled to ultraviolet detector, fluorescence detector, ion-trap mass spectrometer for short peptide amino acid sequence determination with bottom-up strategy. J Chin Chem Soc 65:714–725. https://doi.org/10.1002/jccs.201700380
8. Pandohee J, Stevenson P, Zhou X-R et al (2015) Multi-dimensional liquid chromatography and metabolomics, are two dimensions better than one? Curr Metabolomics 3:10–20. https://doi.org/10.2174/2213235X03666150403231202
9. Kalili KM, De Villiers A (2013) Systematic optimisation and evaluation of on-line, off-line and stop-flow comprehensive hydrophilic interaction chromatography × reversed phase liquid chromatographic analysis of procyanidins, Part I: theoretical considerations. J Chromatogr A 1289:58–68. https://doi.org/10.1016/j.chroma.2013.03.008
10. Bedani F, Kok WT, Janssen HG (2006) A theoretical basis for parameter selection and instrument design in comprehensive size-exclusion chromatography × liquid chromatography. J Chromatogr A 1133:126–134. https://doi.org/10.1016/j.chroma.2006.08.048
11. Striegel AM (2001) Longitudinal diffusion in size-exclusion chromatography: a stop-flow size-exclusion chromatography study. J Chromatogr A 932:21–31. https://doi.org/10.1016/S0021-9673(01)01214-6
12. Donato P, Rigano F, Cacciola F et al (2016) Comprehensive two-dimensional liquid chromatography–tandem mass spectrometry for the simultaneous determination of wine polyphenols and target contaminants. J Chromatogr A 1458:54–62. https://doi.org/10.1016/j.chroma.2016.06.042
13. Vivo G, Van Der Wal S, Schoenmakers PJ (2010) Comprehensive study on the optimization of online two-dimensional liquid chromatographic systems considering losses in theoretical peak capacity in first- and second-dimensions: a pareto-optimality approach. Analysis Anal Chem 82(20):3090–3100. https://doi.org/10.1021/ac101420f
14. Pirok BWJ, Pous-Torres S, Ortiz-Bolsico C et al (2016) Program for the interpretive optimization of two-dimensional resolution. J Chromatogr A 1450:29–37. https://doi.org/10.1016/j.chroma.2016.04.061
15. Stoll DR, Shoykhet K, Petersson P, Buckenmaier S (2017) Active solvent modulation: a valve-based approach to improve separation compatibility in two-dimensional liquid chromatography. Anal Chem 89:9260–9267. https://doi.org/10.1021/acs.analchem.7b02046
16. Vonk RJ, Gargano AFG, Davydova E et al (2015) Comprehensive two-dimensional liquid chromatography with stationary-phase-assisted modulation coupled to high-resolution mass spectrometry applied to proteome analysis of saccharomyces cerevisiae. Anal Chem 87:5387–5394. https://doi.org/10.1021/acs.analchem.5b00708
17. Tian H, Xu J, Xu Y, Guan Y (2006) Multidimensional liquid chromatography system with an innovative solvent evaporation interface. J Chromatogr A 1137:42–48. https://doi.org/10.1016/j.chroma.2006.10.005
18. Navarro-Reig M, Jaumot J, Tauler R (2018) An untargeted lipidomic strategy combining comprehensive two-dimensional liquid chromatography and chemometric analysis. J Chromatogr A 1568:80–90. https://doi.org/10.1016/j.chroma.2018.07.017
19. Navarro-Reig M, Jaumot J, Baglai A et al (2017) Untargeted comprehensive two-dimensional liquid chromatography coupled with high-resolution mass spectrometry analysis of rice metabolome using multivariate curve resolution. Anal Chem 89:7675–7683. https://doi.org/10.1021/acs.analchem.7b01648
20. Donato P, Micalizzi G, Oteri M et al (2018) Comprehensive lipid profiling in the Mediterranean mussel (Mytilus galloprovincialis) using hyphenated and multidimensional chromatography techniques coupled to mass spectrometry detection. Anal Bioanal Chem 410:3297–3313. https://doi.org/10.1007/s00216-018-1045-3
21. Berkecz R, Tömösi F, Körmöczi T et al (2018) Comprehensive phospholipid and sphingomyelin profiling of different brain regions in mouse model of anxiety disorder using online two-dimensional (HILIC/RP)-LC/MS method. J Pharm Biomed Anal 149:308–317. https://doi.org/10.1016/j.jpba.2017.10.043
22. Sun C, Zhao YY, Curtis JM (2015) Characterization of phospholipids by two-dimensional liquid chromatography coupled to in-line ozonolysis-mass spectrometry. J Agric Food Chem 63:1442–1451. https://doi.org/10.1021/jf5049595
23. Holčapek M, Ovčačíková M, Lísa M et al (2015) Continuous comprehensive two-dimensional liquid chromatography-

electrospray ionization mass spectrometry of complex lipidomic samples. Anal Bioanal Chem 407:5033–5043. https://doi.org/10.1007/s00216-015-8528-2

24. Bang DY, Moon MH (2013) On-line two-dimensional capillary strong anion exchange/reversed phase liquid chromatography-tandem mass spectrometry for comprehensive lipid analysis. J Chromatogr A 1310:82–90. https://doi.org/10.1016/j.chroma.2013.08.069
25. Yang L, Lv P, Ai W et al (2017) Lipidomic analysis of plasma in patients with lacunar infarction using normal-phase/reversed-phase two-dimensional liquid chromatography–quadrupole time-of-flight mass spectrometry. Anal Bioanal Chem 409:3211–3222. https://doi.org/10.1007/s00216-017-0261-6
26. Baglai A, Gargano AFG, Jordens J et al (2017) Comprehensive lipidomic analysis of human plasma using multidimensional liquid- and gas-phase separations: two-dimensional liquid chromatography–mass spectrometry vs. liquid chromatography–trapped-ion-mobility–mass spectrometry. J Chromatogr A 1530:90–103. https://doi.org/10.1016/j.chroma.2017.11.014
27. Brouwers JF (2011) Liquid chromatographic-mass spectrometric analysis of phospholipids. Chromatography, ionization and quantification. Biochim Biophys Acta Mol Cell Biol Lipids 1811:763–775. https://doi.org/10.1016/j.bbalip.2011.08.001
28. D'Attoma A, Grivel C, Heinisch S (2012) On-line comprehensive two-dimensional separations of charged compounds using reversed-phase high performance liquid chromatography and hydrophilic interaction chromatography. Part I: orthogonality and practical peak capacity consideratio. J Chromatogr A 1262:148–159. https://doi.org/10.1016/j.chroma.2012.09.028
29. Yang Q, Shi X, Gu Q et al (2012) On-line two dimensional liquid chromatography/mass spectrometry for the analysis of triacylglycerides in peanut oil and mouse tissue. J Chromatogr B Anal Technol Biomed Life Sci 895–896:48–55. https://doi.org/10.1016/j.jchromb.2012.03.013
30. Yang L, Cui X, Zhang N et al (2015) Comprehensive lipid profiling of plasma in patients with benign breast tumor and breast cancer reveals novel biomarkers. Anal Bioanal Chem 407:5065–5077. https://doi.org/10.1007/s00216-015-8484-x
31. Sun C, Zhao YY, Curtis JM (2014) Elucidation of phosphatidylcholine isomers using two dimensional liquid chromatography coupled in-line with ozonolysis mass spectrometry. J Chromatogr A 1351:37–45. https://doi.org/10.1016/j.chroma.2014.04.069
32. Li M, Tong X, Lv P et al (2014) A not-stop-flow online normal-/reversed-phase two-dimensional liquid chromatography-quadrupole time-of-flight mass spectrometry method for comprehensive lipid profiling of human plasma from atherosclerosis patients. J Chromatogr A 1372:110–119. https://doi.org/10.1016/j.chroma.2014.10.094
33. Li M, Feng B, Liang Y et al (2013) Lipid profiling of human plasma from peritoneal dialysis patients using an improved 2D (NP/RP) LC-QToF MS method. Anal Bioanal Chem 405:6629–6638. https://doi.org/10.1007/s00216-013-7109-5
34. Krastanov A (2010) Metabolomics - the state of art. Biotechnol Biotechnol Equip 24:1537–1543. https://doi.org/10.2478/V10133-010-0001-y
35. Woiwode U, Reischl RJ, Buckenmaier S et al (2018) Imaging peptide and protein chirality via amino acid analysis by chiral × chiral two-dimensional correlation liquid chromatography. Anal Chem 90:7963–7971. https://doi.org/10.1021/acs.analchem.8b00676
36. Toro-Uribe S, Montero L, López-Giraldo L et al (2018) Characterization of secondary metabolites from green cocoa beans using focusing-modulated comprehensive two-dimensional liquid chromatography coupled to tandem mass spectrometry. Anal Chim Acta 1036:204–213. https://doi.org/10.1016/j.aca.2018.06.068
37. Muller M, Tredoux AGJ, de Villiers A (2018) Predictive kinetic optimisation of hydrophilic interaction chromatography × reversed phase liquid chromatography separations: Experimental verification and application to phenolic analysis. J Chromatogr A 1571:107–120. https://doi.org/10.1016/j.chroma.2018.08.004
38. Sommella E, Ismail OH, Pagano F et al (2017) Development of an improved online comprehensive hydrophilic interaction chromatography × reversed-phase ultra-high-pressure liquid chromatography platform for complex multiclass polyphenolic sample analysis. J Sep Sci 40:2188–2197. https://doi.org/10.1002/jssc.201700134
39. Montero L, Ibáñez E, Russo M et al (2016) Metabolite profiling of licorice (Glycyrrhiza glabra) from different locations using comprehensive two-dimensional liquid chromatography coupled to diode array and tandem mass spectrometry detection. Anal Chim Acta

913:145–159. https://doi.org/10.1016/j.aca.2016.01.040

40. Dugo P, Herrero M, Kumm T et al (2008) Comprehensive normal-phase × reversed-phase liquid chromatography coupled to photodiode array and mass spectrometry detection for the analysis of free carotenoids and carotenoid esters from mandarin. J Chromatogr A 1189:196–206. https://doi.org/10.1016/j.chroma.2007.11.116
41. Baglai A, Blokland MH, Mol HGJ et al (2018) Enhancing detectability of anabolic-steroid residues in bovine urine by actively modulated online comprehensive two-dimensional liquid chromatography – high-resolution mass spectrometry. Anal Chim Acta 1013:87–97. https://doi.org/10.1016/j.aca.2017.12.043
42. Wang Y, Lu X, Xu G (2008) Development of a comprehensive two-dimensional hydrophilic interaction chromatography/quadrupole time-of-flight mass spectrometry system and its application in separation and identification of saponins from Quillaja saponaria. J Chromatogr A 1181:51–59. https://doi.org/10.1016/j.chroma.2007.12.034
43. Blokland MH, Zoontjes PW, Van Ginkel LA et al (2018) Multiclass screening in urine by comprehensive two-dimensional liquid chromatography time of flight mass spectrometry for residues of sulphonamides, beta-agonists and steroids. Food Addit Contam Part A Chem Anal Control Expo Risk Assess 35:1703–1715. https://doi.org/10.1080/19440049.2018.1506160
44. Wong YF, Cacciola F, Fermas S et al (2018) Untargeted profiling of Glycyrrhiza glabra extract with comprehensive two-dimensional liquid chromatography-mass spectrometry using multi-segmented shift gradients in the second dimension: Expanding the metabolic coverage. Electrophoresis 39:1993–2000. https://doi.org/10.1002/elps.201700469
45. Yan X, Wang LJ, Wu Z et al (2016) New on-line separation workflow of microbial metabolites via hyphenation of analytical and preparative comprehensive two-dimensional liquid chromatography. J Chromatogr B Anal Technol Biomed Life Sci 1033–1034:1–8. https://doi.org/10.1016/j.jchromb.2016.07.053
46. Hájek T, Jandera P, Staňková M, Česla P (2016) Automated dual two-dimensional liquid chromatography approach for fast acquisition of three-dimensional data using combinations of zwitterionic polymethacrylate and silica-based monolithic columns. J Chromatogr A 1446:91–102. https://doi.org/10.1016/j.chroma.2016.04.007
47. Corgier A, Sarrut M, Crétier G, Heinisch S (2016) Potential of online comprehensive two-dimensional liquid chromatography for micro-preparative separations of simple samples. Chromatographia 79:255–260. https://doi.org/10.1007/s10337-015-3012-x
48. Hájek T, Jandera P, Staňková M, Česla P (2016) Automated dual two-dimensional liquid chromatography approach for fast acquisition of three-dimensional data using combinations of zwitterionic polymethacrylate and silica-based monolithic columns. J Chromatogr A 1446:91–102. https://doi.org/10.1016/j.chroma.2016.04.007

Chapter 4

Metabolomics of Small Numbers of Cells Using Chemical Isotope Labeling Combined with Nanoflow LC-MS

Xian Luo and Liang Li

Abstract

Performing metabolome analysis using a small number of cells is desirable and required in many areas of biological research. However, it presents a major analytical challenge to detect and quantify a large number of metabolites for untargeted metabolome analysis due to the low amount of metabolites in a few cells. In this chapter, we describe a highly sensitive workflow for the analysis of thousands of metabolites in hundreds or thousands of cultured mammalian cells, such as those in cancer cell lines. A rationally designed chemical labeling reagent is used to target a chemical-group-based submetabolome to improve both metabolite separation and ionization. The labeled samples are subjected to nanoflow LC-MS for sensitive detection.

Key words Chemical labeling, Isotope reagent, Relative quantification, LC-MS, Metabolomics, Cells

1 Introduction

Untargeted metabolomics involves the detection of as many metabolites as possible, ideally the whole set of metabolites or the metabolome. It also requires the accurate measurement of concentration differences of individual metabolites between comparative samples. This requirement of high-coverage quantitative metabolome analysis presents a huge analytical challenge. Our group has been working on the development of a high-performance chemical isotope labeling (CIL) LC-MS platform to address this challenge. It is based on a divide-and-conquer strategy whereby the whole metabolome is divided into several chemical-group-based submetabolomes and in-depth analysis of each is performed separately.

To analyze a specific submetabolome (e.g., amine submetabolome), a high-performance derivatization chemistry is used to react with metabolites containing the common chemical group (e.g., amine), followed by LC-MS analysis of the labeled products. By high-performance, we mean the structure of the labeling reagent is rationally designed in order to alter the chemical and physical properties of the metabolites, after labeling, to such an extent that

Paul L. Wood (ed.), *Metabolomics*, Neuromethods, vol. 159, https://doi.org/10.1007/978-1-0716-0864-7_4,

we can separate and detect the labeled metabolites with high efficiency and sensitivity. Over the years, we have reported reagents of different chemistries that meet these criteria, including dansyl chloride for amine and phenol submetabolome profiling [1], base-activated dansyl chloride for hydroxyl submetabolome profiling [2], dansyl hydrazine for carbonyl submetabolome profiling [3], and DmPA bromide for carboxylic acid submetabolome profiling [4]. In each case, accurate relative quantification is achieved using differential isotope labeling where a light-isotope-reagent (e.g., ^{12}C-) is used to label the individual samples and a heavy-isotope-reagent (e.g., ^{13}C-) is used to label a reference sample (e.g., a pooled sample of the same type), followed by equal mole mixing of light-labeled-sample and heavy-labeled-reference and LC-MS analysis of the individual mixtures.

Although CIL LC-MS has been successfully applied for metabolome analysis of many different types of samples such as plasma [5], urine [6], sweat [7] and cell cultures [8, 9], there is a need for developing and applying even more sensitive methods to handle samples with limited amounts of starting material such as metabolite extracts from a small number of cells (e.g., $<10,000$ cells). To achieve this, we have combined CIL with nanoflow LC-MS to reduce the amount of sample needed for metabolome analysis. Nanoflow LC-MS is an analytical platform widely used in proteomics and thus may be readily accessible to many researchers. In shotgun proteomics, researchers inject a relatively large volume of sample (several to tens of microliters) prepared from a protein or proteome digest. In order to do this, a short reversed-phase (RP) trap column is often used to focus the sample before flowing into the analytical column for separation and detection. However, because of the presence of ionic, hydrophilic, and hydrophobic metabolites in a metabolome sample, using the RP trap, only the hydrophobic metabolites will be retained. This limitation is largely overcome by applying dansyl or DmPA labeling as the labeled metabolites are all relatively hydrophobic and therefore can be trapped in the RP trap column. The combined benefits of chemical labeling for improving metabolite detectability and nanoLC-MS for handling small amounts of sample make CIL nanoLC-MS a viable platform for performing metabolome analysis of samples with a limited amount of starting material. We have recently demonstrated its applications in metabolomic profiling of sweat [10], small numbers of cells [11], and serum exosomes [12].

In this Chapter, we describe the experimental protocol for metabolome analysis of small numbers of cells, based on the CIL nanoLC-MS workflow as shown in Fig. 1, with a focus of using dansylation for labeling the amine/phenol submetabolome. In a typical workflow, the cells are harvested from cell cultures and lysed, followed by metabolite extraction. Each individual extract or sample is labeled by ^{12}C-dansyl chloride, and a pooled sample is labeled

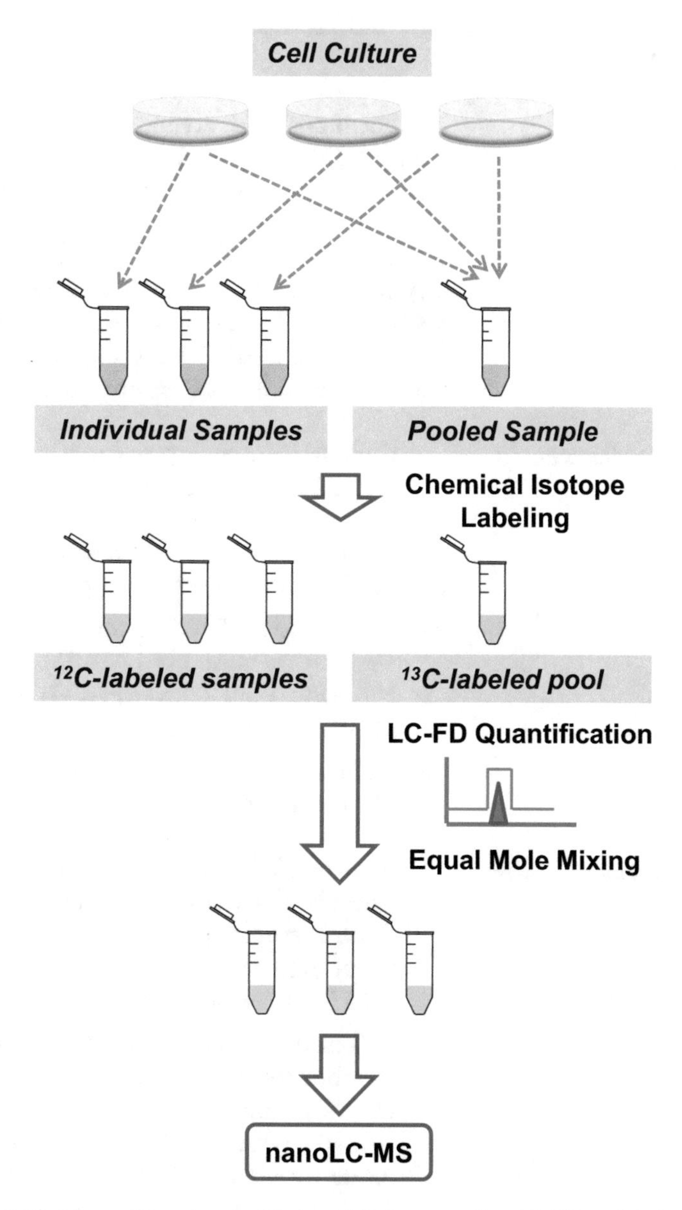

Fig. 1 Workflow of the chemical isotope labeling nanoflow LC-MS metabolomics platform

by ^{13}C-dansyl chloride. The labeled cell lysates are quantified by liquid chromatography equipped with a fluorescent detector (LC-FD) for sample normalization [13]. Equal moles of ^{12}C-labeled sample and ^{13}C-labeled pool are mixed together and injected onto nanoLC-MS for analysis.

2 Materials

2.1 Chemicals and Reagents

All chemicals and reagents are purchased from Sigma-Aldrich Canada (Markham, ON, Canada) unless otherwise indicated. Vials used for dansylation reaction are 0.6 mL Eppendorf Safe-Lock microcentrifuge tubes (Mississauga, ON, Canada). The MS-Pure grade ^{12}C- and ^{13}C-dansyl chloride reagents are purchased from Nova Medical Testing Inc. (Edmonton, AB, Canada). LC-MS grade water, acetonitrile, methanol, and formic acid are purchased from Fisher Scientific Canada (Ottawa, ON, Canada). All solutions are prepared in LC-MS grade solvent.

2.2 Cell Culture

1. Cells: MCF-7 breast cancer cells (ATCC HTB-22) are cultured in DMEM growth medium, supplemented with 10% fetal bovine serum (FBS). The culture plates are incubated at 37 °C in a humidified atmosphere with 5% CO_2. The growth medium is renewed every 2 days.
2. Phosphate-buffered saline (PBS): 1× PBS is prepared for cell washing. It contains 137 mM NaCl, 2.7 mM KCl, 10 mM Na_2HPO_4, and 2 mM KH_2PO_4 with pH 7.4. PBS is autoclaved for sterilization before using.
3. Cell lysis solvent: 1:1 (v/v) water: methanol is used for cell lysis and metabolite extraction.

2.3 Chemical Isotope Labeling

1. 0.25 mg/mL dansyl chloride solution: Weigh 10 mg dansyl chloride in a vial using an analytical balance and dissolve it in 1 mL acetonitrile. Transfer 10 μL 10 mg/mL dansyl chloride solution to another vial, and add 390 μL acetonitrile to prepare a 0.25 mg/mL dansyl chloride solution.
2. 250 mM Na_2CO_3–$NaHCO_3$ buffer: Dissolve 16.8 g sodium bicarbonate and 5.4 g anhydrous sodium carbonate in 800 mL water in a beaker. Transfer the solution to a volumetric flask and add water till 1 L. The final pH of Na_2CO_3–$NaHCO_3$ buffer should be around 9.4.
3. 250 mM NaOH solution: Transfer 25 mL 1 M NaOH standard solution in 100 mL volumetric flask. Dilute NaOH solution to 100 mL by water and transfer it to a plastic bottle.
4. 425 mM formic acid solution: Dilute 160 μL of formic acid to 10 mL by 1:1 (v/v) acetonitrile–water.

2.4 LC-FD

1. Mobile Phase A: 0.1% formic acid in 95:5 (v/v) water–acetonitrile (*see* **Notes 1** and **2**).
2. Mobile Phase B: 0.1% formic acid in 5:95 (v/v) water–acetonitrile (*see* **Note 3**).

3. LC Column: Agilent Poreshell 120 EC-C18 column (2.1 × 30 mm, 2.7 μm) (Santa Clara, CA).
4. Amino acid standard (AAS) solution: AAS solution is available from Sigma-Aldrich Canada. It contains 17 amino acids with individual concentrations of 2.5 mM, with the exception of 1.25 mM for L-cystine.

2.5 Nanoflow LC-MS

1. Mobile Phase A: 0.1% formic acid in 95:5 (v/v) water–acetonitrile (*see* **Note 1**).
2. Mobile Phase B: 0.1% formic acid in 5:95 (v/v) water–acetonitrile (*see* **Note 1**).
3. LC Trap Column: Thermo Scientific Acclaim PepMap 100 trap column (75 μm × 20 mm, 3 μm) (Sunnyvale, CA).
4. LC Analytical column: Thermo Scientific Acclaim PepMap RSLC C18 column (75 μm × 150 mm, 2 μm) (Sunnyvale, CA).

3 Method

3.1 Cell Harvest, Lysis, and Metabolite Extraction

The workflow of handling cell cultures for CIL nanoLC-MS metabolomics is shown in Fig. 2.

1. Remove growth medium by aspiration.
2. Wash cell culture two times by precooled PBS (4 °C) to remove residual growth medium and extracellular metabolites.
3. Add precooled (−20 °C) methanol to quench cellular metabolism.
4. Scrape and detach cells from culture dish by cell scrapers (*see* **Note 4**).
5. Transfer detached cells and solvent into vials and dry down. Store cell pellets in −80 °C freezer if they are not processed immediately.
6. For cell lysis, add 50 μL lysis solvent into cell pellets. Freeze cells in liquid nitrogen and thaw them on ice-water bath for five rounds to disrupt cells and release intracellular metabolites completely.
7. Centrifuge vials at 16,000 × *g* at 4 °C to remove cell debris, and transfer the supernatant to a new vial and dry down.

3.2 Dansylation Labeling

The dansylation labeling scheme is shown in Fig. 3.

1. Add 7.5 μL Na_2CO_3–$NaHCO_3$ buffer to redissolve metabolite extracts.
2. Add 7.5 μL ^{12}C-dansyl chloride solution for light labeling or ^{13}C-dansyl chloride for heavy labeling.

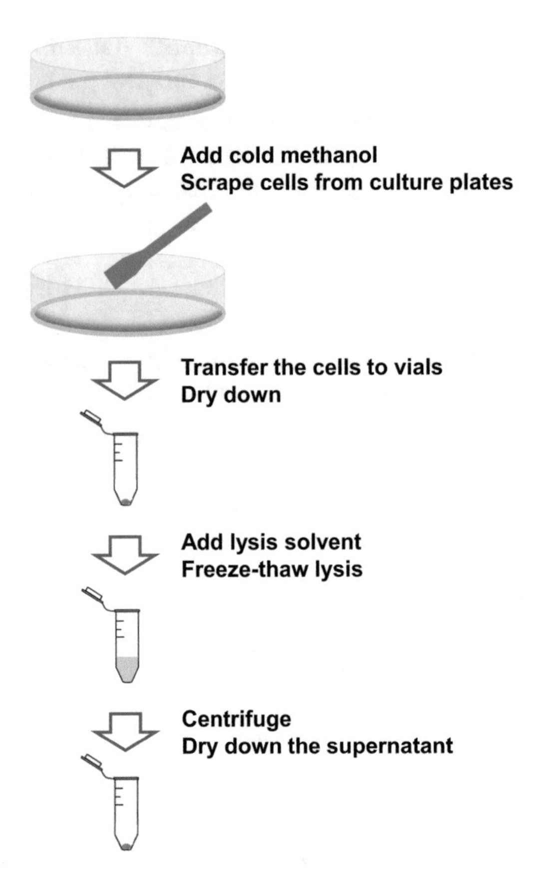

Fig. 2 Workflow of handling cell cultures for CIL nanoLC-MS metabolomics

3. Incubate the mixture for 1 h at 40 °C.
4. Cool down the reaction mixture on ice-water bath. Add 1 μL NaOH and then incubate the vials for another 10 min to quench the reaction by consuming excess dansyl chloride.
5. Add 5 μL formic acid solution to consume excess NaOH.

3.3 Sample Normalization

3.3.1 LC-FD

Agilent 1220 HPLC system equipped with a fluorescent detector is used for measuring the total concentration of labeled metabolites for sample normalization. A step gradient is used to rapidly separate the dansyl hydroxyl and dansyl labeled metabolites (*see* Fig. 4). Chromatographic condition: $t = 0$ min, 100% A; $t = 0.5$ min, 100% A; $t = 0.51$ min, 100% B; $t = 2.5$ min, 100% B; $t = 2.51$ min, 100% A; $t = 6$ min, 100% A. Flow rate: 0.45 mL/min; Excitation wavelength: 250 nm; Emission wavelength: 520 nm; Detector gain: 18.

3.3.2 Calibration Curve from Amino Acid Standard

1. Prepare a series of amino acid standard solutions with different concentrations (*see* **Note 4**).
2. Transfer 5 μL of the amino acid standard solution to a vial and add 2.5 μL Na_2CO_3–$NaHCO_3$ buffer.

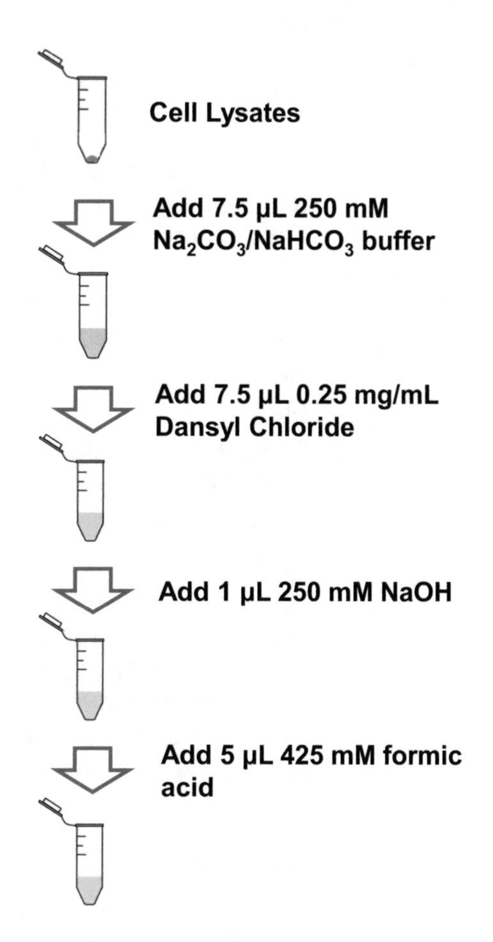

Fig. 3 Dansylation labeling scheme

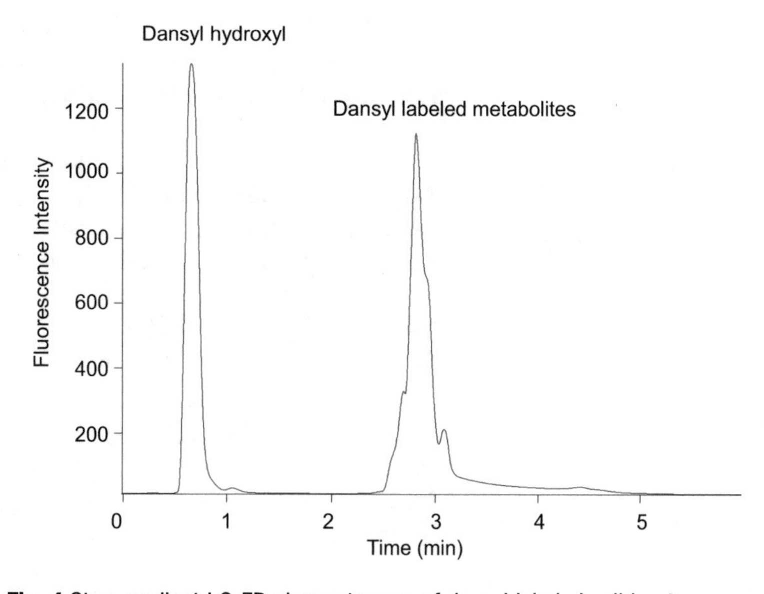

Fig. 4 Step-gradient LC-FD chromatogram of dansyl labeled cell lysates

3. Label the AAS solutions by following **steps 2–5** of Subheading 3.2.
4. Inject 1 μL labeled AAS solutions onto LC-FD, and integrate the peak areas of the labeled amino acids. Plot peak areas as a function of total concentration of amino acids in AAS to create a calibration curve.

3.3.3 Calibration Curve from Cell Lysates

1. Prepare metabolite extracts from a large number of cells (e.g., 10^5 cells) by following Subheading 3.1.
2. Redissolve the metabolite extracts by Na_2CO_3–$NaHCO_3$ buffer.
3. The extracts are diluted to the equivalents of 10,000, 5000, 2000, 1000, 500, 200, 100 cells.
4. Label each individual sample by following **steps 2–5** of Subheading 3.2.
5. Inject 1 μL labeled cell lysates onto LC-FD, and integrate the peak areas of labeled metabolites. The total concentration of 10,000 cell lysates is calculated by using the AAS calibration curve created in Subheading 3.3.2, and the concentrations of other cell lysates is calculated by multiplying their dilution factors. Plot peak areas as a function of concentration of cell lysates to create the calibration curve for cell lysates.

3.3.4 Sample Normalization

1. Inject 1 μL ^{12}C- or ^{13}C- labeled cell lysates onto LC-FD.
2. Integrate the peak area and calculate the concentration of each sample based on the calibration curve created in Subheading 3.3.3 (*see* **Note 5**).
3. Mix the ^{12}C- and ^{13}C-labeled cell lysates by equal mole amount. Dry down and redissolve by adding 9:1 (v/v) water–acetonitrile before injection onto nanoLC-MS.

3.4 Nanoflow LC-MS

1. Waters NanoAcquity UPLC system (Milford, MA, USA) is used for separation. The sample is loaded onto a trap column connected to an analytical column. Trapping condition: 2 min, 5 μL/min with 98% mobile phase A. The chromatographic condition: $t = 0$ min, 15% B; $t = 2.0$ min, 15% B; $t = 4.0$ min, 25% B; $t = 24$ min, 60% B; $t = 28$ min, 99% B, $t = 45$ min, 99% B. The flow rate is 350 nL/min.
2. A Bruker Daltonics Impact HD Q-TOF mass spectrometer (Billerica, MA, USA) equipped with a CaptiveSpray nanoBooster ion source is used for sample analysis. The MS condition: dry temperature, 200 °C; dry gas, 3 L/min; capillary voltage, 1400 V; nanoBooster, 0.2 bar; dopant gas, pure ACN.

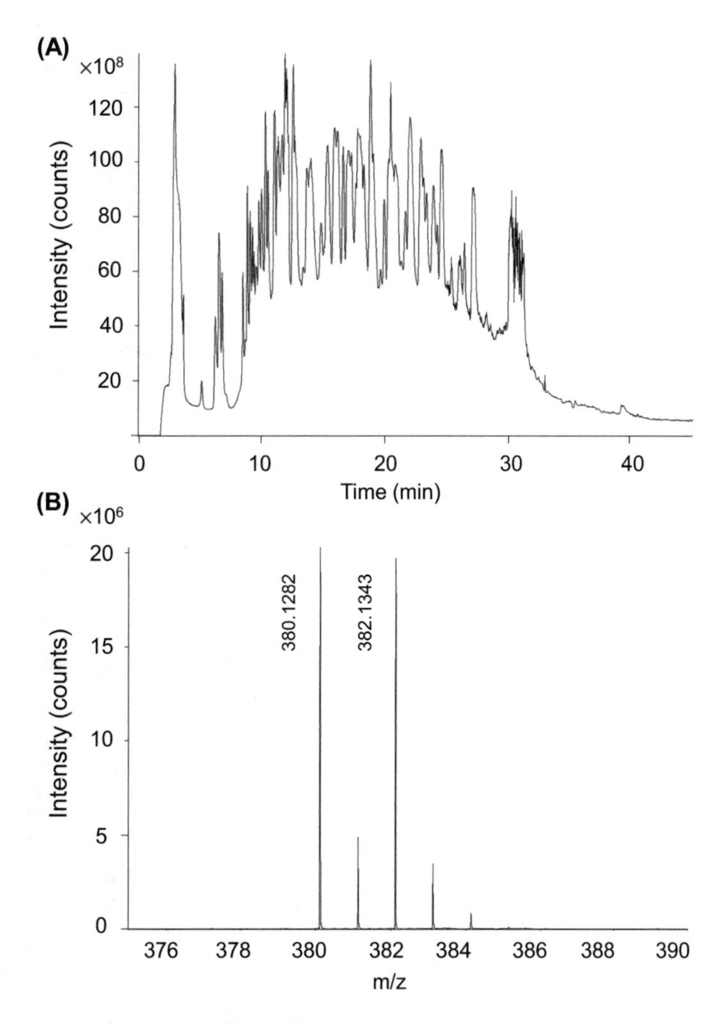

Fig. 5 (**a**) Total ion chromatogram of dansyl labeled cell lysates. (**b**) Representative mass spectrum showing the molecular ion region of labeled glutamine

3. 5 μL of dansyl labeled cell lysates was injected into nanoLC-MS for analysis. Figure 5a shows the total ion chromatogram from dansyl labeled cell lysates. The peak pair of dansyl labeled glutamine is shown in Fig. 5b as an example for a representative mass spectrum.

3.5 Data Processing

The workflow for data processing is shown in Fig. 6 (*see* **Note 6**).

1. The raw LC-MS data is converted to .csv file by using Bruker Daltonics Data Analysis.
2. IsoMS is used to extract peak pairs from raw data file [14]. The peak pairs are filtered by removing redundant peaks and the peak-pair intensity ratios are calculated.
3. The multiple LC-MS data files are aligned together based on accurate mass and retention time.

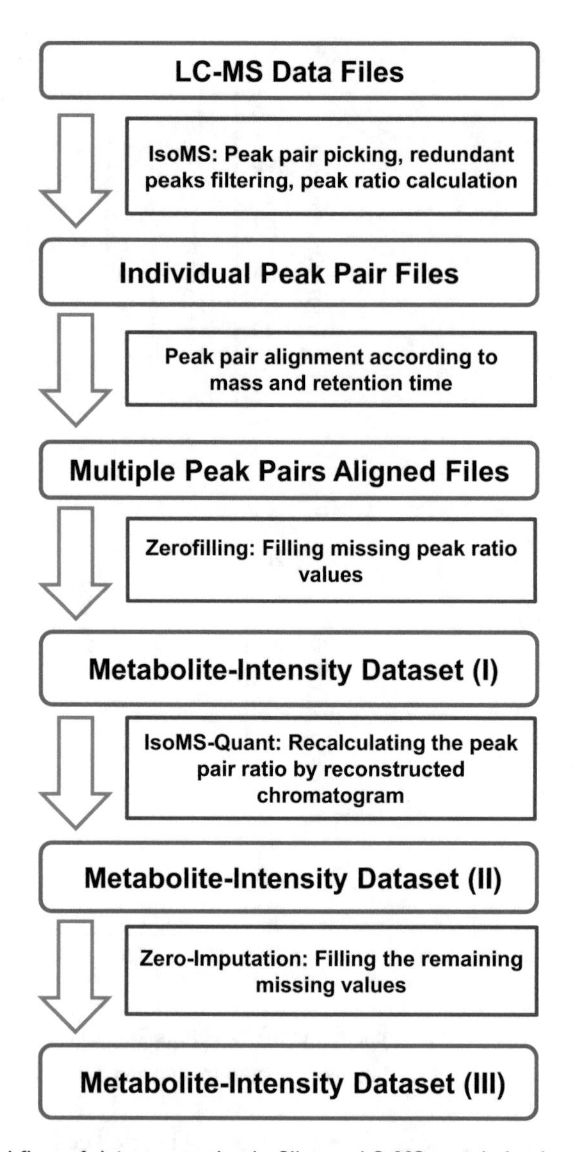

Fig. 6 Workflow of data processing in CIL nanoLC-MS metabolomics

4. The missing values in the data matrix are filled by Zerofill [15].
5. IsoMS-Quant is used to reconstruct the chromatographic peaks and recalculate the ratios based on peak areas of reconstructed chromatograms [16].
6. The rest of missing values are filled by ZeroImputation.

3.6 Metabolite Identification

1. The metabolites are identified by searching against the DnsID Library [17]. This library contains 315 entries from 273 dansylated metabolites (*see* **Note 7**).

2. Other metabolites can be matched by searching against MyCompoundID Zero Reaction Library (8021 known human metabolites) or One Reaction Library (375,809 predicted human metabolites) [18].
3. Univariate and multivariate statistical analysis are performed by using a website based statistical tool, Metaboanalyst. (https://www.metaboanalyst.ca/) (*see* **Note 8**).

4 Notes

1. The mobile phases are filtered by 0.45 μm Nylon 66 filter prior to use.
2. Adding acetonitrile into mobile phase A can prevent the growth of microbes.
3. Adding 5% water into mobile phase B can prevent acetonitrile and other chemical polymerizations.
4. Harvesting cells by scrapping method can reduce the change of concentrations for some metabolites. However, if controlling the number of cells is desirable in the experiment, the trypsinization harvest method as described in the publication is recommended [9].
5. After labeling, the final total concentration of AAS solution should be between 0.001 and 0.13 mM. The total concentration of labeled metabolites from a small number of cells (up to 10,000 cells) falls within this range.
6. The data processing package is available on MyCompoundID website (http://www.mycompoundid.org).
7. The metabolite identification library is available on MyCompoundID website.
8. Other statistical tools, such as Simca-P and SPSS, can also be used for data analysis.

Acknowledgments

This work was supported by the Natural Sciences and Engineering Research Council of Canada, the Canada Research Chairs program, Canada Foundation for Innovation, Genome Canada, and Alberta Innovates.

References

1. Guo K, Li L (2009) Differential C-12/C-13-isotope dansylation labeling and fast liquid chromatography/mass spectrometry for absolute and relative quantification of the metabolome. Anal Chem 81:3919–3932
2. Zhao S, Luo X, Li L (2016) Chemical isotope labeling LC-MS for high coverage and quantitative profiling of the hydroxyl submetabolome in metabolomics. Anal Chem 88:10617–10623
3. Zhao S, Dawe M, Guo K, Li L (2017) Development of high-performance chemical isotope labeling LC–MS for profiling the carbonyl submetabolome. Anal Chem 89:6758–6765
4. Guo K, Li L (2010) High-performance isotope labeling for profiling carboxylic acid-containing metabolites in biofluids by mass spectrometry. Anal Chem 82:8789–8793
5. Chen D, Su X, Wang N, Li Y, Yin H, Li L, Li L (2017) Chemical isotope labeling LC-MS for monitoring disease progression and treatment in animal models: plasma metabolomics study of osteoarthritis rat model. Sci Rep 7:40543
6. Su X, Wang N, Chen D, Li Y, Lu Y, Huan T, Xu W, Li L, Li L (2016) Dansylation isotope labeling liquid chromatography mass spectrometry for parallel profiling of human urinary and fecal submetabolomes. Anal Chim Acta 903:100–109
7. Hooton K, Li L (2017) Nonocclusive sweat collection combined with chemical isotope labeling LC–MS for human sweat metabolomics and mapping the sweat metabolomes at different skin locations. Anal Chem 89:7847–7851
8. Luo X, Zhao S, Huan T, Sun D, Friis RMN, Schultz MC, Li L (2016) High-performance chemical isotope labeling liquid chromatography–mass spectrometry for profiling the Metabolomic reprogramming elicited by ammonium limitation in yeast. J Proteome Res 15:1602–1612
9. Luo X, Gu X, Li L (2018) Development of a simple and efficient method of harvesting and lysing adherent mammalian cells for chemical isotope labeling LC-MS-based cellular metabolomics. Anal Chim Acta 1037:97–106
10. Li Z, Tatlay J, Li L (2015) Nanoflow LC–MS for high-performance chemical isotope labeling quantitative metabolomics. Anal Chem 87:11468–11474
11. Luo X, Li L (2017) Metabolomics of small numbers of cells: metabolomic profiling of 100, 1000, and 10000 human breast cancer cells. Anal Chem 89:11664–11671
12. Luo X, An M, Cuneo KC, Lubman DM, Li L (2018) High-performance chemical isotope labeling liquid chromatography mass spectrometry for exosome metabolomics. Anal Chem 90:8314–8319
13. Luo X, Li L (2020) Rapid liquid chromatography flourescene detection for normalization of samples of limited amounts in metabolomics. Manuscript in preparation
14. Zhou R, Tseng CL, Huan T, Li L (2014) IsoMS: automated processing of LC-MS data generated by a chemical isotope labeling metabolomics platform. Anal Chem 86:4675–4679
15. Huan T, Li L (2015) Counting missing values in a metabolite-intensity data set for measuring the analytical performance of a metabolomics platform. Anal Chem 87:1306–1313
16. Huan T, Li L (2015) Quantitative metabolome analysis based on chromatographic peak reconstruction in chemical isotope labeling liquid chromatography mass spectrometry. Anal Chem 87:7011–7016
17. Huan T, Wu Y, Tang C, Lin G, Li L (2015) DnsID in MyCompoundID for rapid identification of dansylated amine- and phenol-containing metabolites in LC–MS-based metabolomics. Anal Chem 87:9838–9845
18. Li L, Li R, Zhou J, Zuniga A, Stanislaus AE, Wu Y, Huan T, Zheng J, Shi Y, Wishart DS, Lin G (2013) MyCompoundID: using an evidence-based Metabolome library for metabolite identification. Anal Chem 85:3401–3408

Chapter 5

Untargeted Metabolomics to Interpret the Effects of Social Defeat: Integrating Chemistry and Behavioral Approaches in Neuroscience

Allen K. Bourdon and Brooke N. Dulka

Abstract

Combining behavioral approaches with innovative chemistry techniques is an important first step toward characterizing neurochemicals that could potentially serve as biomarkers related to the prevention, treatment, or diagnosis of psychopathologies such as post-traumatic stress disorder (PTSD). Here, we describe the integration of acute social defeat procedures with untargeted metabolomics techniques in order to identify such biomarkers.

Key words Untargeted metabolomics, Ultra high-performance liquid chromatography, High-resolution mass spectrometry, Social defeat, Biomarkers

1 Introduction

While there is a growing interest in identifying biomarkers that could potentially relate to the prevention, treatment, or diagnosis of stress-induced psychopathologies such as post-traumatic stress disorder (PTSD) [1], until recently it has not been feasible to associate levels of large numbers of neurochemicals and metabolites to stress-related phenotypes [2]. Metabolomics is the quantitative analysis of small molecules (<1500 Da) present in biological systems and has been increasingly used for the discovery of biomarkers [3]. A discovery-based chemistry approach, such as untargeted metabolomics, allows for thousands of metabolites to be detected with little-to-no prior knowledge of the metabolite composition of a given sample [4, 5]. Social defeat, on the other hand, is a behavioral paradigm that has increasingly been used to investigate individual differences in stress resilience [6]. Through the integration of chemistry techniques, such as untargeted metabolomics, and behavioral procedures, such as social defeat, it is possible that we

Paul L. Wood (ed.), *Metabolomics*, Neuromethods, vol. 159, https://doi.org/10.1007/978-1-0716-0864-7_5,

can make strides in the field of neuroscience toward identifying novel biomarkers of stress-related mental illness.

We have previously utilized such an integrated approach to identify functional groups of neurochemicals that distinguish stress-resistant and stress-susceptible individuals in two rodent models of acute social defeat stress, which was accomplished using ultra high-performance liquid chromatography coupled with high resolution mass spectrometry (UPLC-HRMS) [2]. Here, we will focus on one of these rodent models (C57BL/6 mice) and how untargeted metabolomics procedures can be used to distinguish resistant, susceptible, and control mice based on their complete metabolic profile using both known and unknown neurochemicals.

2 Materials

2.1 Animals

1. Subjects: Obtain male C57BL/6 mice (7–8 weeks old, 20–27 g) from Envigo (Indianapolis, IN, USA). Maintain mice on a 12:12 light–dark cycle with ad libitum access to food and water in a temperature-controlled room (21 ± 2 °C). Animals are individually housed in polycarbonate cages (18.4 cm × 29.2 cm × 12.7 cm) with corncob bedding, cotton nesting materials, and wire mesh tops. Limit access to environmental enrichment such as toys or plastic tubes (*see* **Note 1**).
2. Resident aggressors/stimulus animals: Adult male Hsd:ICR (CD1) mice that are individually housed to maximize aggression (35–40 g, Envigo).

2.2 Untargeted Metabolomics

1. All samples are stored in a −80 °C freezer prior to extraction. Tissue extraction is done in 5 °C cold room. All solvents are HPLC grade, or higher, and purchased from Fisher Scientific (Atlanta, GA, USA).
2. Sample analysis is conducted on an ultra high-performance liquid chromatography-high-resolution mass spectrometry (UPLC-HRMS) instrument purchased from Thermo Fisher Scientific (Waltham, MA, USA). The UPLC is constructed with an ultimate 3000 LC pump, which was coupled to an Exactive Plus Orbitrap™ MS. All mobile phase solvents are HPLC grade (Fisher Scientific). A single Synergi 2.5 μm Hydro RP 100, 100 × 2.0 mm column (Phenomenex) is used to analyze all samples to maintain consistency.
3. Data processing is carried out using MAVEN software, which allows for the identification of all detected small molecules. Statistical analyses are done using various R software scripts, Cluster, and Java Treeview software.

3 Methods

3.1 Behavioral Procedures

Carry out all behavioral procedures during the first 3 h of the dark phase of the light–dark cycle.

1. Handle subjects gently several times 1 week prior to social defeat to habituate them to the stress of human handling.
2. Subject mice to acute social defeat, as described previously [7]. Briefly, acute social defeat stress consists of three, 2-min aggressive encounters in the home cage of a novel CD1 resident aggressor mouse with 2-min intertrial intervals in the subjects' home cage, for a total duration of 10 min (*see* **Note 2**). Expose nondefeated control animals to the empty home cage of three separate CD1 mice for 2-min with 2-min intertrial intervals in their home cage.
3. Twenty-four hours later, test all subjects in a social interaction test in order to identify those animals susceptible and resilient to the effects of acute social defeat stress. Testing should be performed in an open field arena (43.2 cm × 43.2 cm × 43.2 cm) under dim light conditions. Social interaction testing consists of two 5-min trials: CD1 mouse target absent and CD1 mouse target present. During the target absent trial, subjects are habituated to an empty perforated plastic box that should be positioned against one of the four walls. The target present trial occurs immediately following the target absent trial, and a novel CD1 mouse should be placed inside the perforated plastic box. An observer blind to experimental conditions will quantify the duration of time the subject spent in the interaction zone, which is defined as the 3 cm area surrounding the plastic box (*see* **Note 3**).
4. To investigate defeat-induced changes in neurochemical activity, mice receive a second bout of social defeat stress or empty cage exposure 1 week following the social interaction test.

3.2 Tissue Collection

1. Immediately following the final social defeat encounter, animals are sacrificed with isoflurane and rapidly decapitated. A brain matrix is then used to generate 1 mm thick brain slices which should be rapidly frozen on glass microscope slides. Tissue punches (1 mm diameter) can then collected bilaterally from brain regions of interest, such as the nucleus accumbens (NAc).
2. Tissue punches are flash-frozen in liquid nitrogen and stored at −80 °C until metabolite extraction.

3.3 Extraction Protocol

Prior to extraction, samples are removed from the −80 °C freezer into the 5 °C cold room for 20 min to allow thawing. The extraction solvent is a 40:40:20 acetonitrile–methanol–water solution containing 0.1 M formic acid. To the tissue punches, 500 μL of extraction solvent is added, vortexed, then transferred to a −20 °C freezer for 20 min. Samples are then centrifuged for 5 min at 13,300 × *g* and the resulting supernatant is collected. To the remaining pellet, 100 μL of extraction solvent is added, vortexed, placed in −20 °C freezer for 20 min, and centrifuged. This wash step is done twice for each pellet. The combined supernatants are then dried under nitrogen, followed by resuspension with 150 μL of Millipore water.

3.4 Ultra high-Performance Liquid Chromatography–High-Resolution Mass Spectrometry

Untargeted metabolomics is carried out using a UPLC-HRMS instrument, as explained in Subheading 2. For UPLC, mobile phase A is 97:3 water–methanol containing 10 mM tributylamine, as a proton acceptor, and mobile phase B is 100% methanol. The following multistep gradient is used for analyte separation: 0–5 min, 0% B; 5–13 min, 20% B; 13–15.5 min, 55% B; 15.5–19 min, 95% B; and 19–25 min, 0% B. The 10 μL sample injection is kept at a constant flow rate of 200 μL/min and the column temperature is kept constant at 25 °C. Analytes are detected in negative mode, full scan with the eluent being introduced to the MS by an electrospray ionization source. The instrument analyzes at a resolving power of 140,000 and a scan window of 72–800 *m/z* from 0 to 9 mins, and 100–1000 *m/z* from 9 to 25 min. Samples should be randomized to reduce instrumental error.

3.5 Data Processing

1. Raw files from the Xcaliber software are converted to an open source format, mxML via MSConvert software. The converted files are then uploaded into the Metabolomic Analysis and Visualization Engine (MAVEN) software package. Using MAVEN, known metabolites are selected from a preexisting list of 270 compounds with annotated biological activity. Unknown analytes are selected automatically using a peak identification function that allows the user to adjust parameters such as signal-to-noise ratio and minimum area. Common adducts and isotopomers are removed using XCMS [8].
2. Using the known and unknown datasets, separately, several different statistical approaches are utilized. One of the most common statistical tools in metabolomics is the use of heatmaps. For our heatmaps, we use Cluster software, v3.0, to log2 transform and cluster analytes by average linkage, which adjusts the location of each metabolite to aid in visual identification of patterns. The data are then displayed into a heatmap using Java Treeview software, where a ±3 scale is used to visualize an

increase in concentration as red and decrease in concentration as blue. Lastly, overlay p-values generated from a one-way analysis of variance (ANOVA) followed by Tukey's post hoc tests.

3. Another major statistical tool that should be used is the multivariate technique, partial-least squares discriminant analysis (PLS-DA). This widely accepted tool enables the end user to highlight metabolite clusters and focus on the interpretation of an analyte subset. For PLS-DA figures, use the R software package, DiscriMiner. Prior to visualization, data should be normalized by autoscaling, which attempts to center the data by dividing each value by the metabolite's standard deviation. After inputting the normalized data into DiscriMiner, the PLS-DA scores values are extracted for visualization, which is carried out with the ggplot package. Colored ellipses identify a 95% confidence interval for each data group (Fig. 1a). An integral part of the PLS-DA is the extraction of variable importance of projection (VIP) scores to assist in identification of important features. This tool allows one to identify metabolites that were most significant to the observed separation in the PLS-DA plot.

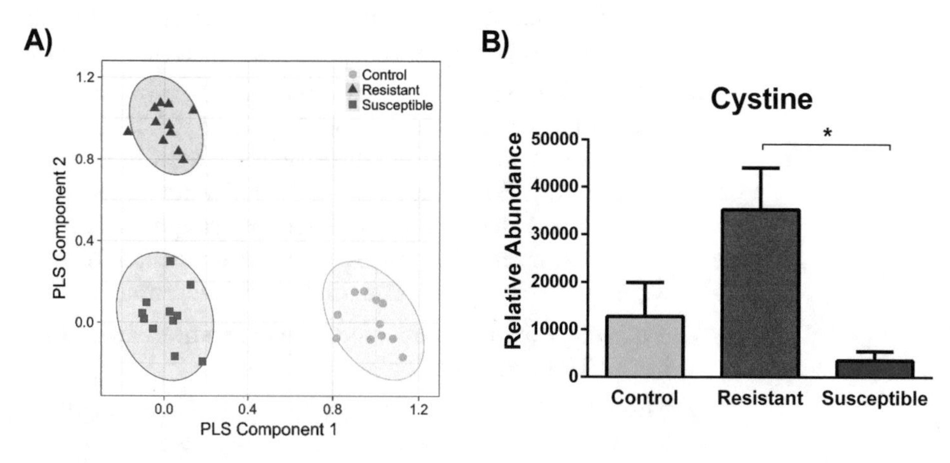

Fig. 1 (**a**) Metabolite patterns from untargeted metabolomic approach in the NAc. Qualitative profiling for control (yellow), resistant (red), and susceptible (blue) mice was conducted with PLS-DA. PLS component 1 (X-axis) and PLS component 2 (Y-axis) represent the highest two X-scores of dimensions 1 and 2 for matrix X of metabolite ion abundances, respectively. Ellipses express a 95% confidence interval. Plots exhibit separation and clustering of samples. (**b**) Effects of social defeat on the relative abundance of neurochemical metabolites in mice. In the NAc, resistant mice had significantly higher levels of cystine. (Figures adapted from [2; https:/doi.org/10.1016/J.YNSTR.2017.08.001]. This article and its contents are available for reproduction under the terms of a Creative Commons Attribution License (CC BY 4.0; https:/creativecommons.org/licenses/by/4.0/))

4 Integration and Interpretation

1. Identify which brain regions display separation in the PLS-DA plots.
2. Use the VIP scores for a given brain region to identify the most important features (*see* **Note 4**) that are contributing to the separation of clusters.
3. Cross reference selected features with the ANOVA values derived from the heatmaps.
4. If a given feature has the required VIP score and a significant ANOVA, check the post hoc analyses to determine which groups are significantly different from one another, this is how representative metabolites (Fig. 1b) that distinguish susceptible, resistant, and control mice are identified (*see* **Notes 5** and **6**).

5 Notes

1. Environmental enrichment buffers mice against the effects of social defeat stress [9] and therefore should be minimized or eliminated.
2. To correct for potential variation in the amount of aggression subjects received, the first defeat episode should not begin until the subject submits to an attack from the resident aggressor.
3. An interaction ratio has been used to categorize mice as either susceptible or resistant to social defeat [10]. The interaction ratio is calculated as (time investigating target present) / (time investigating target absent). Defeated mice with interaction ratios of less than 1.0 are defined as susceptible and defeated mice with interaction ratios equal to or greater than 1.0 are defined as resistant or resilient to social defeat.
4. Typically, a feature is deemed relevant to the PLS-DA plot if it has a VIP score ≥ 1.0.
5. This high-throughput method can be used to characterize both known and unidentified spectral features.
6. Cross-reference statistically significant representative metabolites with the academic literature in order to better understand the biological significance of the data.

6 Conclusions

In closing, a combination of untargeted metabolomics, social defeat, and statistical techniques can be used to identify representative metabolites (such as cystine, *see* Fig. 1b) that may serve as central, mechanistic biomarkers of stress resilience. The authors suggest that future studies focus on peripheral, predictive biomarkers of stress resilience using a variation of the paradigm described above. Altogether, a better understanding of the biological basis of stress-related mental illness, including metabolomics data and their integration with other approaches, might identify new molecular targets and provide improved therapeutic approaches for disorders such as PTSD [11].

Acknowledgments

This work was supported by National Institute of Health grant MH107007, NARSAD grant 19260 from the Brain and Behavior Research Foundation, and a seed grant from the NeuroNET Research Center. Partial funding for AKB was received from the University of Tennessee–Oak Ridge National Lab, Joint Institute of Biological Sciences.

References

1. Yehuda R, Neylan TC, Flory JD, McFarlane AC (2013) The use of biomarkers in the military: from theory to practice. Psychoneuroendocrinology 38(9):1912–1922
2. Dulka BN, Bourdon AK, Clinard CT, Muvvala MB, Campagna SR, Cooper MA (2017) Metabolomics reveals distinct neurochemical profiles associated with stress resilience. Neurobiol Stress 7:103–112
3. Kaddurah-Daouk R, Kristal BS, Weinshilboum RM (2008) Metabolomics: a global biochemical approach to drug response and disease. Annu Rev Pharmacol Toxicol 48:653–683
4. Dunn WB (2008) Current trends and future requirements for the mass spectrometric investigation of microbial, mammalian and plant metabolomes. Phys Biol 5(1):011001
5. Dunn WB, Erban A, Weber RJ, Creek DJ, Brown M, Breitling R, Hankemeier T, Goodacre R, Neumann S, Kopka J (2013) Mass appeal: metabolite identification in mass spectrometry-focused untargeted metabolomics. Metabolomics 9(1):44–66
6. Krishnan V, Han M-H, Graham DL, Berton O, Renthal W, Russo SJ, LaPlant Q, Graham A, Lutter M, Lagace DC, Ghose S, Reister R, Tannous P, Green TA, Neve RL, Chakravarty S, Kumar A, Eisch AJ, Self DS, Lee FS, Tamminga CA, Cooper DC, Gershenfeld HK, Nestler EJ (2007) Molecular adaptations underlying susceptibility and resistance to social defeat in brain reward regions. Cell 131(2):391–404
7. Dulka BN, Ford EC, Lee MA, Donnell NJ, Goode TD, Prosser R, Cooper MA (2016) Proteolytic cleavage of proBDNF into mature BDNF in the basolateral amygdala is necessary for defeat-induced social avoidance. Learn Mem 23(4):156–160
8. Tautenhahn R, Patti GJ, Rinehart D, Siuzdak G (2012) XCMS online: a web-based platform to process untargeted metabolomic data. Anal Chem 84(11):5035–5039
9. Lehmann ML, Herkenham M (2011) Environmental enrichment confers stress resiliency to social defeat through an infralimbic cortex-dependent neuroanatomical pathway. J Neurosci 31(16):6159–6173
10. Golden SA, Covington HE III, Berton O, Russo SJ (2011) A standardized protocol for repeated social defeat stress in mice. Nat Protoc 6(8):1183
11. Konjevod M, Tudor L, Strac DS, Erjavec GN, Barbas C, Zarkovic N, Perkovic MN, Uzun S, Kozumplik O, Lauc G (2019) Metabolomic and glycomic findings in posttraumatic stress disorder. Prog Neuro-Psychopharmacol Biol Psychiatry 77:181–193

Chapter 6

Flow Infusion Electrospray Ionization High-Resolution Mass Spectrometry of the Chloride Adducts of Sphingolipids, Glycerophospholipids, Glycerols, Hydroxy Fatty Acids, Fatty Acid Esters of Hydroxy Fatty Acids (FAHFA), and Modified Nucleosides

Paul L. Wood and Randall L. Woltjer

Abstract

Ceramides are generally monitored via either their anions or cations formed with electrospray ionization (ESI). While these methods appear adequate for chromatographic assays, they offer limited sensitivity with flow infusion ESI ("shotgun lipidomics"). We therefore evaluated the chloride adducts $[M + Cl]^-$ of a number of lipid classes and found 10- to 50-fold increases in assay sensitivity for ceramides, hexosylceramides, lactosylceramides, phytoceramides, dihydroceramides, and ceramide phosphoethanolamines. Chloride adducts of glycerophosphocholines, sphingomyelins, hydroxyl fatty acids, and modified nucleotides also were characterized. Chloride adducts of glycerophosphocholines were particularly useful for the MS/MS analysis of the constituent fatty acids at sn-1 and sn-2.

Key words Negative ion electrospray mass spectrometry, Chloride adducts, Ceramides

1 Introduction

A number of ion adducts have been characterized with electrospray ionization mass spectrometry (ESI-MS) and are utilized to improve the sensitivity of assays for a diversity of biomolecules. Chloride ion adducts are one such example with reports of the attachment to electrophilic hydrogens of hydroxyl groups in steroids [1], mono- and oligosaccharides [2], xyloglucans [3], ceramides [4], lactosylceramides [5], ceramide phosphoethanolamines [5], gangliosides [4], and phenols [4]. In addition, chloride attachment to tertiary and quaternary ammonium compounds, including glycerophosphocholines [4, 5], sphingomyelins [5], nitrated sugar alcohols [6], and nitrate explosives [7], has been reported.

Historically, our flow infusion ESI lipidomics analytical platform [8] has been limited by low sensitivity for ceramides and

Paul L. Wood (ed.), *Metabolomics*, Neuromethods, vol. 159, https://doi.org/10.1007/978-1-0716-0864-7_6,

ceramide derivatives by weak $[M + H]^+$, $[M\text{-}H]^-$, and $[M + HCOO]^-$ ions. We therefore explored chloride adducts of ceramides and monitored 10- to 50-fold increases in sensitivity. We also expanded our evaluations to a number of other candidate biomolecules and found that the chloride adducts demonstrate increased sensitivity with:

- Ceramides.
- Hydroxyceramides.
- Phytoceramides.
- Dihydroceramides.
- Hexosylceramides.
- Lactosylceramides.
- Ceramide phosphoethanolamines.

In the cases of hydroxy fatty acids, sphingomyelins, and N-acylethanolamides we did not detect any increased sensitivity but a significant decrease in mass error for a number of lipids in each of these lipid families. With DAGs and MAGs we also did not detect any alterations in sensitivity, but this still represents an alternate quantitation method in cases where ion suppression of the $[M + NH_3]^+$ ions occurs.

2 Materials

2.1 Buffers and Standards

- Lipid Infusion Buffer: 2-propanol–methanol–dichloromethane (8:4:4) + 5 mM ammonium chloride (*see* **Note 1**).
- Small Molecular Weight Metabolite Infusion Buffer: acetonitrile–methanol–dichloromethane–water (100:100:100:10); the water contains 83 mg. of NH_4Cl (*see* **Note 2**).
- Standards and Internal Standards are dissolved in methanol.
- Stable isotope standards are purchased from CDN Isotopes, Cambridge Isotope Labs., and Avanti Polar Lipids.
- Cold small molecule extraction buffer: acetonitrile–methanol–formic acid (800:200:2.5).

2.2 Analytical Equipment and Supplies

- Q Exactive benchtop Orbitrap (Thermo Fisher).
- Centrifugal dryer (Eppendorf Vacufuge Plus).
- Refrigerated microfuge capable of 30,000 × *g* (Eppendorf 5430R).
- Sonicator (Thermo Fisher FB5).
- 1.5 mL screw-top microfuge tubes (Thermo Fisher).

3 Methods

3.1 Lipid Extraction

- Plasma or serum (100 μL), in 7 mL tubes, is vortexed with 1 mL of methanol and 50 μL of the Stable Isotope Internal Standard solution followed by the addition of 1 mL of water and 2 mL of t-butylmethylether at room temperature.
- Tissues (500–600 mg wet weight) are sonicated in 1 mL of methanol, 50 μL of the Stable Isotope Internal Standard solution, and 1 mL of water. Next 2 mL of t-butylmethylether is added.
- The samples are then shaken at room temperature for 30 min using a Fisher Multitube Vortexer.
- Next samples are centrifuged at 4000 × *g* for 30 min at room temperature.
- One milliliter of the supernatant is transferred to the wells of a 96-deep-well microplate.
- The samples are next dried by vacuum centrifugation and then dissolved in 200 μL of Lipid Infusion Buffer before sealing the plate.

3.2 Small Molecular Weight Metabolite Extraction

- Plasma or serum (100 μL), in 1.5 mL microfuge tubes, is vortexed with 1 mL of cold Extraction Buffer and 50 μL of the Internal Standard solution.
- Tissues (500–800 mg), in 1.5 mL microfuge tubes, are sonicated in 1 mL of cold Extraction Buffer 1 mL of cold Extraction Buffer.
- Next samples are centrifuged at 30,000 × *g* and 4 °C for 30 min.
- Seven-hundred and fifty microliters of the supernatant is transferred to a clean 1.5 mL microfuge tube.
- The samples are next dried by vacuum centrifugation.
- One-hundred and fifty microliters of Small Molecular Weight Metabolite Infusion Buffer is added to the dried samples with vortexing.
- The tubes are centrifuged at 30,000 × *g* for 10 min to precipitate any potential suspended particulates.

3.3 Flow Injection Analysis

The mass axis is calibrated as discussed in Chapter 1 of this volume.

- Samples are infused at a flow rate of 12 μL/min for ion electrospray ionization (ESI).
- The syringe and tubing are flushed between samples with acetonitrile–methanol (1:1).
- The $[M + Cl]^-$ anions are monitored.

Table 1
Cations of the chloride adducts of analytical standards of ceramide subfamilies

Ceramide	Exact	Calc [M + Cl]$^-$	ppm	NCE	Fragment	Calc Cation	ppm
Cer 16:0 [D31]	567.6991	602.6685	1.04	25	HCl loss	566.6918	0.59
Cer 16:0	537.5121	572.4815	0.58	25	HCl loss	536.5048	0.24
Cer 18:0	565.5433	600.5127	0.65	40	HCl loss	564.5361	0.85
Hydroxy Cer 18:0	581.5383	616.5077	0.32	30	HCl loss	580.5310	0.64
Glu-Cer 18:1 [D5]	730.6119	765.5813	1.1	25	[M-(Glu-H_2O)]	567.5518	0.61
Gal-Cer 16:0	699.5649	734.5343	0.86	25	[M-(Gal-H_2O)]	536.5048	1.9
Gal-Cer 24:1	809.6744	844.6438	0.99	25	[M-(Gal-H_2O)]	646.6143	2.2
Lac-Cer	861.6177	896.5871	0.61				
Sphingosine	299.2824	334.2518	1.2				

Table 2
Cations of the chloride adducts of standards of hydroxy fatty acids

Lipid	Exact	Calc [M + Cl]$^-$	ppm	NCE	Fragment	Calc Cation	ppm
17-Hydroxy-DHA	344.2351	379.2049	0.23	25	$C_{16}H_{22}O_2$	245.1547	0.40
14-Hydroxy-DHA	344.2351	379.2049	0.23	25	$C_{13}H_{18}O_2$	205.1234	0.87
7-Hydroxy-DHA	344.2351	379.2049	0.23	25	$C_6H_{10}O_2$	113.0608	0.88
[$^2H_4$]PGE2	356.2500	391.2199	1.8	25	$C_{20}H_{24}O_3D_4$	319.2212	1.2
[$^2H_4$]PGF2α	358.2657	393.2355	1.6	50	$C_{12}H_{14}O_2D_4$	197.1484	0.5
RvD5	360.2300	395.1994	2.3	25	$C_{15}H_{19}$	199.1492	2.5
RvD2	376.2249	411.1943	2.2	25	$C_{16}H_{20}O_3$	259.1339	0.46
RvE1	350.2093	385.1787	2.4	25	$C_{11}H_{15}O_3$	195.1027	1.5
PGE2	352.2249	387.1943	2.4	25	$C_{20}H_{28}O_3$	315.1965	1.0
13-HETE	320.2351	355.2045	2.7	35	$C_{20}H_{29}O_2$	301.2173	3.1

- Characteristic product anions (MS^2) were determined for metabolites of interest (Tables 1, 2, and 3). MS^2 also always generated the [M]$^-$ anions, as a result of loss of the chloride adduct.

3.4 Data Reduction

- Data are transferred to an Excel spreadsheet for analyses of ppm mass error and mass intensity for each lipid of interest.

3.5 Biological Analytical Example

In our continuing analysis of lipid alterations in age-related brain proteinopathies [9–11], we performed a pilot analysis of individual

Table 3
Cations of the chloride adducts of standards of modified nucleotides

Nucleoside	Exact	Calc $[M + Cl]^-$	ppm	NCE	Fragment	Calc Cation	ppm
$[^{13}C_5]$Adenosine	272.1135	307.0833	0.5	45	$C_5H_5N_5$	134.0467	0.41
Adenosine	267.0967	302.06654	0.4	45	$C_5H_5N_5$	134.0467	0.6
2-Hydroxyadenine	151.0494	186.01924	0.7	25	$C_5H_5N_5O$	150.0421	0.4
Uridine	244.0695	279.0393	0.6	35		110.0144	0.6
Methyluridine	258.0851	293.0550	0.5	35	$C_5H_6N_2O_2$	125.0351	2.1
Formylcytidine	271.0844	306.0542	0.2	35		137.0357	1.0
Guanosine	283.0916	282.0843	0.3	35	$C_5H_5N_5O$	150.0421	1.1
Hydroxydeoxyguanosine	283.0916	282.0843	0.4	45	$C_7H_7N_5O_2$	192.0521	0.5
Methylthioadenosine	297.0895	332.0594	0.8	35	$C_5H_5N_5$	134.0467	0.8
Methylguanosine	297.1073	332.0771	1,0	35	$C_6H_7N_5O$	164.0577	1.5
Dimethylguanosine	311.1229	346.0928	0.8	45	$C_7H_9N_5O$	178.0734	0.56

human brain cell populations to determine the potential anatomical relevance of data we have obtained previously with postmortem human brain tissue and for our future studies. For these analyses we purchased human cell pellets from ScienCell (Carlsbad, CA). These included ten million cells for hippocampal astrocytes, Schwann cells, brain vascular pericytes, and brain microvascular endothelial cells. Relative metabolite levels were determined and normalized to protein prior to ranking each cell type relative to neuronal levels (value of 1 assigned to neuronal levels; three million cells).

These data demonstrate the sensitivity (Table 4) monitored with chloride adducts of sphingolipids (Fig. 1) that are weak $[M-H]^-$ signals.

3.6 Glycerophosphocholines (GPC)

Positive ion tandem MS of GPC generates a product ion from phosphocholine (184.0739) and does not allow for the identification of the sn-1 and sn2 fatty acid substituents. Lithium adducts have been reported to be useful for this function, however, in our hands, while this approach works well with standards it lacks sensitivity for biological samples. In contrast, we have found that the chloride adducts of GPC [4, 5] provide significant sensitivity with biological samples for identifying the fatty acid substituents (Fig. 2).

Table 4
Lipidomics of human cells. Cellular relative metabolite levels were normalized to the protein level and neurons assigned a value of 1

Lipid	Exact	$[M + Cl]^-$	ppm	Astro	Schwann	Pericyte	Endo
Sphingosine	299.2824	334.2522	1.9	9.4	10.0	7.6	3.9
Cer d18:1/16:0	537.5121	572.4809	0.58	1.6	6.7	2.9	2.8
Cer d18:1/24:0	649.6373	684.6061	0.17	11.5	25.5	21.5	13.8
Cer d18:1/24:1	647.6216	682.5905	0.36	7.5	19.5	8.8	9.9
H-Cer d18:1/24:0	665.6322	700.6010	0.18	4.2	10.6	5.9	4.0
P-Cer t18:0/24:0	667.6478	702.6167	0.38	40.9	94.5	34.3	573.9
DHC d18:1/16:0	539.5277	574.4975	0.99	1.2	2.8	1.5	0.7
Hex-Cer d18:1/18:0	727.5962	762.5650	0.62	3.2	4.9	2.2	3.2
Hex-Cer d18:1/24:0	811.6901	846.6589	0.43	32.0	52.4	71.1	23.4
Hex-Cer d18:1/24:1	809.6744	844.6433	0.80	13.2	24.7	15.3	9.3
Lac-Cer d18:1/16:0	861.6177	896.5866	1.5	1[a]	0.49	0.65	ND
Lac-Cer d18:1/24:1	971.7273	1006.6961	0.89	1[a]	0.21	0.37	ND
Cer-PE d18:1/22:2	740.5832	775.5520	2.0	642.8	613.8	386.7	549.5
Cer-PE d18:1/24:1	770.6301	805.5990	0.89	10.6	18.8	8.9	71.1
SM d18:1/24:1	812.6771	847.6460	0.33	11.3	23.7	14.2	8.6
H-SM d18:1/24:0	830.6877	865.6565	1.4	1.4	0.9	0.8	0.12
FAHFA 16:0/18(OH)	538.4961	573.4649	1.5	3.2	6.7	3.9	11.5
PC 40:6	833.5934	868.5633	0.16	1.3	3.8	0.8	4.2
PCp 40:6	817.5985	852.5683	0.71	2.6	6	12.2	2.7
Lipid	Exact	$[M-H]^-$	ppm	Astro	Schwann	Pericyte	Endo
PE 36:2	743.5465	742.5392	0.22	2.6	2.8	1.9	4.2
PEp 36:2	727.5515	716.5443	1.3	1.4	1.5	0.91	0.89
PI 38:4	886.5571	885.5498	0.43	0.94	3.6	2.3	1.2
PS 36:1	789.5519	788.5447	0.73	1.4	1.2	1.7	1.1
BMP 44:12	866.5098	865.5021	1.3	ND	2.9	ND	2.3

Astro hippocampal astrocytes, *Schwann* peripheral nerve Schwann cells, *Pericyte* brain vascular pericytes, *Endo* brain microvascular endothelial cells. *BMP* lysobisphosphatidic acid (lysosomal biomarker), *Cer* ceramide, *DHC* dihydroceramide, *FAHFA* fatty ester of hydroxyl fatty acid, *H-Cer* hydroxyceramide, *H-SM* hydroxysphingomyelin, *Hex* hexosyl, *Lac* lactosyl, *PC* phosphatidylcholine, *PCp* choline plasmalogen, *P-Cer* phytoceramide, *PE* phosphatidylethanolamine, *PEp* ethanolamine plasmalogen, *PI* phosphatidylinositol, *PS* phosphatidylserine, *SM* sphingomyelin

[a]Neuronal levels were below the detection limit, therefore, a value of 1 was assigned to astrocytes

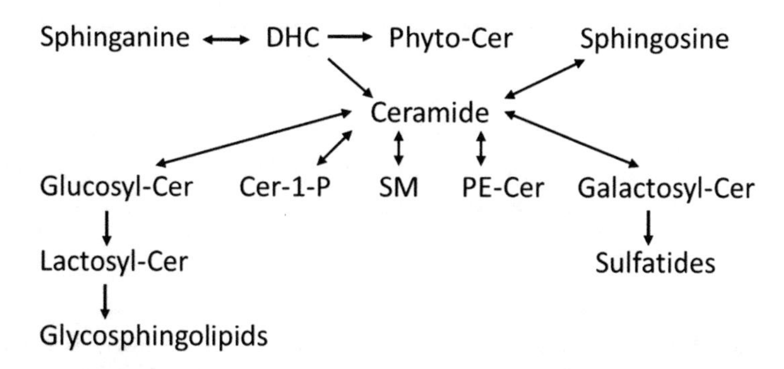

Fig. 1 Schematic overview of sphingolipid metabolism that can be assessed via high-resolution mass analyses of the chloride adducts of these lipids. *Cer* ceramide, *DHC* dihydroceramide, *P* phosphate, *PE* phosphatidylethanolamine, *SM* sphingomyelin

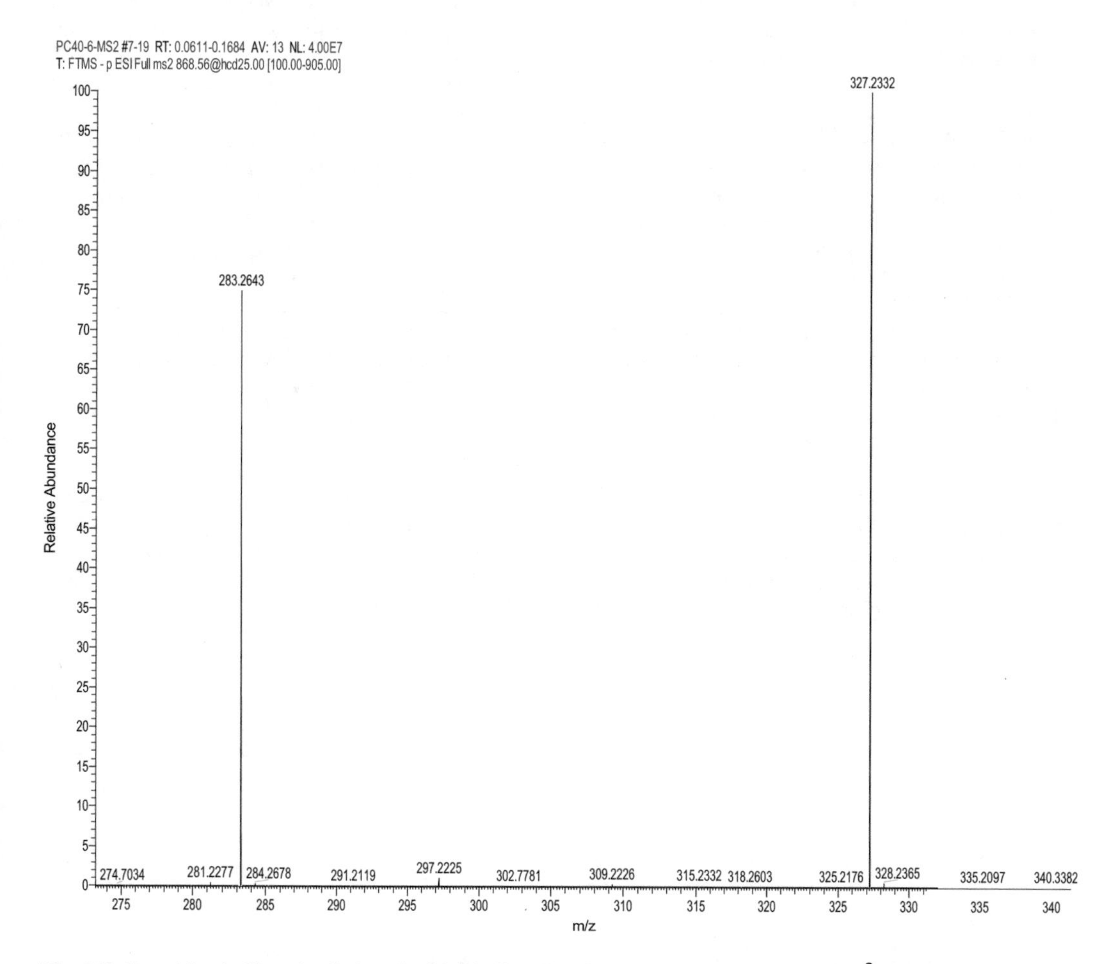

Fig. 2 Fatty acid substituents of phosphatidylcholine 40:6 in egg yolk determined by MS^2 of the parent chloride adduct 868.56. The 2 major fatty acids are 22:6 (327.233; 2.7 ppm) at sn-2 and 18:0 (283.264; 0.35 ppm) at sn-1

4 Notes

1. All organic solvents should be of LC-MS grade.
2. The Small Molecular Weight Metabolite Extraction Buffer must be cold to stabilize metabolites from further metabolism.

References

1. Rannulu NS, Cole RB (2012) Novel fragmentation pathways of anionic adducts of steroids formed by electrospray anion attachment involving regioselective attachment, regiospecific decompositions, charge-induced pathways, and ion-dipole complex intermediates. J Am Soc Mass Spectrom 23:1558–1568
2. Zhu J, Cole RB (2001) Ranking of gas-phase acidities and chloride affinities of monosaccharides and linkage specificity in collision-induced decompositions of negative ion electrospray-generated chloride adducts of oligosaccharides. J Am Soc Mass Spectrom 12:1193–1204
3. Vinueza NR, Gallardo VA, Klimek JF, Carpita NC, Kenttämaa HI (2013) Analysis of xyloglucans by ambient chloride attachment ionization tandem mass spectrometry. Carbohydr Polym 98:1203–1213
4. Zhu J, Cole RB (2000) Formation and decompositions of chloride adduct ions. J Am Soc Mass Spectrom 11:932–941
5. Han X, Yang K, Gross RW (2012) Multidimensional mass spectrometry-based shotgun lipidomics and novel strategies for lipidomic analyses. Mass Spectrom Rev 31:134–178
6. Ostrinskaya A, Kelley JA, Kunz RR (2017) Characterization of nitrated sugar alcohols by atmospheric-pressure chemical-ionization mass spectrometry. Rapid Commun Mass Spectrom 31:333–343
7. Bridoux MC, Schwarzenberg A, Schramm S, Cole RB (2016) Combined use of direct analysis in real-time/Orbitrap mass spectrometry and micro-Raman spectroscopy for the comprehensive characterization of real explosive samples. Anal Bioanal Chem 408:5677–5687
8. Wood PL (2017) Non-targeted lipidomics utilizing constant infusion high resolution ESI mass spectrometry. In: Wood PL (ed) Springer protocols, Neuromethods: lipidomics, vol 125, pp 13–19
9. Wood PL, Tippireddy S, Feriante J, Woltjer RL (2018) Augmented frontal cortex diacylglycerol levels in Parkinson's disease and Lewy body disease. PLoS One 13:e0191815
10. Wood PL, Medicherla S, Sheikh N, Terry B, Phillipps A, Kaye JA, Quinn JF, Woltjer RL (2015) Targeted lipidomics of fontal cortex and plasma diacylglycerols (DAG) in mild cognitive impairment and Alzheimer's disease: validation of DAG accumulation early in the pathophysiology of Alzheimer's disease. J Alzheimers Dis 48:537–546
11. Wood PL, Barnette BL, Kaye JA, Quinn JF, Woltjer RL (2015) Non-targeted lipidomics of CSF and frontal cortex grey and white matter in control, mild cognitive impairment, and Alzheimer's disease subjects. Acta Neuropsychiatr 27:270–278

Chapter 7

High Accurate Mass Gas Chromatography–Mass Spectrometry for Performing Isotopic Ratio Outlier Analysis: Applications for Nonannotated Metabolite Detection

Yunping Qiu and Irwin J. Kurland

Abstract

Identification of unknown metabolites is one of the major issues for untargeted metabolomics studies. Isotopic ratio outlier analysis (IROA) technique uses specially designed isotopic patterns to reveal the carbon number of each metabolite and distinguish the biological originated metabolites from artifacts. This chapter describes the procedure to utilize the IROA technique with high accurate mass GC-MS to identify unknown metabolites.

Key words Isotopic ratio outlier analysis, High accurate mass GC-MS, Metabolomics, Metabolite identification, Yeast

1 Introduction

In general, mass spectrometry–based metabolomics has higher sensitivity than nuclear magnetic resonance–based profiling and therefore generates a much larger metabolic feature data set. However, only a small portion of this metabolic feature data set can be credentialed against known standards [1, 2]. The identification of features as non-annotated metabolites of biological origin, rather than being an artifact, cannot be resolved even with ultrahigh-resolution MS, as resolution alone cannot identify whether the metabolite, annotated or nonannotated, was generated in the biological system examined. For identification of small metabolites even with unit Dalton resolution, gas chromatography/mass spectrometry (GC/MS) has the advantage of the electron ionization (EI) fragmentation mode, which generates relatively stable fragmentation patterns and enables the confidence of library search, with over one million EI spectra in the commercial libraries such as NIST, WILEY, and FiehnLib for metabolite identification [3, 4]. However, a typical unit Dalton GC/MS based

Paul L. Wood (ed.), *Metabolomics*, Neuromethods, vol. 159, https://doi.org/10.1007/978-1-0716-0864-7_7,

metabolomics can only annotate a small part of all features (less than 15%)[5]. The reasons for low annotation rate for untargeted metabolites are (1) limited availability of authentic metabolite standards, (2) co-eluting metabolites, and (3) that low concentration metabolites can be hard to be distinguished from noise signals. In order to enhance the identification of small metabolites, we have recently developed a high-resolution, accurate mass GC-MS–based isotopic ratio outlier analysis (IROA) metabolite identification procedure [6, 7].

IROA is a specific stable isotope labeling technology to generate randomized ^{13}C labeled biomolecules for the whole cell population [2, 8, 9]. Compared with conventional stable isotope labeling methods using natural abundance (1.1% of ^{13}C isotopomer seen in carbon atoms in nature) and 98–99% enrichment ^{13}C as carbon source [10–14], IROA uses randomized 95% ^{12}C (5% ^{13}C), and randomized 95% ^{13}C (5% ^{12}C) glucose as carbon sources. The promise of IROA for metabolic phenotyping in model organisms is seen in recent studies including for prototrophic *S. cerevisiae* [6, 7] and *C. elegans* [9].

These randomized 95% ^{12}C and 95% ^{13}C glucose labeling experiments, when extracted and combined, show distinctive IROA patterns for metabolites of biological origin: ^{12}C-derived molecules and ^{13}C-derived molecules can be distinguished from artifacts which lack IROA patterns. Only metabolites of biological origin will have mirrored ^{12}C and ^{13}C metabolite peaks at the same retention time. The mass difference between the light isotopic ^{12}C monoisotopic peak (M_0), and the heavy ^{13}C monoisotopic peak (M_n), reveals the total number of carbons (n) in the metabolite (Fig. 1). Furthermore, the abundance of the heavy isotopologs in the 95% ^{12}C samples (M_{0+1}, M_{0+2}, etc.; the ^{12}C envelope) or light isotopologs in the 95% ^{13}C samples (M_{n-1}, M_{n-2}, etc.; the ^{13}C envelope) follows the binomial distribution based on the initial substrate enrichment. Not only do the molecular ions show

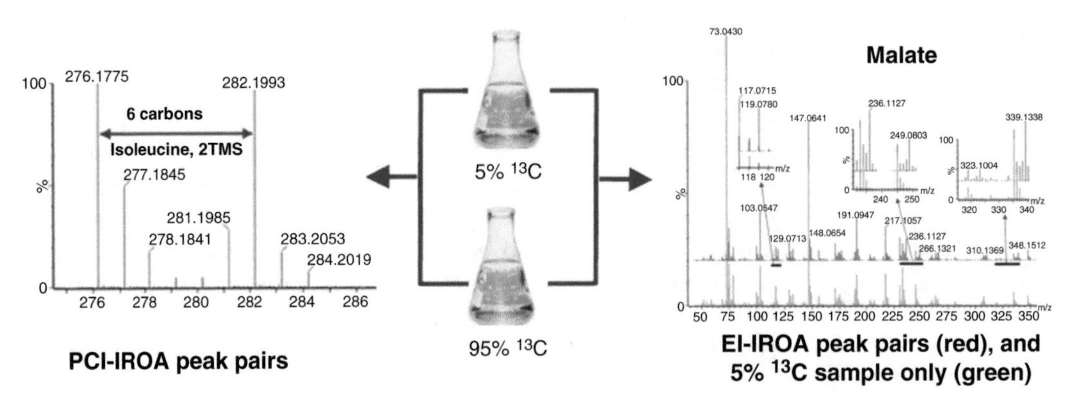

Fig. 1 Demonstration of typical GC-MS based IROA peak pairs with silylation (isoleucine), in positive chemical ionization (PCI) mode (left side), and electron impact (EI) mode (right side) (adapted from our publication [6])

IROA patterns, but the EI fragmentation spectra also does allows one to see silylation artifacts easily, as well as the structure/identity of the true fragments of the IROA metabolites.

When IROA patterns are generated, we can easily distinguish true metabolites from artificial peaks even for low concentration signals. In addition, M_n generated from the heavy ^{13}C monoisotopic peaks can be used as internal standards for each individual metabolite to normalize the M_0 peaks generated the light isotopic ^{12}C monoisotopic peaks, which reduces the analytical variations [9].

The chemical derivatization is performed prior to GC-MS base metabolomics analysis. With the capability of reacting with a wide range of metabolites, methoximation followed by silylation has been used as the most popular derivatization for metabolomics studies. The silylation reaction will add silicon and additional carbons onto the metabolite, which will affect the isotopic distributions for IROA peaks (based on the nature abundance distribution of silicon (92.2% for ^{28}Si, 4.7% for ^{29}Si, and 3.7% for ^{30}Si) and carbon (98.9% of ^{12}C, and 1.1% of ^{13}C)). The add-on isotopic enrichment from derivatization reagents can be calculated and used for the determination of active functional groups for unknown metabolites. For the ^{13}C monoisotopic peak (M_n) generated form 95% ^{13}C glucose, the binominal distribution will generate isotopic peaks of M_{n-x} ($x = 1,2,3...$). The derivatized groups will generate isotopic peaks of M_{n+x} ($x = 1,2,3...$, *see* Sect. 3.5.3(b)), which can be used to calculate the number of derivatization groups on the metabolites. The number of derivatization groups can be verified by deuterated silylation reagents (such as BSTFA_d9, [6]). In addition, ethoximation can be used to determine ketone groups in the unknown compounds. However, deuterated silylation reagents and ethoximation will result in retention time shift from the originally derivatized compounds (derivatized with methoximation and silylation reagents of natural abundance isotopic distribution) which need careful examination to match with the right IROA pairs.

The in silico fragmentation methodologies which predict fragmentations for MS spectra have been developed to facilitate metabolite identification [15]. In silico fragmentation has been reported by Blaženović et al. to increase metabolite identification accuracy with the Critical Assessment of Small Molecule Identification (CASMI) data [16]. The in silico fragmentation software packages include CFM-ID [17], MassFrag [18], MS-Finder [19], and Mass Frontier [20], which mainly focus on LC-MS/MS data. Recently, this technique has also been used for high-resolution GC/MS data [21], and IROA aids in the confirmation of CFM-ID predictions (*see* Sect. 3.5.3(h) and Fig. 3).

In this chapter, we describe a procedure of an IROA based widely unknown metabolite identification procedure for metabolomics analysis with high-resolution GC-MS. In this procedure, we used 95% randomized ^{12}C glucose (5% ^{13}C) and 95% randomized ^{13}C glucose (5% ^{12}C) as carbon sources for prototrophic yeast culture, and generated IROA patterns for all biological originated metabolites. Ethoximation versus methoximation and deuterated silylation reagents helped to confirm the existence and the number of active carbonyl groups and the number of silylation groups, separately. Positive chemical ionization (PCI) and electron impact (EI) modes were used to generate molecular ions and fragmentations, separately.

2 Materials

1. Prepare the IROA Yeast medium (50 mL for each): Weigh 0.67 g YNB medium (BD Difco PN#291940), no unlabeled carbon sources, *see* **Note 1**) into a clean cylinder, and dissolve in 100 mL of distilled water. Weigh 1 g of 5% randomized $U^{13}C$ glucose (IROAtech Inc) dissolved in 50 mL of YNB medium to prepare 5% labeled IROA medium. Weigh 1 g of 95% randomized $U^{13}C$ glucose (IROAtech Inc) dissolved in 50 mL of YNB medium to prepare 95% labeled IROA medium.
2. Yeast cell harvest: Prewet with H_2O, a 25 mm 0.45 μm nylon filter in the side arm of the clamp filtration apparatus (Magna Nylon 0.45um (Maine PN# 1213775) with a 15 mL Pyrex Millipore flask).
3. Extraction solution: 40% acetonitrile (HPLC grade), 40% methanol (HPLC grade), and 20% H_2O (HPLC grade). Prepare no more than 1 hour before use to minimize water condensation in the solvent. Store at −20 °C to cool down and gently shake to mix the solvents before use.
4. Methoxyamine hydrochloride (MOA) solution: Before the experiment freshly make 15 mg/mL methoxyamine hydrochloride (MOA) in pyridine (*see* **Note 2**).
5. GC-MS analysis: Use the DB-5MS column with a 10 m guard column (0.25 mm × 0.25 μM, Agilent PN#122-5532G).

3 Methods

3.1 Metabolite Extraction

1. 1. Prototrophic Yeast strains (*Saccharomyces cerevisiae*) are grown in YNB IROA medium (see above) (95% randamized $U^{12}C$ or $U^{13}C$ glucose medium, no unlabeled glucose or amino acid carbon sources) from OD 0.2 to OD 2.5. The yeast cell suspension (5 mL) is then vacuum filtered, and the yeast adherent to the nylon filter, is further washed with 5 mLs distilled H_2O using vacuum flitration.

2. Take the nylon filter with the adherent yeast out of the vacuum apparatus, and place the filter with yeast side down into a 35 mm petri dish (cooled with using a metal/ice block) containing 700 μL prechilled (−20 °C) extraction solvent. Use 1 mL pipette to pipette extraction solvent multiple times to release cells from the filter.
3. Collect all solvent from plate and filter plus cells into a 2.0 mL Eppendorf screw cap tube, containing 500 μL glass beads. Keep on dry ice. Rinse the filter with an additional 500 μL extraction solvent, and combine with the extract in the previous Eppendorf tube.
4. Break yeast cells in the Eppendorf tube with the glass beads in the Biospec beads beater, cycling for 30 s, and 1 min rest on ice for 4 cycles. Collect the supernatant after centrifugation at 18,400 × *g* for 5 min at 4 °C.
5. Finally, back extract the cell debris and glass beads by vortexing with 250 μL of extraction solution. Combine the supernatant with previous extracts. Perform the back extraction twice. The pooled supernatants are centrifuged again to discard any possible debris at 18,400 × *g* for 5 min. The supernatant is then pipetted into a new screw top tube and stored at −80 °C.

3.2 Metabolites Derivatization

1. It is recommended that five sets of samples are prepared and derivatized: (1) the combined 95% ^{12}C and 95% ^{13}C metabolite extracts, derivatized with MOX and unlabeled BSTFA; (2) The 95% ^{12}C metabolite extract derivatized with MOX and unlabeled BSTFA; (3) the 95% ^{13}C metabolite extract which is identically derivatized with MOX and unlabeled BSTFA. If confirmation of silylation is desired then prepare (4) the 95% ^{12}C metabolite extract derivatized with MOX and BSTFA_d9. If the number of methoximations are to be determined, then prepare (5) the 95% ^{12}C metabolite extract derivatized with EOX and unlabeled BSTFA. These samples are used as follows:
   (a) The pooled derivatized extracts are combined as an IROA sample. The IROA sample consists of the combined 95% ^{12}C and 95% ^{13}C metabolite extracts, derivatized with MOX and unlabeled BSTFA. The 95% ^{12}C metabolite extract alone, and the 95% ^{13}C metabolite extract alone, are derivatized with MOX and unlabeled BSTFA, and are used to confirm the two mirror halves of the IROA pattern. The IROA sample, the 95% ^{12}C sample, and the 95% ^{13}C sample, are then analyzed, in both EI and PCI modes;
   (b) For confirmation of the number of silylations on a given metabolite, 95% ^{12}C metabolite extract derivatized with MOX and unlabeled BSTAFA is compared to 95% ^{12}C metabolite extract derivatized with MOX and BSTFA_d9, and analyzed in the PCI mode;

(c) For estimating the number of methoximations, the 95% ^{12}C metabolite extract derivatized with MOX and unlabeled BSTFA is compared to the 95% ^{12}C metabolite extract derivatized with EOX and unlabeled BSTFA, and analyzed in the PCI mode.

2. Derivatization Procedure: Pipette a volume of 200 μL yeast extract into a high recovery cramp vial (glass vial, Agilent PN#5183-4497), and dry under gentle nitrogen flow. Make sure samples are totally dried.
3. Add 50 μL of methoxyamine hydrochloride or ethoxyamine hydrochloride (15 mg/mL in pyridine) and crimp-cap the vial. Perform the reaction at 30 °C for 90 min.
4. Add 50 μL of N,O-bis(trimethylsilyl)trifluoroacetamide (BSTFA containing 1% TMCS) or BSTFA_d9 and keep the samples at 70 °C for 60 min.
5. Keep the samples in room temperature to cool down, pending for instrumental analysis.

3.3 GC-MS Analysis

GC settings:

The oven program was set as following: initiating the temperature gradient at 60 °C and keep for 1 min, then allowed to rise by 10 °C/min to 320 °C and then held at 320 °C for 3 min. The injection temperature and the transfer line temperature were set to 270 °C and 250 °C, separately. Helium is used as a carrier gas at a flow rate of 1 mL/min.

MS settings for Waters GC-TOF/MS:

For the EI mode: the data was collected with the scan time of 0.3 s and inter scan delay of 0.1 s. The mass range was set to 35–650 Da. Heptacosa was used as the EI reference gas with a lock mass of 218.9856 ± 0.2 Da.

For the Positive CI mode: the data was collected at the same speed as in the EI mode. The mass range was set to 80–1200 Da for PCI. Tris (trifluoromethyl) triazine was used as reference gas with lock mass of 286.0027 ± 0.2 Da. 5% ammonium in methane was used as the regent gas; however, 10% ammonium in methane can be used as the regent gas, depending on availability.

MS settings for Thermo GC-QExactive:

Orbitrap was run in a full scan mode with the resolution of 60,000 relative to 200 Da (at this resolution setting, the maximum rate is 7 Hz). The AGC gain should be set to 10e6, with a maximum injection time of 119 ms. The mass range was set to

80–1000 Da for PCI, and 30–600 Da for EI. Pure methane or either 5% or 10% ammonia in methane (depending on availability) was used as a reagent gas in the PCI mode at a flow rate of 1.5 mL/min (*see* **Note 3**).

3.4 Data Analysis

Raw data files (.raw) generated from Waters GC-TOF/MS in the PCI and EI modes were transferred to CDF files with the DataBridge option in MassLynx software. The CDF files were analyzed in Refiner MS (Genedata Expressionist, Basel, Switzerland) software, to generate the deconvoluted data. Peaks that elute within the same retention time window, signifying a metabolite signature (cluster) were outputted from Refiner MS. IROA peak pairs were selected based on the characteristics of mirrored isotopolog pairs.

3.5 Unknown Metabolite Identification

The GC-IROA based metabolite identification flowchart is shown in Fig. 2. The detailed steps are as following:

1. The metabolites of biological origin are distinguished from artifactual metabolites, which manifest as single peaks unpaired at a given retention time, with the pooled IROA sample analyzed from the PCI data (as described in Subheading 3.2, the IROA sample is the combined 95% ^{12}C and 95% ^{13}C metabolite extracts, derivatized with MOX and unlabeled BSTFA).

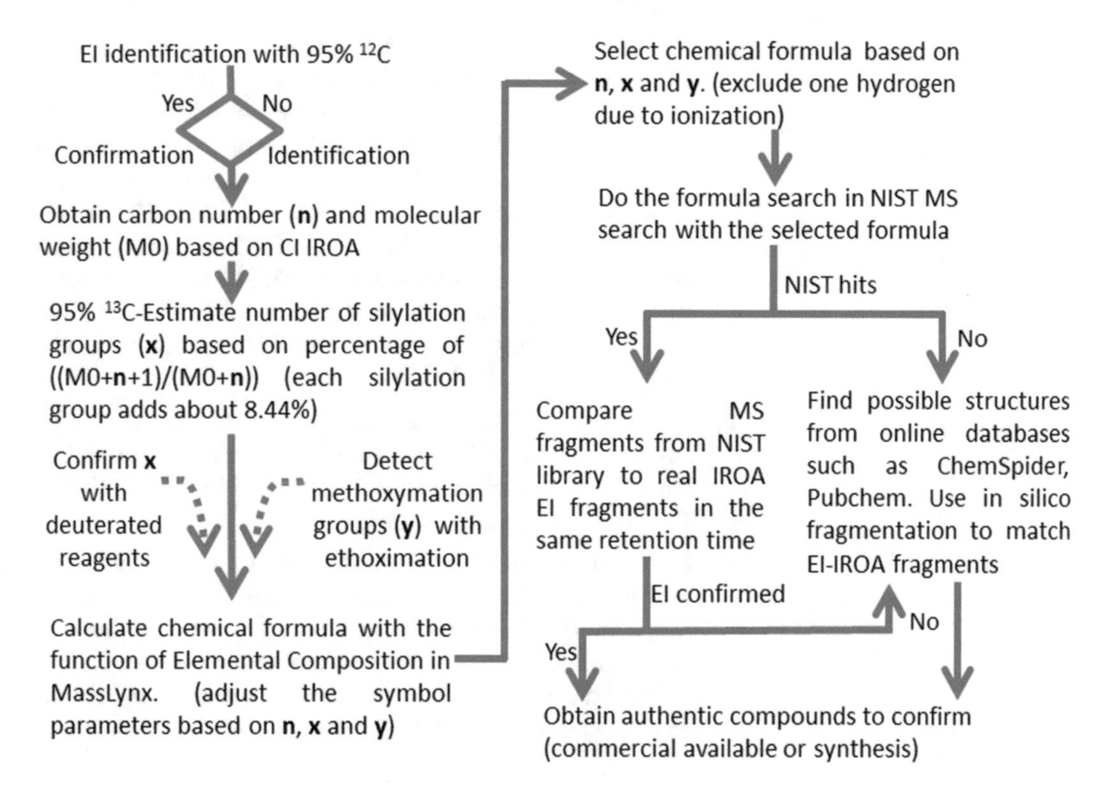

Fig. 2 The GC-IROA based metabolites identification flow chart (adapted from our publication [6])

Metabolites of biological origin in the pooled IROA sample will generate unique mass spectral mirror patterns, due to the higher intensity isotopologs (as compared with basal enrichment). With the unique mirrored IROA pattern, it is easy to distinguish metabolites of biological origin from artifactual peaks.

2. Annotated metabolite identification can be done with the 95% $U^{12}C$ labeled samples in EI mode. With the EI spectrum from 95% $U^{12}C$ sample, metabolite identification can be performed using AMDIS software with our in-house libraries, or the commercial libraries such as NIST, Wiley, and Fiehnlib (*see* **Note 4**).

3. The non-annotated metabolite identification procedure is dependent on characterization of the IROA pattern. The non-annotated peaks are those not confirmed with retention time of authentic standards, or there is no annotation with a confident fit in the libraries. These non-annotated peaks with IROA patterns require further characterization for identification.

   (a) The CI data (molecular ions) from pooled IROA samples will be used to identify total carbon numbers in the backbone of the compound based on the distance between M_0 and M_n peak (carbon number = n) (*see* **Note 5**).

   (b) Calculate the intensity ratio between the isotopolog of M_{n+1} to M_n to estimate the number of silylation groups (x). As described previously, the isotopolog Mn + 1 mainly generated from the natural abundance of ^{13}C, ^{2}H and ^{29}Si from silylation groups. The natural abundance of the isotopes of hydrogen, silicon and carbon are 0.015% for ^{2}H, 4.67% for ^{29}Si, and 1.1% for ^{13}C. Each silylation group ($-SiC_3H_9$) would contribute about 8.54% of the intensity of $(M_{n+1})/(M_n)$ (3 × 1.1% × (0.989)2/(0.989) 3 + 4.67%/0.9223 + 9 × 0.015% × (0.9985)8/ (0.9985) × 9 = 8.54%).

   (c) If desired, validate the number of silylation groups (x) with BSTFA_d9 derivatized 5% ^{13}C metabolite extract (sample 4 in Subheading 3.2) by searching through the PCI chromatography (*see* **Note 6**).

   (d) Detect methoximation groups (y) by comparing to samples treated with ethoximation. In **step 3** of Subheading 3.2, o-methoxylamine can be replaced by o-ethoxylamine, which will result in one carbon difference for each methoximated carbony group. In this way, the carbonyl group can be recognized and used for metabolite identification.

(e) Based on the carbon number (n), number of silylation groups (x), and methoximation information (y), the elements compositions can be precisely determined, rather than just relying on mass differences, and reduces the number of possible chemical formula(e) calculated from the detected molecular ion, which can be done using the Elemental Composition function in MassLynx (for data generated from the Waters GC-TOFMS).

(f) Input the leading possible formula into NIST MS search to find possible chemical structures, and the EI spectrum of the structure.

(g) Compare the MS spectrum from the NIST library to the detected EI-IROA pairs that have been matched with the CI-IROA peak (also strongly considering its retention time).

(h) If there is no potential hits from NIST MS search, there are two ways to find potential hits with in silico fragmentation from their light isotopic sample (95% ^{12}C only). The possible hits are compared with EI-IROA spectrum for their *n*, *x*, and *y* (as shown in Fig. 3).

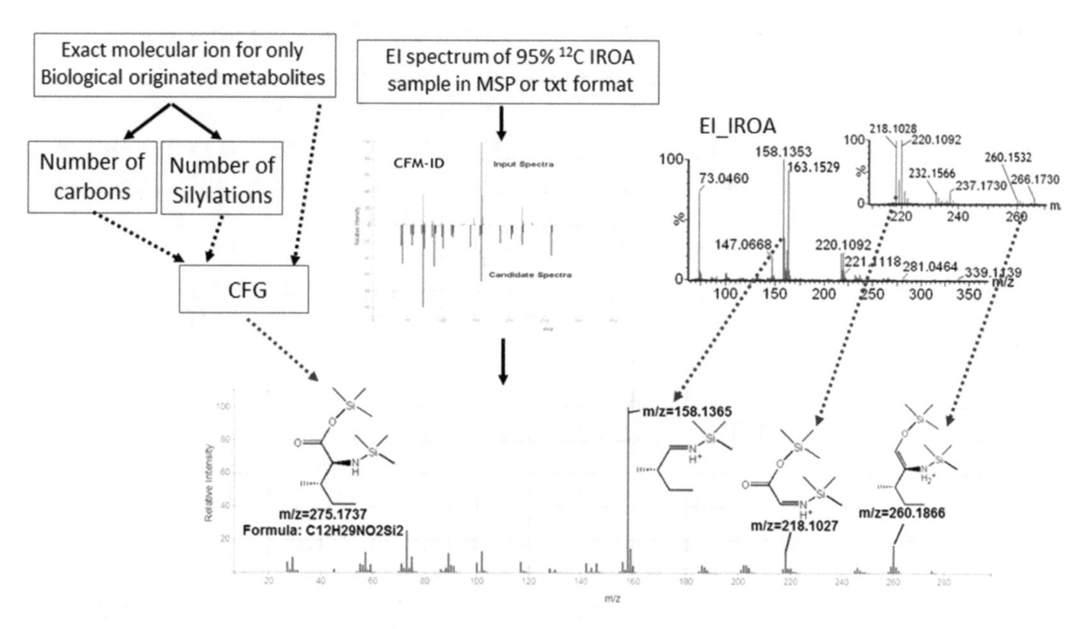

Fig. 3 Expended IROA metabolites identification procedure with *in silico* fragmentation, scheme 1. Isoleucine is used as an example here. Positive CI-IROA distinguishes unknown metabolites of biological origin from any artificial peaks, and identifies the molecular ion, number of carbons (n) in the metabolite from the dalton difference of the M_0 and M_{0+n} IROA peaks, and number of silylation groups attached to the metabolites. The exact mass of M0 molecular ion is combined with the EI mass spectrum found from the 95% ^{12}C IROA sample as inputs into CFM-ID to search for potential IDs. The structural identity and its molecular ion can be validated/confirmed from the CI-IROA molecular ion, and the CFM-ID in silico EI fragmentation pattern can be confirmed from the IROA EI pattern (adapted from our publication [7])

- The EI spectrum can then be input to the Wishart lab's Competitive Fragmentation Modeling program, for metabolite identification (CFM-ID, http://cfmid.wishartlab.com/), to compare with HMDB structures derivatized with the in silico library (Fig. 3) (*see* **Note 7**).
- Another way is to search the formula(e) online such as ChemSpider, Pubchem, or Metabolic In Silico Network Expansion Databases (MINE)) to search the possible chemical structures. The possible hits can then be input into MetaboloDerivatizer (http://prime.psc.riken.jp/Metabolomics_Software/MetaboloDerivatizer/index.html) software to obtain in silico derivatized structure information (*see* **Note 8**).

(i) The potential hits from NIST, or in silico searches with matched EI-IROA peak pairs, can be verified with authentic standards with their retention time and mass spectrum.

4 Notes

1. To insure glucose is the solo carbon source, YNB (BD Difco PN#291940) is purchased glucose and amino acid free.
2. Add some sodium sulfate anhydrous to absorb possible moisture in the solution. Make fresh MOA solution every week.
3. 5% ammonium in methane is preferred for the PCI reagent gas, and 10% ammonium in methane also acceptable. 10% ammonium in methane was used for GC-Orbitrap experiment because of its availibility.
4. The library match score with the 95% $U^{12}C$ sample will be slightly lower than the nature abundance sample due to altered isotopic pattern. Retention time, or retention index, is additionally helpful here for metabolite identification.
5. For the CI data, most of the adducts of molecular ions dominate with the form of MH^+, the adducts for some molecular ions (e.g., sugars) can take the form of $M(NH_4)^+$.
6. There will be a retention time deviation of the deuterated derivatives from the unlabeled metabolites (M_0). Deuterated derivatives have a faster retention time than unlabeled ones. The more the deuterium added on to the metabolite being derivativized (via BSTFA_d9), the larger the retention time shift.
7. This scheme shown in Fig. 3 works with the potential hits within the in silico libraries such as HMDB derivatized library in CFM-ID.

8. Chemical formula(e) calculated from Masslynx contains derivatization groups. You need to subtract the derivatization groups from the molecular formula before submitting the formula to online databases.

Acknowledgments

This work was supported by the Stable Isotope and Metabolomics Core Facility of the Diabetes Research and Training Center (DRTC) of the Albert Einstein College of Medicine (NIH P60DK020541).

References

1. Mahieu NG et al (2014) Credentialing features: a platform to benchmark and optimize untargeted metabolomic methods. Anal Chem 86(19):9583–9589
2. de Jong FA, Beecher C (2012) Addressing the current bottlenecks of metabolomics: isotopic ratio outlier analysis, an isotopic-labeling technique for accurate biochemical profiling. Bioanalysis 4(18):2303–2314
3. Stein S (2012) Mass spectral reference libraries: an ever-expanding resource for chemical identification. Anal Chem 84(17):7274–7282
4. Kind T et al (2009) FiehnLib: mass spectral and retention index libraries for metabolomics based on quadrupole and time-of-flight gas chromatography/mass spectrometry. Anal Chem 81(24):10038–10048
5. Hummel J et al (2010) Decision tree supported substructure prediction of metabolites from GC-MS profiles. Metabolomics 6 (2):322–333
6. Qiu Y et al (2016) Isotopic ratio outlier analysis of the S. cerevisiae metabolome using accurate mass gas chromatography/time-of-flight mass spectrometry: a new method for discovery. Anal Chem 88(5):2747–2754
7. Qiu Y et al (2018) Enhanced isotopic ratio outlier analysis (IROA) peak detection and identification with ultra-high resolution GC-Orbitrap/MS: potential application for investigation of model organism metabolomes. Metabolites 8(1):9
8. Clendinen CS et al (2015) An overview of methods using (13)C for improved compound identification in metabolomics and natural products. Front Plant Sci 6:611
9. Stupp GS et al (2013) Isotopic ratio outlier analysis global metabolomics of Caenorhabditis elegans. Anal Chem 85(24):11858–11865
10. Giavalisco P et al (2009) 13C isotope-labeled metabolomes allowing for improved compound annotation and relative quantification in liquid chromatography-mass spectrometry-based metabolomic research. Anal Chem 81 (15):6546–6551
11. Wu L et al (2005) Quantitative analysis of the microbial metabolome by isotope dilution mass spectrometry using uniformly 13C-labeled cell extracts as internal standards. Anal Biochem 336(2):164–171
12. Bennett BD et al (2008) Absolute quantitation of intracellular metabolite concentrations by an isotope ratio-based approach. Nat Protoc 3 (8):1299–1311
13. Weindl D et al (2015) Isotopologue ratio normalization for non-targeted metabolomics. J Chromatogr A 1389:112–119
14. Blank LM et al (2012) Analysis of carbon and nitrogen co-metabolism in yeast by ultrahigh-resolution mass spectrometry applying 13C- and 15N-labeled substrates simultaneously. Anal Bioanal Chem 403(8):2291–2305
15. Wolf S et al (2010) In silico fragmentation for computer assisted identification of metabolite mass spectra. BMC Bioinformatics 11:148
16. Blaženović I, Kind T, Torbašinović H, Obrenović S, Mehta SS, Tsugawa H et al (2017) Comprehensive comparison of in silico MS/MS fragmentation tools of the CASMI contest: database boosting is needed to achieve 93% accuracy. J Cheminform 9:32

17. Allen F et al (2014) CFM-ID: a web server for annotation, spectrum prediction and metabolite identification from tandem mass spectra. Nucleic Acids Res 42(Web Server issue): W94–W99
18. Ruttkies C et al (2016) MetFrag relaunched: incorporating strategies beyond in silico fragmentation. J Cheminform 8:3
19. Tsugawa H et al (2016) Hydrogen rearrangement rules: computational MS/MS fragmentation and structure elucidation using MS-FINDER software. Anal Chem 88 (16):7946–7958
20. Zhou J et al (2014) HAMMER: automated operation of mass frontier to construct in silico mass spectral fragmentation libraries. Bioinformatics 30(4):581–583
21. Lai Z, Kind T, Fiehn O (2017) Using accurate mass gas chromatography-mass spectrometry with the MINE database for epimetabolite annotation. Anal Chem 89(19):10171–10180

Chapter 8

Enantioseparation and Detection of (R)-2-Hydroxyglutarate and (S)-2-Hydroxyglutarate by Chiral Gas Chromatography–Triple Quadrupole Mass Spectrometry

Shinji K. Strain, Morris D. Groves, and Mark R. Emmett

Abstract

(R)-2-hydroxyglutarate (R-2-hg) has gained significant interest as a circulating biomarker for patients with cancers that harbor mutations in the isocitrate dehydrogenase (IDH) gene. R-2-hg detection is challenging as it must be separated from its enantiomer, (S)-2-hydroxyglutarate (S-2-hg). Here, we present a novel gas chromatography–tandem mass spectrometry method (GC-MS/MS) assay that can be used to separate and detect 2-hg enantiomers for clinical use.

Key words 2-hydroxyglutarate enantiomer, Gas chromatography–tandem mass spectrometry, Chiral Chromatography, Serum, Urine, Isocitrate dehydrogenase, Glioma, Acute Myeloid Leukemia, Cholangiocarcinoma

1 Introduction

The isocitrate dehydrogenase (IDH) gene has recently been identified to be mutated in several human cancers, including glioma, acute myeloid leukemia (AML), cholangiocarcinoma, including others. The mutant IDH (mutIDH) produces the oncometabolite, (R)-2-hydroxyglutarate (R-2-hg), from alpha-ketoglutarate (α-kg) at a rate above its metabolism by R-2-hydroxyglutarate dehydrogenase (R-2HGDH) (Fig. 1) [1]. The result is a significant increase of intracellular R-2-hg which has been shown to exert proliferative and tumorigenic effects on cells leading to its label as an oncometabolite [2, 3]. Blood levels of R-2-hg as a biomarker of early detection, treatment response and recurrence has garnered significant interest since deadly cancers, such as glioma and cholangiocarcinoma, generally portend a poor prognosis, even with treatment [4–8]. However, in order to properly utilize R-2-hg levels as a biomarker, it must be distinguished from the enantiomer (S)-2-hydroxyglutarate (S-2-hg) [9–11]. S-2-hg is made normally

Paul L. Wood (ed.), *Metabolomics*, Neuromethods, vol. 159, https://doi.org/10.1007/978-1-0716-0864-7_8,

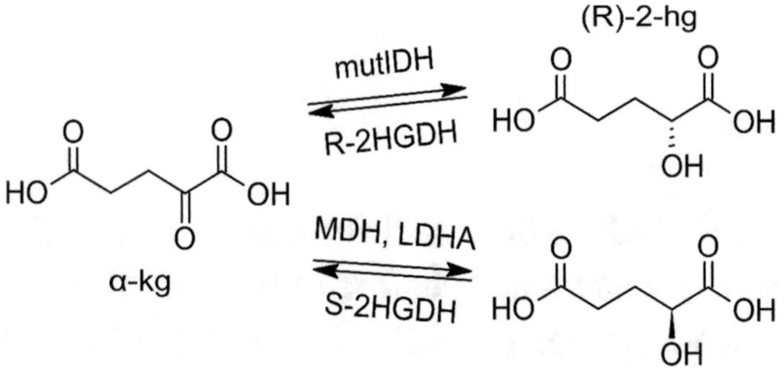

Fig. 1 Structures of (R)-2-hydroxyglutaric acid (R-2-hg) and (S)-2-hydroxyglutaric acid (S-2-hg) and their chief metabolic enzymes. Hydroxyacid-oxoacid transhydrogenase (HOT) is another enzyme that is responsible for basal levels of R-2-hg (not shown). Alpha-ketoglutarate (α-kg), mutantIDH (mutIDH), R-2-hg dehydrogenase (R-2HGDH), malate dehydrogenase (MDH), lactate dehydrogenase A (LDHA), S-2-hg dehydrogenase (S-2HGDH)

in the body during metabolism of α-ketoglutarate (α-KG) by malate dehydrogenase (MDH), degraded by S-2-hydroxyglutarate dehydrogenase (S-2HGDH), keeping basal circulation levels comparable to R-2-hg basal levels [1, 12]. However, although not affected by IDH mutations, S-2-hg concentrations increase during hypoxia through upregulation of MDH and lactate dehydrogenase A (LDHA) [13, 14]. Since hypoxic conditions are commonly found within tumors, it is imperative that R-2-hg and S-2-hg be detected separately in order to use R-2-hg as a biomarker for management of patients with mutIDH cancers [15].

2-hg is a dicarboxylic acid with a hydroxyl group at the alpha or second carbon (Fig. 1). The configuration of the asymmetric carbon differentiates between the R- form (or the D- form using D/L notation) and the S- (L-) form, and the chirality is used as the basis of separation of the enantiomers. A widely utilized methodology converts the enantiomers into diastereomers by use of a chiral derivatizing agent followed by separation over an achiral stationary phase. Both gas chromatography–mass spectrometry (GC-MS) and liquid chromatography–tandem mass spectrometry (LC-MS/MS) assays have been established using such stereochemistry [16–18]. However, chiral derivatizations can result in racemization which is the interconversion of one enantiomer to the other, reducing sensitivity and specificity. Although the LC-MS/MS agents have reported lower racemization (1% or less vs. 5–12% for GC/MS methods), any racemization can skew measured levels of 2-hg enantiomers, especially when monitoring blood levels of 2-hg from brain tumor patients for surveillance, where sensitivity and accuracy are vital.

The methodology reported herein is a gas chromatography–tandem mass spectrometry (GC-MS/MS) assay that utilizes a simple nonchiral derivatization and a chiral column to separate 2-hg enantiomers. Chiral columns separate enantiomers through the chirality of the stationary phase. Though once regarded as expensive and application specific, chiral GC has made significant advances in the separation of metabolites from complex matrices for biological applications [19]. Furthermore, chiral GC columns are commercially available across a wide number of vendors with a multitude of chiral stationary phases (CSP).

The GC-MS/MS assay described extracts 2-hg metabolites using an ethyl acetate liquid–liquid extraction that was optimized for clinical use. The metabolites are derivatized for volatization using (trimethylsilyl)diazomethane and then separated over a β-cyclodextrin chiral GC column. The 2-hg enantiomers are detected using MS/MS and quantitated using a stable-isotope internal standard, disodium (RS)-2-hydroxy-1,5-pentanedioate-2,3,3-d3,OD, which is a racemic mixture of deuterated 2-hg. The assay is used to detect levels of R-2-hg and S-2-hg from blood or urine samples obtained from both healthy or patients with IDH mutant cancers. The methodology is optimized to detect physiologic serum concentrations with near-baseline resolution while maintaining high sensitivity as demonstrated by a signal-to-noise ratio greater than 20:1. This is significant as many prior assays have optimized and demonstrated 2-hg enantiomer separation in urine where basal concentrations of 2-hg are much higher per unit volume. Thus, the GC-MS/MS methodology in this book chapter is optimal for detecting any small changes in serum levels of 2-hg enantiomers that result from a mutIDH brain tumor.

2 Materials and Reagents

All solvents should be mass spectrometry grade or the highest grade obtainable. Reagents were stored at room temperature unless otherwise indicated. Handling of human fluids, such as blood, should be carried out in a biosafety cabinet. Accordingly, all consumables utilized for measurement of 2-hg enantiomers from human samples are discarded into biological hazard waste containers.

2.1 2-hg Standards and Reagents

1. D-α-hydroxyglutaric acid disodium salt (≥98%), L-α-hydroxyglutaric acid disodium salt (≥98%) (Millipore Sigma, St. Louis, MO).
2. Deuterated internal standard (IS), disodium (RS)-2-hydroxy-1,5-pentanedioate-2,3,3-d3,OD (≥95%) (CDN Isotopes, Pointe-Claire, Québec).
3. Mass spectrometry grade H_2O (J.T. Baker, Phillipsburg, NJ).

2.2 Biological Sample Materials and Reagents

1. Serum and urine samples can be stored at −20 °C or lower. Freeze–thaw cycles should be kept at a minimum. Samples are stable at room temperature for several hours but immediate extraction after thawing is optimal. The normal human serum and urine sample used in this chapter are obtained from Innovative Research (Novi, MI, USA), and the serum samples from brain tumor patients are obtained from The Austin Brain Tumor Center (Texas Oncology, Austin, TX) under an IRB-approved study.
2. Mass spectrometry grade ethyl acetate (Millipore Sigma, St. Louis, MO).
3. Hydrochloric acid solution, 6.0 N (Millipore Sigma, St. Louis, MO).
4. Sodium chloride, NaCl (≥99%) (Millipore Sigma, St. Louis, MO).
5. 1.5 mL Microcentrifuge Tube, Screw Cap (Millipore Sigma, St. Louis, MO)
6. 11 mm Micro-sampling 1.5 mL Vial, <15 μL reservoir (Thermo Fisher Scientific, Waltham, MA).

2.3 Derivatization Reagents

1. (Trimethylsilyl)diazomethane ((TMS)DAM) solution (2.0 M) in Hexanes (Millipore Sigma St. Louis, MO).
2. Mass spectrometry grade methanol (J.T. Baker, Phillipsburg, NJ).
3. Glacial acetic acid (≥99%) (Millipore Sigma St. Louis, MO).

2.4 Gas Chromatography–Tandem Mass Spectrometry (GC-MS/MS) Equipment and Supplies

1. CP-Chirasil-Dex CB Column (0.25 mm × 25 m × 25 μm) CP7502 (Agilent, Santa Clara, CA).
2. Shimadzu GCMS-TQ8050 Triple Quadrupole Mass Spectrometer equipped with an AOC-20i/20s autoinjector and sampler.

3 Methods

As compared to current methods used to detect 2-hg enantiomers, the GC-MS/MS assay presented herein is advantageous as it improves the efficiency of the extraction procedure, avoids racemization, and utilizes a commercially available chiral column. Thus, the assay is optimized for the extraction of 2-hg from biological fluids for clinical use.

Liquid–liquid extraction using methanol or ethyl acetate as the organic solvent is commonly used to extract small molecule metabolites like 2-hg. An ethyl acetate–based extraction is advantageous

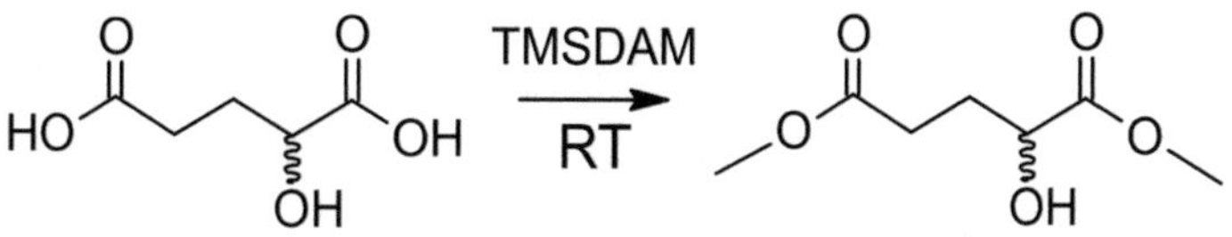

Fig. 2 Derivatization scheme for methyl esterification of 2-hg. The reaction is carried out at room temperature (RT) using (TMS)DAM ((trimethysilyl)diazomethane)

as it is selective for organic acids and does not require the use of dry ice temperatures (−80 °C) or an incubation period [20, 21]. Prior studies using ethyl acetate have utilized organic solvent volumes of up to 5 mL [22, 23]. However, the method described herein reduces the volume to 1 mL to allow for the derivatization to take place in the same glass vial that the entire organic layer is freeze-dried in.

The sample is derivatized for the purposes of volatilization and not for separation which chiral derivatization achieves. Methyl esterification with (TMS)DAM can be done at room temperature for 30 min and is ready for analysis immediately (Fig. 2). The derivatization avoids racemization since only the carboxyl groups are methylated. Chiral derivatization reagents such as diacetyl-L-tartaric anhydride (DATAN) and N-(o-toluenesulfonyl)-L-phenylalanyl chloride (TSPC) chemically modify the chiral center, resulting in a low but nonnegligible percentage of racemization [17, 18]. Thus, methods that utilize chiral derivatization may not be suitable for applications that require high sensitivity, such as the detection of changes in serum 2-hg enantiomer levels resulting from a mutIDH brain tumor.

After methylation, 2-hg enantiomers are separated over a β-cyclodextrin capillary GC column, which is a common, commercially available CSP that is now comparable in cost to the majority of reverse phase LC columns. Electron ionization (EI) results in fragments that are monitored by tandem MS. The resulting areas of each enantiomer fragments are integrated, and the concentration of each 2-hg enantiomer is determined relative to the internal standard.

3.1 Preparation of Standards

Stock solutions of R-2-hg, S-2-hg, and the deuterated R/S 2-hg are made by dissolving 1 mg of powder in the appropriate amount of H_2O to make a 1 mM solution. The molecular weight of the sodium salts for each compound must be accounted for when making the stock solution. Stock solutions can be aliquoted into 0.5 mL or 1.5 mL Eppendorf tubes and are stable at −80 °C for several years.

3.2 Instrument Response and Validation

Characterization of instrument response can be used for calibration purposes or for the validation of the GC-MS/MS assay. The following recommended steps for validation are taken from the US

Food and Drug Administration (FDA) guidelines for Bioanalytical Method Validation Assays but can be skipped if assay has been previously validated elsewhere (*see* **Note 1**) [24].

3.2.1 Calibration Curve

Serial dilutions of a stock solution can be used to characterize the instrument response and to build a calibration curve for an external calibration if an internal standard is not utilized. For validation, the calibration curve must include the upper and lower limits of quantitation, a blank, and should be linear within the range of the levels of the analyte of interest. A comprehensive range with 9 concentrations from 100 nM to 1 mM (100 fmol/μL to 1 nmol/μL) can be prepared as follows:

1. Label eight (8) 0.5 mL Eppendorf tubes as 100 nM, 500 nM, 1 μM, 5 μM, 10 μM, 50 μM, 100 μM, 500 μM and ten 12 × 32 mm clear crimp top vials w/ glass insert as Blank, 100 nM, 500 nM, 1 μM, 5 μM, 10 μM, 50 μM, 100 μM, 500 μM, and 1 mM.
2. A 500 μM solution is created by adding 100 μL of 1 mM R-2-hg and 100 μL of 1 mM S-2-hg to the "500 μM" glass vial, and a 100 μM solution by adding 800 μL H_2O and 100 μL of R-2-hg stock solution and 100 μL of S-2-hg stock solution to the "100 μM" glass vial. Vortex for 10 s (*see* **Note 2**).
3. Successive serial 1:10 dilutions of the 500 μM and 100 μM solutions in water are then performed to make up the rest of the 6 concentrations.
4. Add 10 μL of each solution (100 nM to 1 mM) to its corresponding glass vial and freeze-dry.
5. Derivatize (*see* Subheading 3.4) and analyze by GC-MS/MS (*see* Subheading 3.5).
6. Define the ULoQ and LLoQ of the assay (*see* **Note 3**) and complete validation studies if necessary.

3.3 Extraction of Biological Samples

Label the appropriate amount of microcentrifuge tubes according to the number of biological samples to be analyzed with the addition of a "blank extraction" tube. The blank has the internal standard added in but not a biological sample and is extracted along with the rest of the samples (*see* **Note 4**).

3.3.1 Serum

1. In a 1.5 mL microcentrifuge tube, add 250 μL of serum.
2. Add 3 μL of 100 μM deuterated 2-hg as an internal standard followed by 10 μL of 6 N HCl.
3. Saturate the solution with approximately 100–300 mg NaCl.
4. Screw cap tightly and vortex each sample for approximately 30 s or until the sample turns into a white milk-like solution (*see* **Note 5**).

5. To each sample then add 1 mL of mass spec grade ethyl acetate and vortex for another 30 s.
6. Centrifuge the sample at 5000 × *g* for 5 min and while the samples are spinning, label the appropriate amount of 1.5 mL 12 × 32 mm micro-v glass vials.
7. The spun down samples will have an organic (top) and aqueous layer (bottom) separated by a thick, white solid layer. Pipet 900 μL of the organic layer into the micro-v glass vial.
8. Dry the organic layer in a vacuum concentrator and repeat **steps 5–7** for a second extraction if desired (*see* **Note 6**).

3.3.2 Urine

1. In a 1.5 mL microcentrifuge tube, add 100 μL of urine.
2. Add 12 μL of 100 μM deuterated 2-hg as an internal standard followed by 10 μL of 6 N HCl.
3. Saturate the solution with approximately 100–300 mg NaCl.
4. To each sample then add 1 mL of mass spec grade ethyl acetate and vortex for 30 s.
5. Centrifuge the sample at 5000 × *g* for 5 min and while the samples are spinning, label the appropriate amount of 1.5 mL 12 × 32 mm micro-v glass vials.
6. The spun down samples will have an organic (top) and aqueous layer (bottom). Pipet 900 μL of the organic layer into the micro-v glass vial.
7. Dry the organic layer in a vacuum concentrator and repeat **steps 5–7** for a second extraction if desired (*see* **Note 6**).

3.4 Derivatization

1. To each dried sample is added 100 μL of MeOH followed by 20 μL of 2.0 M (TMS)DAM in hexanes.
2. The sample is then sealed with a crimp cap and placed on a shaker for 30 min at room temperature (~25 °C).
3. After derivatization, the reaction is quenched with 20 μL of acetic acid and the sample is ready for analysis.

3.5 GC-MS/MS Analysis

1. A 1 μL splitless injection of each sample is performed using a Shimadzu AOC-20i injector and AOC-20s autosampler.
2. The chiral GC column (25 m × 0.25 mm × 0.2 5 μm) used to separate 2-hg enantiomers contains β-cyclodextrin bonded directly to dimethylpolysiloxane and conditioned according to manufacturer recommendations.
3. Helium was used as the carrier gas with a flow rate of 1.5 mL/min. After injection, the GC oven was held at 80 °C for 1 min and subsequently ramped at 4 °C/min to 180 °C for 4 min for a total run time of 30 min for each sample.

Table 1
GC-MS/MS parameters

Parameter	Setting
Inlet temperature	220 °C
Injection pressure	250 kPa
Sampling time	1 min
Flow control	Linear velocity
Column flow	1.50 mL/min
Total flow	25 mL/min
Purge flow	3.0 mL/min
Linear velocity	49.4 cm/s
Interface temperature	200 °C
Ion source temperature	200 °C
Ionization energy	70 eV
MRM event time	0.040 s
117 > 85 *m/z* collision energy	13.00 V
85 > 29 *m/z* collision energy	6.00 V
Q1/Q3 resolution	Unit/unit

4. 2-hg enantiomer compounds elute from the column into the ion volume where they are ionized and fragmented by electrons at 70 eV. Multiple reaction monitoring (MRM) is used to detect two transitions unique to 2-hg, 117 *m/z* > 85 *m/z* and 85 *m/z* > 29 *m/z*, where the former transition is used for quantitation and the later as a qualifier transition. The internal standard contains three deuterated carbons atoms so the quantifier transition is 120 *m/z* > 88 *m/z* and the qualifier transition, 88 *m/z* > 32 *m/z*.
5. All samples were run with the GC-MS/MS parameters listed in Table 1.
6. R-2-hg and S-2-hg elute between 15 and 16 min with R-2-hg eluting prior to S-2-hg with near-baseline resolution. Figure 3 demonstrates a typical chromatogram of 2-hg enantiomers as measured from serum (A) and urine (B).

3.6 Quantification of 2-hg Enantiomers

1. Integration of the endogeneous and spiked 2-hg peaks is done automatically using the Shimadzu Postrun software. The slope and the width usually must be adjusted accordingly to <5/min and 5 s, respectively, in order for the software to integrate the full peak.

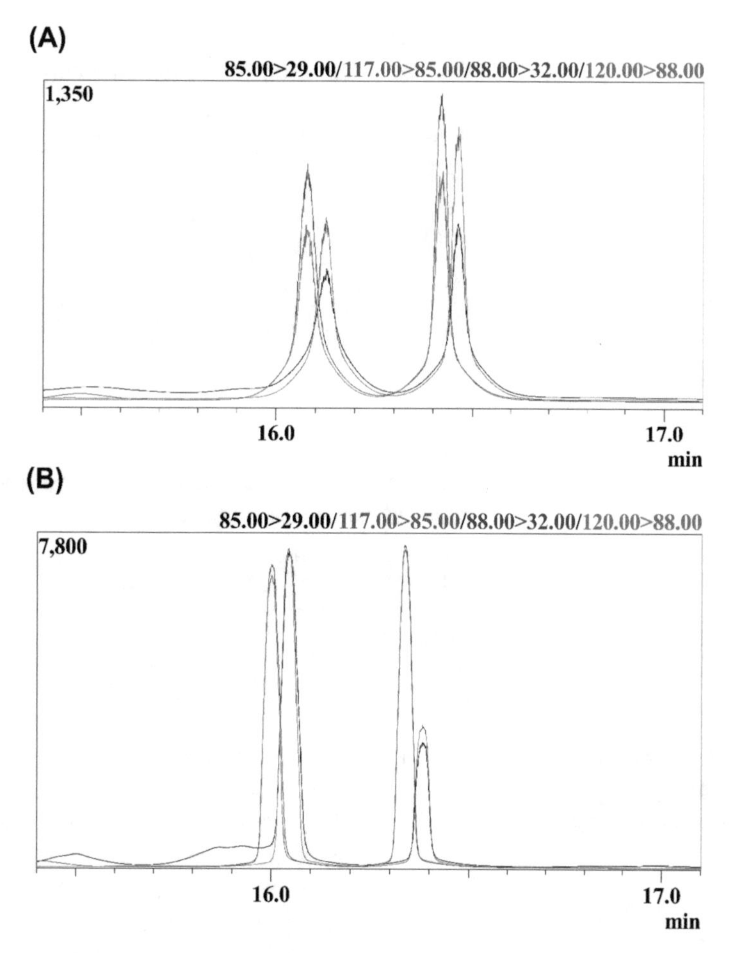

Fig. 3 MRM Chromatograms demonstrating separation of detection of R-2-hg and S-2-hg in normal serum (**a**) and urine (**b**). R-2-hg elutes prior to S-2-hg. Endogenous 2hg: 85 > 29 *m/z* (black) and 117 > 85 *m/z* (magneta), Internal Standard: 88 > 32 *m/z* (blue) and 120 > 88 *m/z* (green)

2. To calculate the concentration of endogeneous 2-hg, the following equation is used: $2\text{hg} = \frac{\text{AUC}_{\text{endo2hg}}}{\text{AUC}_{\text{IS}}} \times \text{Conc}_{\text{IS}}$ where AUC = Area Under the Curve, IS = Internal Standard, Endo2hg = endogenous 2-hg, and Conc_{IS} = the concentration of the spiked deuterated 2-hg (*see* **Note 7**).

3.7 Normal Concentrations for Serum

Expected values for R-2-hg, S-2-hg, Total 2-hg, and the ratio of R-2-hg:S-2-hg, as measured in serum, is shown in Table 2. Basal levels of each 2-hg enantiomer are comparable individually but have a wide range. Thus, measuring Total 2-hg or reporting only the 2-hg Ratio could mask any small but significant changes in each individual enantiomer which is important when monitoring R-2-hg levels as a biomarker of treatment response, assessment of residual tumor, and recurrence.

Table 2
Normal serum 2-hg values reported in μM

	Mean (Min – Max)
R-2-hg	1.09 (0.32 – 2.30)
S-2-hg	1.11 (0.48 – 2.00)
Total 2-hg	2.20 (1.08 – 3.86)
2-hg ratio (R-2-hg:S-2-hg)	1.07 (0.26 – 2.70)

Table 3
2-hg enantiomer serum levels from a patient with and without a mutIDH brain tumor. Values are report in μM

IDH status	R-2-hg	S-2-hg	Total 2-hg	2-hg ratio
Mutant	2.94	1.24	4.18	2.37
WT	1.61	1.32	2.93	1.22

3.8 Changes in Serum 2-hg Levels from mutIDH Glioma

Serum 2-hg enantiomer levels taken from one patient with mutIDH glioma is shown in Table 3. The patient's R-2-hg, Total 2-hg, and 2-hg ratio values are increased relative to the normal values reported in Table 2. However, the S-2-hg levels are normal which is expected since a mutated tumor only makes the (R) enantiomer. A patient with wild-type IDH tumor is shown in Table 3 where all values are normal. The results shown here only demonstrate feasibility and the application to patients with mutIDH brain tumors. Further validation studies are required to prove efficacy of R-2-hg as a biomarker of disease.

4 Notes

1. FDA Bioanalytical Method Validation parameters include but are not limited to assessment of Quality Controls (QCs), Calibration Curves, Accuracy & Precision, LLoQ and ULoQ, Carryover, Selectivity/Specificity, and Stability. Values for all validation parameters are not displayed in this chapter for the sake of brevity.
2. If only testing the instrument response for the deuterated R/S 2-hg compound, instead add 100 μL of H_2O and 100 μL of 1 mM deuterated 2-hg to the "500 μM" glass vial, and 900 μL of H_2O and 100 μL of 1 mM deuterated 2-hg to the "100 μM" glass vial since the vial contains a racemic mixture. Furthermore, at this step, other 1:2 and 1:4 dilutions can be made from each of the 1:10 serial dilutions to build a set of QCs for validation purposes.

3. The ULoQ is defined as the highest concentration on the linear calibration curve and LLoQ should be a concentration where the amount injected does not vary more than 20%. For this GC-MS/MS assay, the instrument response is linear between 35 fmol and 35 pmol (amount injected) and also corresponded to the LLoQ and ULoQ, respectively.
4. PBS can be added as another control to approximate a biological matrix with a similar salt content as that which is found in the body. To our knowledge, no perfect biological matrix mimicking serum or urine without 2-hg exists.
5. If the solution does not turn "milk-white," add more NaCl, cap, and vortex once more until color change has been achieved.
6. A reextraction with a second milliliter of ethyl acetate step will result in a higher amount of 2-hg extracted from the sample. However, the signal-to-noise of 2-hg enantiomer peaks produced with only one extraction is adequate, especially when extracting from urine. Interassay variability (relative standard deviation) is approximately 9–10% on average using only one ethyl acetate extraction.
7. When calculating urine 2-hg levels, it is recommended to normalize to urine creatinine levels to accommodate for any changes in creatinine clearance that could be altered during any cancer therapies.

Acknowledgments

This work was supported by The Shimadzu Corporation and The Head for The Cure Foundation. SKS was supported by The Institute for Translational Sciences at The University of Texas Medical Branch, supported, in part, by a Clinical and Translational Science Award NRSA (TL1) Training Core (TL1TR001440) from the National Center for Advancing Translational Sciences, National Institutes of Health.

References

1. Losman JA, Kaelin WG (2013) What a difference a hydroxyl makes: mutant IDH, (R)-2-hydroxyglutarate, and cancer. Genes Dev 27:836–852
2. Koivunen P et al (2012) Transformation by the (R)-enantiomer of 2-hydroxyglutarate linked to EGLN activation. Nature 483:484–488
3. Losman J-A et al (2013) (R)-2-hydroxyglutarate is sufficient to promote leukemogenesis and its effects are reversible. Science 339:1621–1625
4. Capper D et al (2012) 2-Hydroxyglutarate concentration in serum from patients with gliomas does not correlate with IDH1/2 mutation status or tumor size. Int J Cancer 131:766–768
5. DiNardo CD et al (2013) Serum 2-hydroxyglutarate levels predict isocitrate dehydrogenase mutations and clinical outcome in acute myeloid leukemia. Blood 121:4917–4924

6. Wang J-H et al (2013) Prognostic significance of 2-hydroxyglutarate levels in acute myeloid leukemia in China. Proc Natl Acad Sci U S A 110:17017–17022
7. Borger DR et al (2014) Circulating oncometabolite 2-hydroxyglutarate is a potential surrogate biomarker in patients with isocitrate dehydrogenase-mutant intrahepatic cholangiocarcinoma. Clin Cancer Res 20:1884–1890
8. Brunner AM et al (2015) Use of 2HG levels in the serum, urine, or bone marrow to predict IDH mutations in adults with acute myeloid leukemia. Blood 126:2597
9. Delahousse J et al (2018) Circulating oncometabolite D-2-hydroxyglutarate enantiomer is a surrogate marker of isocitrate dehydrogenase--mutated intrahepatic cholangiocarcinomas. Eur J Cancer 90:83–91
10. Janin M et al (2014) Serum 2-hydroxyglutarate production in IDH1- and IDH2-mutated de novo acute myeloid leukemia: a study by the acute leukemia French Association group. J Clin Oncol 32:297–305
11. Struys EA (2013) 2-Hydroxyglutarate is not a metabolite; D-2-hydroxyglutarate and L-2-hydroxyglutarate are! Proc Natl Acad Sci U S A 110:E4939
12. Rzem R, Vincent M-F, Van Schaftingen E, Veiga-da-Cunha M (2007) l-2-Hydroxyglutaric aciduria, a defect of metabolite repair. J Inherit Metab Dis 30:681–689
13. Tyrakis PA et al (2016) S-2-hydroxyglutarate regulates CD8+ T-lymphocyte fate. Nature 540:236–241
14. Intlekofer A et al (2015) Hypoxia induces production of L-2-hydroxyglutarate. Cell Metab 22:304–311
15. Hanahan D, Weinberg RA (2011) Hallmarks of cancer: the next generation. Cell 144:646–674
16. Duran M, Kamerling JP, Bakker HD, van Gennip AH, Wadman SK (1980) L-2-hydroxyglutaric aciduria: an inborn error of metabolism? J Inherit Metab Dis 3:109–112
17. Struys EA, Jansen EEW, Verhoeven NM, Jakobs C (2004) Measurement of urinary D- and L-2-hydroxyglutarate enantiomers by stable-isotope-dilution liquid chromatography-tandem mass spectrometry after derivatization with diacetyl-L-tartaric anhydride. Clin Chem 50:1391–1395
18. Cheng Q-Y et al (2015) Sensitive determination of onco-metabolites of D- and L-2-hydroxyglutarate enantiomers by chiral derivatization combined with liquid chromatography/mass spectrometry analysis. Sci Rep 5:15217
19. Patil RA, Weatherly CA, Armstrong DW (2018) Chiral gas chromatography. Chiral Anal 468–505. https://doi.org/10.1016/B978-0-444-64027-7.00012-4
20. Jones PM, Boriack R, Struys EA, Rakheja D (2017) Measurement of oncometabolites D-2-hydroxyglutaric acid and L-2-hydroxyglutaric acid. Methods Mol Biol 1633:219–234. https://doi.org/10.1007/978-1-4939-7142-8_14
21. Oldham W, Loscalzo J (2016) Quantification of 2-hydroxyglutarate enantiomers by liquid chromatography-mass spectrometry. Bio Protoc 6:e1908
22. Kamerling JP, Gerwig GJ, Vliegenthart JF (1977) Determination of the configurations of lactic and glyceric acids from human serum and urine by capillary gas-liquid chromatography. J Chromatogr 143:117–123
23. Sahm F et al (2012) Detection of 2-hydroxyglutarate in formalin-fixed paraffin-embedded glioma specimens by gas chromatography/mass spectrometry. Brain Pathol 22:26–31
24. U. S. Food and Drug Administration (2018) Bioanalytical method validation. Guidance for industry

Chapter 9

Enantioselective Supercritical Fluid Chromatography (SFC) for Chiral Metabolomics

Robert Hofstetter, Andreas Link, and Georg M. Fassauer

Abstract

Supercritical fluid chromatography (SFC) is a minor but significant tool in chiral analysis of drug substances and the greenest semipreparative purification technique known but still awaits its breakthrough in metabolomics. Until recently, sample preparation has been identical for HPLC and SFC. Here, we describe the online solid-phase CO_2 extraction SFE-SFC-MS analysis of a mixture of chiral metabolites of the investigational antidepressant ketamine from urine. Direct transfer of the CO_2 extract to the stationary phase enables fast separation of 8 analytes in a single run.

Key words SFC, SFE, Enantioselective, Chirality, Ketamine, Extraction, Method development, Validation

1 Introduction

While supercritical fluid chromatography (SFC) is mainly applied to industrial-scale preparative separations or chiral chromatography of active pharmaceutical ingredients (API), packed-column SFC has begun to encroach bioanalytical territory, for example enantioselective quantitation of chiral metabolites from organic matrices like urine or plasma. Due to its potentially superior performance and easy hyphenation to mass spectrometry (MS), SFC-MS may invigorate chiral metabolomics on a scale similar to LC-MS technology's revolutionary impact on bioanalysis.

SFC relies on supercritical carbon dioxide ($scCO_2$) as low viscous, nonpolar component of its mobile phase [1]. Although the mobile phase may fluctuate between the super- and subcritical state during analysis, separation power does not suffer from this fact [2] and generally remains at least equivalent to most other forms of chromatography in terms of theoretical plates. Van Deemter curves in SFC are less steep than those in HPLC, indicating higher performance even at increased velocity, and indeed run times in SFC are often only a fraction of those in HPLC [3]. This makes $scCO_2$ a

Paul L. Wood (ed.), *Metabolomics*, Neuromethods, vol. 159, https://doi.org/10.1007/978-1-0716-0864-7_9,

versatile means for rapid analysis, especially on stationary phases containing sub-2 μm particles, and in the near future, in 3 mm columns. Together with low dispersion plumbing and minimal extracolumn variance, pressure drops and resistive heating can be minimized and ultrahigh-performance SFC (UHPSFC) will be achievable, soon. Due to the slower development of sub-5 μm particles for chiral stationary phases (CSP), enantioselective methods relying on 5 μm particles are still common, with most applications using polysaccharide-based CSPs [4, 5]. The combination of cellulose- or amylose-based CSPs and the unique properties of $scCO_2$-containing mobile phases (e.g., low surface tension and high penetration of porous particles) unlocks new possibilities for metabolism studies in terms of efficiency and analysis speed compared to classical or even ultrahigh-performance liquid chromatography (UHPLC) [6].

Similarly, $scCO_2$ can be utilized for supercritical fluid extraction (SFE) of biological samples, a rarely reported but highly auspicious alternative to conventional extraction procedures. Due to the overlap in mobile phase and apparatus, SFE and SFC can easily be interfaced to construct a fully automatable platform (online SFE-SFC), freeing up both time and labour by simplifying sample treatment, one of the most resource-consuming steps in bioanalysis [7].

A prime example of SFC's advantages can be found in the enantioselective quantification of racemic ketamine (KET) and its chiral metabolites (*RS*)-norketamine (NK), (*RS*)-dehydronorketamine (DNK), and (2*R*,6*R*),(2*S*,6*S*)-hydroxynorketamine (HNK) in human urine [8, 9]. The enantioselective elucidation of KET-metabolism is of particular neurological interest, as (*RS*)-KET and (2*R*,6*R*)-HNK are understood to exert almost instantaneous antidepressant effects, requiring further investigation in order to harness KET's full pharmacological potential as a next-generation antidepressant [10–12]. While future clinical trials need to confirm the active metabolite(s) as well as their eudysmic ratio, renally excreted (2*R*,6*R*)-HNK is among the most promising candidates regarding antidepressant action [13]. Here, we present a protocol for quantitation using SFC-MS following either conventional liquid–liquid extraction (offline LLE-SFCMS) or online $scCO_2$ extraction (SFE-SFC-MS) to illustrate three key points: (1) the practicability of SFC-MS in metabolomics, (2) the contribution of sample preparation to chromatographic success, as well as (3) the differences in expenditure of human labour between online and offline approaches (Fig. 1).

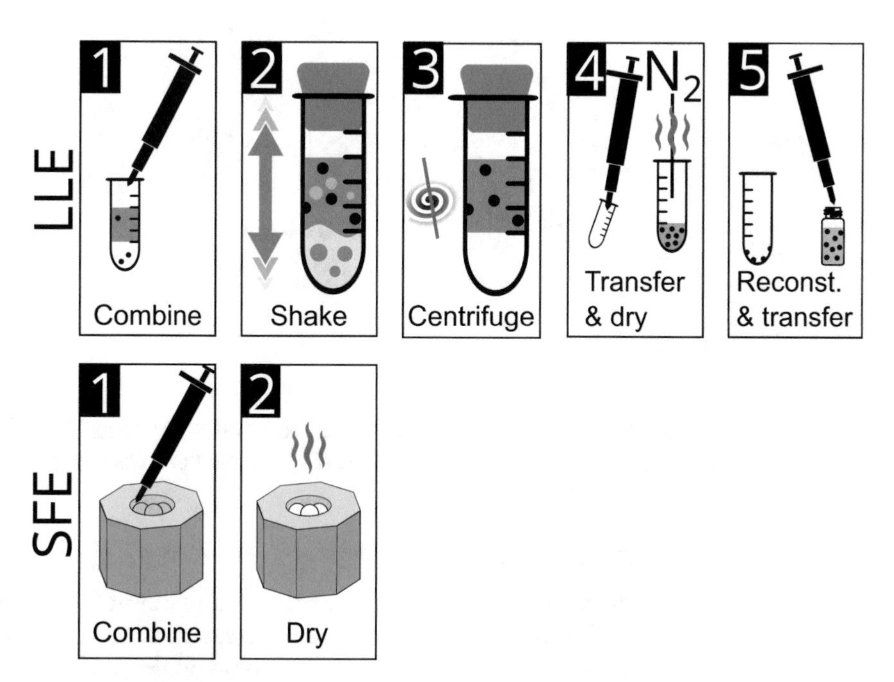

Fig. 1 Workflow chart illustrating the differences in sample preparation for liquid-liquid extraction (LLE) and supercritical fluid extraction (SFE). LLE (top) consists of five steps: (1) introduction of the biosample into a test tube containing an aqueous and organic phase, (2) extended shaking to allow for analyte enrichment in the organic phase, (3) centrifugation for phase separation and sedimentation of solid matrix particles, (4) transfer of the organic phase into a clean tube and evaporation under a stream of nitrogen, and (5) reconstitution in new solvent and transfer into an HPLC vial. In contrast, SFE (bottom) requires only two steps: (1) introduction of the biosample into an extraction vessel containing an adsorbent and (2) removing excess moisture. This can be achieved at 51 °C and 7 mbar (1 h) or at room temperature in a desiccator (overnight)

2 Materials

Prepare the following materials from chemicals of MS grade, unless stated otherwise.

1. Stock solutions: 0.1 mg/mL analyte in acetonitrile (*see* **Note 1**). Weigh (if procured as solid) or dilute (liquids) corresponding amounts to prepare separate solutions of each target analyte, that is, racemic KET, NK, DNK, and HNK (*see* **Note 2**), as well as a suitable internal standard (IS) (e.g., tetra-deuterated racemic KET-d_4 or NK-d_4) (*see* **Note 3**). Store at −20 °C.
2. Working solution: 1 μg/mL per target analyte in 2-propanol. Charge one volumetric flask with 100 μL of each target analyte stock solution and add 2-propanol to yield 10.0 mL. Store at 4 °C.
3. The chromatographic system needs an additional CO_2 pump, a backpressure regulator (BPR), and a supercritical fluid extraction unit besides the standard HPLC components (Fig. 2).

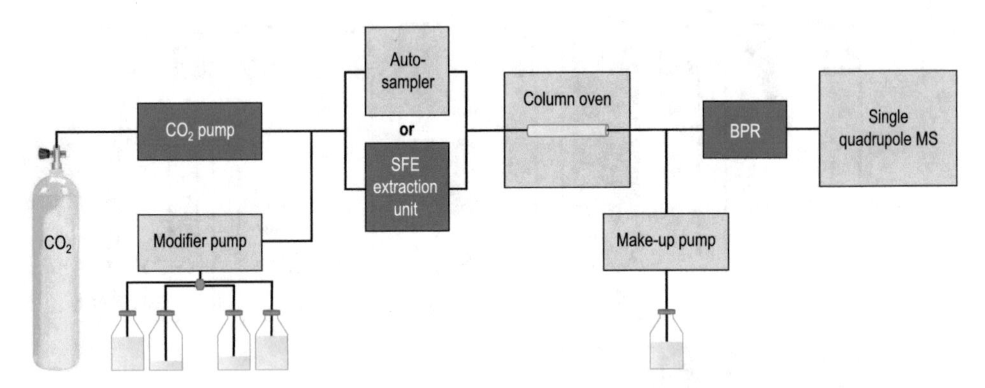

Fig. 2 SFC-MS or SFE-SFC-MS setup: Light blue–colored parts can be found in "classical" HPLC systems. Dark blue parts are additional modules required for SFC and SFE-SFC. Note that the autosampler and the SFE extraction unit cannot be used at the same time

4. Mobile phase A: CO_2 (99.995% purity, **Note 4**).
5. Mobile phase B (Modifier): MeOH containing 0.075% ammonia (*see* **Note 5**). Add 1.5 mL of concentrated ammonia (25% aqueous solution) to 500 mL of MeOH and sonicate for 10 min (*see* **Note 6**).
6. Mobile phase C (Makeup): MeOH containing 0.1% formic acid. Add 0.5 mL formic acid to 500 mL of MeOH and sonicate for 10 min (*see* **Note 7**).
7. Stationary phase: tris(5-chloro-2-methylphenylcarbamate) amylose of 150 × 4.6 mm dimension and particle size of 5 μm (Phenomenex® Lux-Amylose 2). Use the same material as guard column (4 × 3.0 mm).
8. Target biomatrix: urine (*see* **Note 8**).
9. Extraction media and equipment.
   (a) LLE: methyl *tert*-butyl ether and sodium carbonate solution (1:1 dilution of saturated aqueous solution).
   (b) SFE: diatomaceous earth (ISOLUTE® HM-N) and stainless-steel SFE extraction vessels.

3 Methods

3.1 Preparations

1. Set your chromatographic system according to following data: 3 mL/min mobile phase flow rate, 0.3 mL/min makeup flow rate, 8% B for the LLE method and 3% B for the SFE method. Set back pressure regulator to 150 bar and 50 °C, column oven to 30 °C.
2. Set your MS system according to following settings: electrospray ionization (ESI) in positive mode (interface voltage of 4.5 kV), nitrogen as nebulizing (1.5 L/min) and drying gas

(12 L/min). Set heat block temperature to 300 °C, interface temperature to 350 °C and desolvation line to 250 °C (*see* **Note 9**).

3. Analytes are detected in selected ion monitoring (SIM) mode as the molecular ions $[M + H]^+$ (KET: 238, NK: 224, DHK: 222, HNK: 240, NK-d_4: 228, KET-d_4: 242).
4. Flush the system under starting conditions for at least 20 min (*see* **Note 10**).
5. Determine selectivity (retention times) using diluted working solution without urine matrix at a concentration level of about 500 ng/mL.
6. Prepare 1-mL samples of known concentration for calibration curve and quality control (QC). This is achieved by spiking blank human urine with suitable volumes of working solution to yield at least six defined concentration levels (e.g., 5, 10, 25, 50, 100 and 200 ng/mL) (*see* **Note 11**).
7. Prepare 1-mL aliquots of clinical urine samples (if available).
8. To each sample (known and unknown), add 10 μL of the IS stock solution to yield a final concentration of approximately 1 μg/mL IS. Also prepare a blank urine sample with IS (*zero*) and one without IS (*blank*) (*see* **Note 12**).

3.2 LLE-SFC

1. Transfer each sample to a fresh 10-mL tube and add 750 μL of sodium carbonate solution and 4 mL of methyl *tert*-butyl ether (*see* **Note 13**).
2. Fit each tube with a stopper and shake vigorously for 15 min at room temperature.
3. Centrifuge for 5 min at 3200 × *g* to ensure phase separation.
4. Transfer 3 mL of the organic layer to a second tube and evaporate the ether under a gentle nitrogen stream at room temperature (30–60 min, depending on nitrogen flow rate).
5. Reconstitute the residue in 100 μL of MeOH and transfer the solution into an HPLC vial (*see* **Note 14**). Specimen can be stored for up to 96 h in the autosampler set to 6 °C.
6. Begin SFC by injecting 5 μL and running the following gradient:

   0–7 min at 8% B, 7–7.25 min increasing gradually to 20% B, 7.26–11 min at 20% B, and 11–15 min at 8% B.
7. For elution order check Fig. 3.

3.3 SFE-SFC

1. Fill SFE extraction vessels with diatomaceous earth (100 μL) and add 50 μL of IS-containing samples (blanks, calibration curve, QC or clinical samples if available) (*see* **Note 15**).
2. Let dry (*see* **Note 16**).

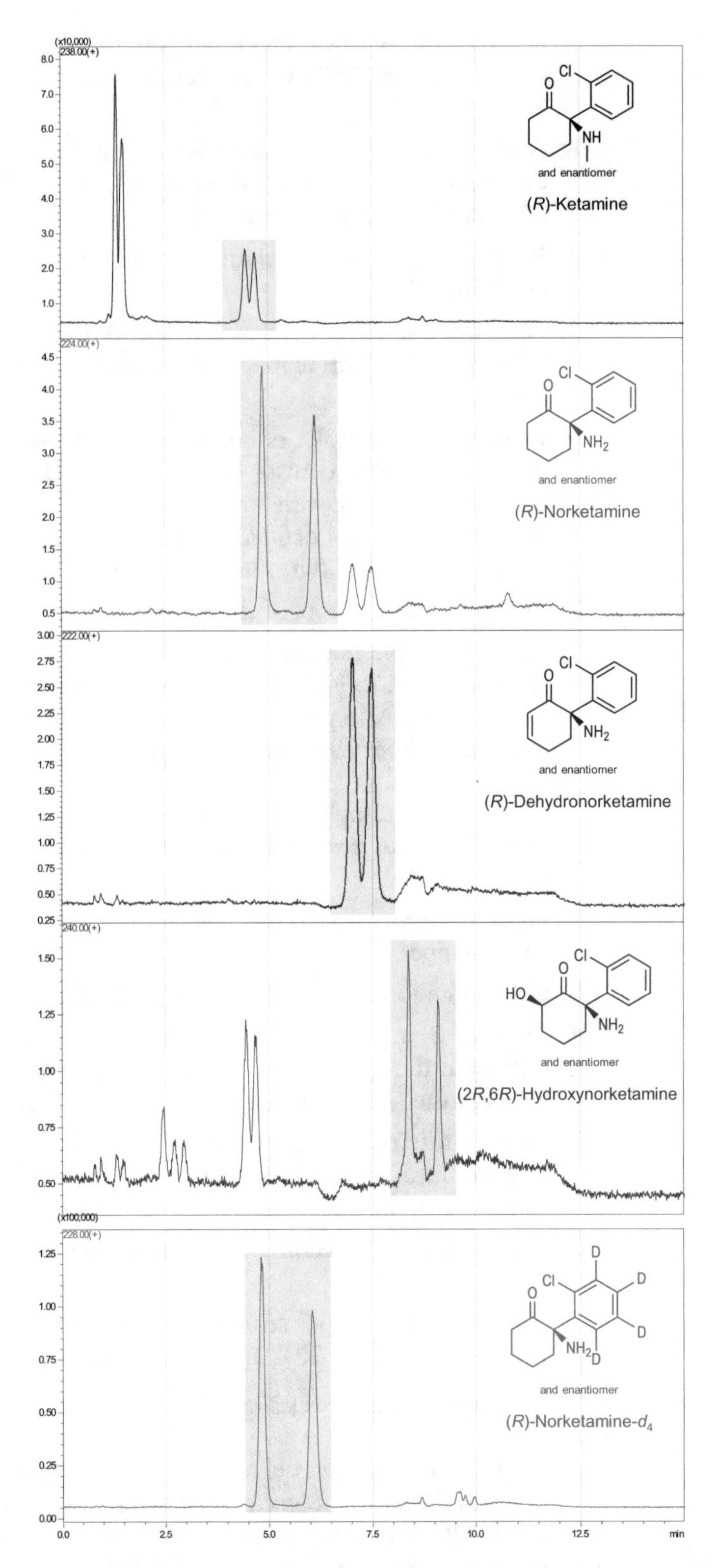

Fig. 3 Representative chromatogram for the SFC-MS protocol with LLE as sample preparation at the level of 5 ng/mL per analyte in the original urine sample (*see* **Note 21**)

3. Close lids and run online SFE-SFC:
   0–3.5 min at 3% B (*static extraction*: supercritical fluid is being introduced to the extraction vessel and becomes saturated with analytes), 3–3.5 min at 3% B (*dynamic extraction*: the mixture is being released onto the chromatographic system, but analytes become trapped at the head of the column due to the low elution power of the mobile phase mixture), 3.5–11 min gradually increasing to 20% B ("SFC"), 11–14 min at 20% B ("column cleaning"), 14–15 min at 3% B ("equilibration" in preparation of the next run).
4. For elution order check Fig. 4 (*see* **Note 17**).

3.4 Processing

1. Integrate the peaks of calibration samples to construct a calibration curve (*see* **Note 20**).
2. Analyze QC samples as part of the validation of the method (*see* **Note 21**).
3. Apply the method to patient samples for clinical results.

4 Notes

1. Although basic analytes such as KET are often distributed as hydrochloride salts, concentrations refer to the free base. For example, 1.15 mg of (*R*)-KET·HCl ($M_r = 274.185$) need to be dissolved in 10.0 mL in order to yield a 0.1 mg/mL KET ($M_r = 237.727$) solution (because the ratio of 274.185 to 237.727 is 1.15).
2. Metabolites of (*rac*)-KET also represent one pair of enantiomers each and therefore yield two peaks after enantioselective separation. There is no general correlation between configuration and elution order (such as "*R* always comes before *S*"), and therefore separate solutions containing only *S*- or only *R*-enantiomers need to be analysed at least once in order to assign each peak to its corresponding enantiomer. Once assigned, racemic mixtures can be used.
3. Methods involving sample preparation (such as an extraction step), complex matrices or detectors of limited robustness (e.g., mass detection) immensely benefit from the use of an internal standard (IS) that reflects variations in extraction ratio or ionization suppression. In order to do so, the IS ought to be similar to the analyte in terms of chemistry (extraction and ionization behavior) and retention time (chromatographic behavior), making deuterated analytes ideal standards for MS detection.

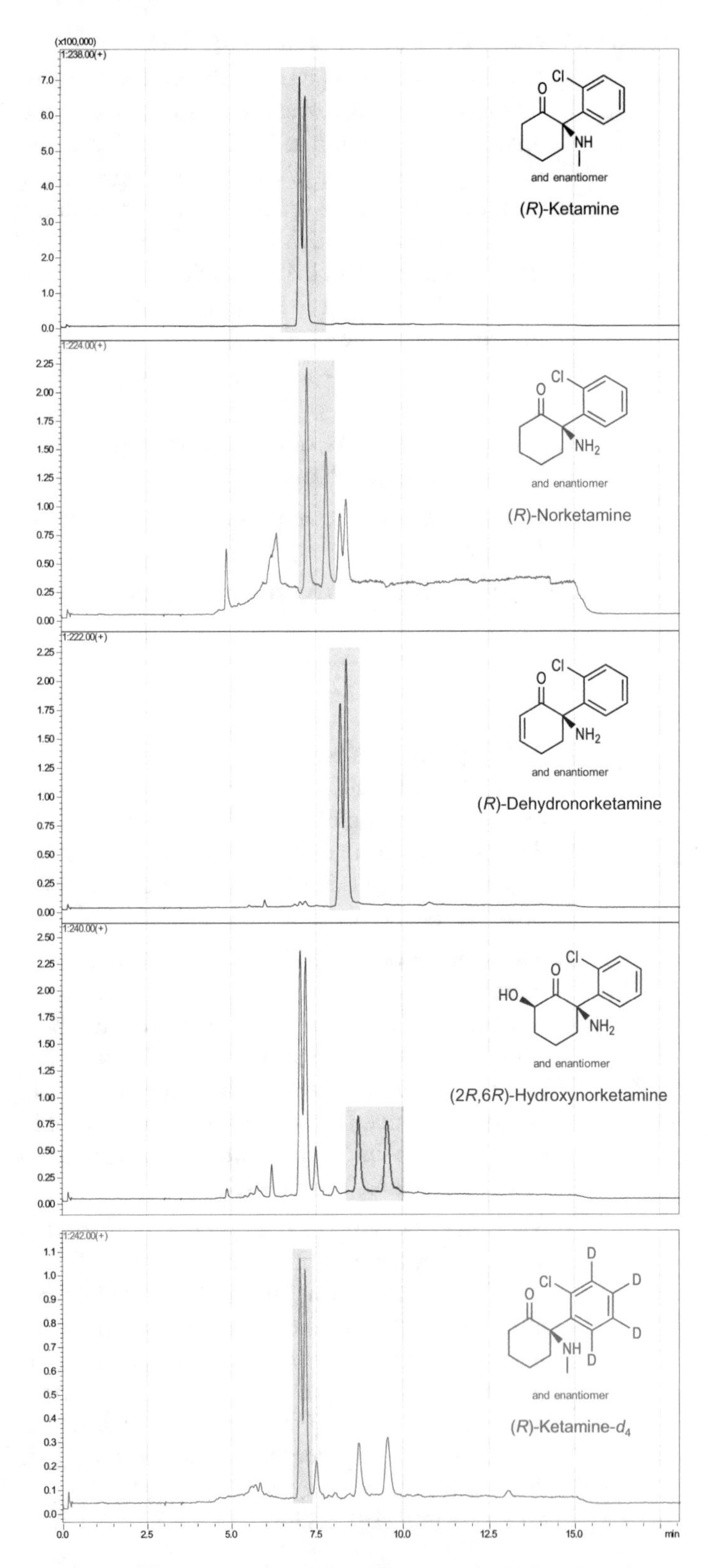

Fig. 4 Representative chromatogram for the SFE-SFC-MS protocol at the 10 ng per analyte on column (*see* **Notes 18** and **19**)

4. Depending on the SFC apparatus, CO_2 can be procured in gaseous or liquid state. Shimadzu's® Nexera UC system uses liquid CO_2, which is compressed without prior expansion to the gaseous state and heated to a supercritical state. In terms of quality, successful separations have been reported using CO_2 of as low as 99.9% purity. Instruments of other manufacturers (Agilent®, Waters®) reliquefy CO_2 that is supplied from the CO_2 cylinder by evaporation by means of a so-called booster compressor.
5. In SFC, the mobile phase is constituted mainly by nonpolar and acidic $scCO_2$ but can be adjusted in terms of polarity (through use of organic solvents, termed "modifier") and pH (by addition of MS-compatible acids or bases, termed "additives").
6. Generally, higher concentrations of basic additives in the modifier improve peak shape and resolution of basic analytes, but alterations in solvent strength are nonlinear. However, we found an increase in column backpressure when queueing longer batches at higher amounts of ammonia. Thus, 0.075% ammonia was chosen as a compromise.
7. The make-up is not participating in the separation itself, as it is added postcolumn with a T-joint prior to the BPR. Its main function is to prevent analyte precipitation when the mobile phase depressurizes and CO_2 evaporates after having left the column. In addition, an acidic makeup assists in ionization of basic analytes, thus increasing signal intensity and subsequently lowering the limit of detection.
8. Collect urine devoid of KET and KET metabolites from volunteers (laboratory staff).
9. The given parameters are customized to Shimadzu's® Single Quadrupole Mass Spectrometer LCMS-2020. Settings may vary or have different names if applied to systems of other manufacturers.
10. Equilibration times in SFC are shorter than in LC. If in doubt, monitor the detector and backpressure output until a constant base-line signal can be observed.
11. Calibration curves should be prepared daily to minimize errors stemming from solvent evaporation (enrichment) or analyte degradation (depletion). The range must cover the expected concentration of clinical samples and should reflect the target biomatrix and dose/route of application (here: 24-h urine obtained after i.v. application of 5 mg (*rac*)-KET). When spanning several orders of magnitude, it is helpful to prepare tenfold dilutions of the working solution in order to prevent handling of volumes ≤ 10 μL.

12. Both *blank* and *zero* demonstrate the absence of interfering matrix and IS components: In the final chromatogram, signals at the specific retention times of analytes must not exceed guideline specifications (20% of lower limit of quantification (LLOQ) for target analytes, 5% for IS, according to FDA and EMA).
13. Sodium carbonate is added to prevent protonation of basic analytes and aid accumulation in the organic phase.
14. Reconstitution is an important step of sample preparation: On the one hand, it allows for the adjustment of analyte concentration by choosing a higher (dilution) or lower reconstitution volume (enrichment) than that of the original sample. On the other hand, the reconstitution solvent can greatly affect peak shape and needs to be optimized experimentally. When reconstituting in less than 200 μL, it is helpful to use microvolume inserts for standard HPLC vials to ensure full immersion of the autosampler needle.
15. Extraction is almost quantitative if the sample is adsorbed onto a highly porous material like diatomaceous earth. Other materials (e.g., filter paper) can also be used, but need to be investigated beforehand to determine extraction yield and absence of interfering background signals. Note the 20-fold difference in sample volume required for LLE-SFC (1 mL) and SFE-SFC (50 μL). Presupposing nigh-complete extraction, the full amount introduced into the SFE-extraction vessels is transferred onto the chromatographic system, while only 5 μL of the LLE-extract are injected in LLE-SFC. On the other hand, reconstitution in one-tenth of the initial sample volume after LLE allows for a tenfold analyte enrichment, resulting in approximately the same LLOQs for both methods.
16. Residual moisture greatly affects the rate of extraction, and must therefore be standardized to obtain reproducible results. For example, store all samples overnight in an evacuated desiccator, or heat for 1 h to 51 °C at 7 mbar.
17. SFE is superior in terms of sample efficiency, preventing artificial contaminations and time-management. However, direct comparison to conventional LLE demonstrates the price for said expedience, that is, a reduction in selectivity: In the given protocol, differences in terms of separation performance following either LLE or SFE can be observed. While both protocols allow for enantioselective quantification of ketamine metabolites, other analytes may well be more suitably addressed by either LLE-SFC-MS (when interference proves an issue) or SFE-SFC-MS (when focus is placed on speedy analysis of labile analytes).

18. In conventional HPLC and also SFC, liquid samples (albeit after treatment) are injected into the chromatographic system, and therefore quantities are generally characterized in terms of concentration (e.g., *ng/mL* pertaining to the original sample). SFE-SFC, on the other hand, is performed after drying of the originally liquid sample (also termed "solid-phase SFE"). Therefore, declaration of absolute analyte amounts introduced into the chromatographic system are easier to handle (e.g., *ng on column*).
19. Analytes with molecular masses differing in 2 amu will appear in the mass trace of the heavier molecule (e.g., KET will show up in HNK's mass trace and DHK will appear in NK's mass trace). This circumstance is due to the existence of two stable isotopes of chlorine, namely, 35Cl and ^{37}Cl. Nevertheless, selectivity is not negatively affected by this phenomenon, since analytes can be easily distinguished on grounds of significantly differing retention times.
20. Depending on the method, the relationship between signal and concentration may best be described by linear or nonlinear models (quadratic, cubic, etc.), but these need to be confirmed statistically before violating the principles of parsimony. For example, a quadratic relationship (nonweighted) was confirmed by Mandel's fitting test after LLE-SFC, whereas the relationship was linear (albeit heteroscedastic) in SFE-SFC.
21. FDA's revised version on Bioanalytical Method Validation (2018) [14] recommends at least three independent accuracy and precision runs, four QC levels per run (at the lower limit of quantification, at a low level, at a medium level and at a high Level close to the upper limit of quantification) with at least five replicates per QC level. Accuracy (i.e., the ratio of measured to calculated concentration) and precision (i.e., the relative standard deviation between independent samples) should fall within 85–115% (except close to LLOQ, where 80–120% are permitted).

References

1. Tarafder A (2016) Metamorphosis of supercritical fluid chromatography to SFC: an overview. TrAC 81:3–10
2. Tarafder A, Hill JF, Baynham M (2014) Convergence chromatography versus SFC - what's in a name? Chromatogr Today 7:34–36
3. Grand-Guillaume Perrenoud A, Veuthey JL, Guillarme D (2012) Comparison of ultra-high performance supercritical fluid chromatography and ultra-high performance liquid chromatography for the analysis of pharmaceutical compounds. J Chromatogr A 1266:158–167
4. West C (2018) Current trends in supercritical fluid chromatography. Anal Bioanal Chem 410:6441–6457
5. Harps LC, Joseph JF, Parr MK (2019) SFC for chiral separations in bioanalysis. J Pharm Biomed Anal 162:47–59
6. Franzini R, Ciogli A, Gasparrini F, Ismail OH, Villani C (2018) Chapter 14 - Recent developments in chiral separations by supercritical fluid

chromatography. In: Polavarapu PL (ed) Chiral analysis, 2nd edn. Elsevier, Netherlands, pp 607–629

7. Hofstetter R, Fassauer GM, Link A (2018) The many (inter-)faces of supercritical fluid chromatography: the present and future prospects of online supercritical fluid extraction–supercritical fluid chromatography. Bioanalysis 10:1073–1076
8. Fassauer GM, Hofstetter R, Hasan M, Oswald S, Modeß C, Siegmund W, Link A (2017) Ketamine metabolites with antidepressant effects: fast, economical, and eco-friendly enantioselective separation based on supercritical-fluid chromatography (SFC) and single quadrupole MS detection. J Pharm Biomed Anal 146:410–419
9. Hofstetter R, Fassauer GM, Link A (2018) Supercritical fluid extraction (SFE) of ketamine metabolites from dried urine and on-line quantification by supercritical fluid chromatography and single mass detection (on-line SFE–SFC–MS). J Chromatogr B 1076:77–83
10. Zanos P, Moaddel R, Morris PJ, Georgiou P, Fischell J, Elmer GI, Alkondon M, Yuan P, Pribut HJ, Singh NS, Dossou KSS, Fang Y, Huang XP, Mayo CL, Wainer IW, Albuquerque EX, Thompson SM, Thomas CJ, Zarate CA Jr, Gould TD (2016) NMDAR inhibition-independent antidepressant actions of ketamine metabolites. Nature 533:481–486
11. Yamaguchi JI, Toki H, Qu Y, Yang C, Koike H, Hashimoto K, Mizuno-Yasuhira A, Chaki S (2018) (2*R*,6*R*)-Hydroxynorketamine is not essential for the antidepressant actions of (*R*)-ketamine in mice. Neuropsychopharmacology 43:1900–1907
12. Zanos P, Gould TD (2018) Mechanisms of ketamine action as an antidepressant. Mol Psychiatry 23:801–811
13. Dinis-Oliveira RJ (2017) Metabolism and metabolomics of ketamine: a toxicological approach. Forensic Sci Res 2:2–10
14. U.S. Department of Health and Human Services, Food and Drug Administration (2018) Guidance for industry: bioanalytical method validation. https://www.fda.gov/downloads/drugs/guidances/ucm070107.pdf

Chapter 10

Capillary Electrophoresis–Mass Spectrometry of Hydrophilic Metabolomics

Masahiro Sugimoto

Abstract

Metabolomics is defined as the comprehensive identification and quantification of small molecules that provide a holistic view of cellular metabolism. The metabolomic network is downstream of the central dogma that transfers regulatory information from the genome. The metabolomic profile, that is, concentration patterns of metabolites on metabolic pathways, can be expected to directly reflect cellular phenotype. Thus, metabolomics has been used for many biomarker discoveries for metabolic diseases, such as diabetes, myocardial infarctions, and cancers. However, there are large limitations in the measurement technologies, compared to the other *omics*. Currently, a single analytical methodology cannot profile whole metabolites and various methods should be used to monitor the wide range of metabolic pathways. Here, we introduce metabolomics using capillary electrophoresis-mass spectrometry which is suitable for hydrophilic metabolite quantifications.

Key words Capillary electrophoresis-mass spectrometry, Metabolomics, Serum, Saliva

1 Introduction

Metabolomics is the newest omics which enables simultaneous profiling of hundreds of metabolites. This method is useful to diagnose a wide range of diseases using low-invasively available biofluid. There are several methodologies are available as measurement technologies. Nuclear magnetic resonance (NMR) [1] and mass spectrometry (MS) [2, 3] are the major analytical techniques in metabolomics. A merit of the use of NMR is noninvasive sample processing. However, the relatively low sensitivity of this method and spectral overlap limits the number and variety of metabolites that can be observed.

MS is another major instrument in metabolomics. MS required destructive sample processing while the sensitivity is quite high compared to NMR. MS is usually used with a separation system prior to MS. Gas chromatography (GC)-MS [4], liquid chromatography (LC)-MS [5], and capillary electrophoresis (CE)-MS [6]

Paul L. Wood (ed.), *Metabolomics*, Neuromethods, vol. 159, https://doi.org/10.1007/978-1-0716-0864-7_10,

are currently the leading analytical platform. The combination of these separation systems with MS provides higher selectivity and sensitivity. However, because of the large variety of physical and chemical properties of the metabolites, no single analytical method can comprehensively measure all metabolites [7]. GC-MS is already conventional, that is, well-established, technology, which is capable of profiling volatile compounds. GC-MS generally requires an initial derivatization procedure for profiling nonvolatile compounds. LC-MS can be used to monitor a wider variety of nonvolatile compounds; however, optimization of sample processing and LC column selection that depends on the target analytes is necessary. In contrast, CE-MS can monitor all charged metabolites within only two (positive/negative) modes, which allows for the simultaneous profiling of many pathways.

We have conducted biomarker discoveries in various biofluid samples. From blood samples, biomarkers for pancreatic cancer and liver diseases were explored [8–10]. Urinary samples were used for colon cancer detection [11]. Saliva samples were used for biomarker discovery for various cancers, such as oral, breast, and pancreatic cancers, and dementia [12–15]. Here, we introduce the protocol for analyzing blood and salivary samples.

2 Materials

2.1 Subjects

All patients have pancreaticobiliary cancers diagnosed histologically. All patients are recently diagnosed with a primary disease and none have received any prior treatment in the form of chemotherapy, radiotherapy, surgery, or alternative therapy. No subjects have a history of prior malignancy.

2.2 Samples

To explore the metabolite biomarkers to discriminate patients with pancreatic cancers from healthy controls and chronic pancreatitis, biofluid samples are collected [8, 13]. Here, we introduce serum sample analyses, while the metabolomic profiles of plasma sample are more reproducible compared to serum samples [16]. For eliminating batch effect, *see* **Note 1**.

3 Methods

3.1 Blood Sample Collection

1. All subjects are not allowed to take any food except water intake after 9:00 p.m. on the previous day.
2. Blood samples is collected at 8:00–11:00 a.m.
3. The collected blood is immediately centrifuged at 3000 × *g* for 10 min at 4 °C, and the serum is transferred to a 1.2 mL

polypropylene serum tube (SUMILON, Sumitomo Bakelite Co., LTD., Tokyo, Japan).

4. Samples are stored at −80 °C until metabolomic analyses.

3.2 Saliva Sample Collection

1. All subjects are not allowed to take any food except water intake after 9:00 p.m. on the previous day.
2. Saliva samples are collected at 8:00–11:00 a.m.
3. The subjects are required to brush their teeth without toothpaste on the day of saliva collection and have to refrain from drinking water, smoking, toothbrushing, and intense exercise 1 h before saliva collection. They are required to gargle with water just before saliva collection.
4. Approximately 400 μL of unstimulated saliva is collected in a 50 cc polypropylene tube. A polypropylene straw 1.1 cm in diameter is used to assist the saliva collection. Saliva are collected on ice. After collection, saliva samples are immediately stored at −80 °C until metabolite measurements. For the storage condition, *see* **Note 2** for the detail.

3.3 Sample Preparation for CE-MS

3.3.1 Preparation of Serum Samples (Fig. 1)

1. Frozen samples are thawed and 50 μL aliquots of serum are mixed with 450 μL of methanol containing internal standards (20 μM each of methionine sulfone and camphor 10-sulfonic acid, described as IS1 in Fig. 1).
2. The samples are vortex-mixed and 500 μL of chloroform and 200 μL of Milli-Q water (Millipore, Billerica, MA, USA) are added.
3. The mixture is centrifuged at 10,000 × *g* for 3 min at 4 °C.
4. The 150 μL of aqueous layer of each sample is filtered through a centrifugal filter tube with a 5-kDa cutoff (Millipore) at 9100 × *g* for 2.5 h at 4 °C to remove large molecules.
5. The filtrates are concentrated by centrifugation for about 2 h at 40 °C using Acid-Resistant CentriVap Benchtop Centrifugal Vacuum Concentrator (Labconco Corp., Kansas, MO, USA) and resuspended in 50 μL of Milli-Q water containing other internal standards (3-aminopyrrolidine and trimesate) prior to CE-MS analyses. These internal standards (IS2 in Fig. 1) are used for correction drift of migration times of each measurement.

3.3.2 Preparation of Saliva Samples (Fig. 2)

1. Frozen saliva samples are thawed at 4 °C for approximately 1.5 h.
2. Samples are completely dissolved using a Vortex mixer at room temperature.

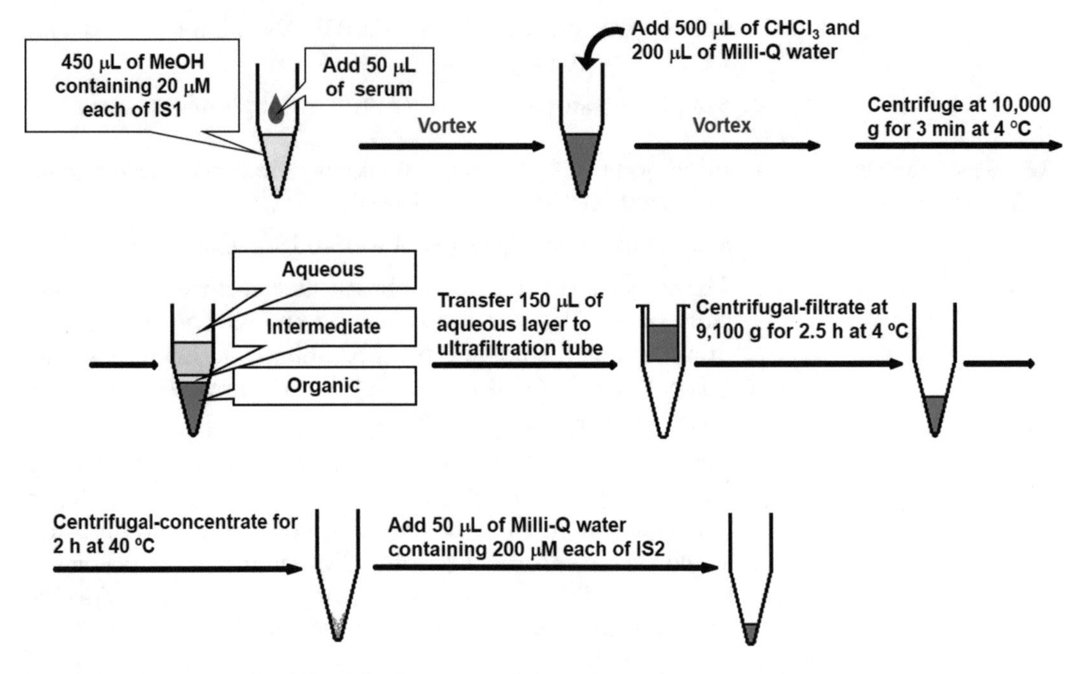

Fig. 1 Sample processing of blood (both serum and plasma) samples

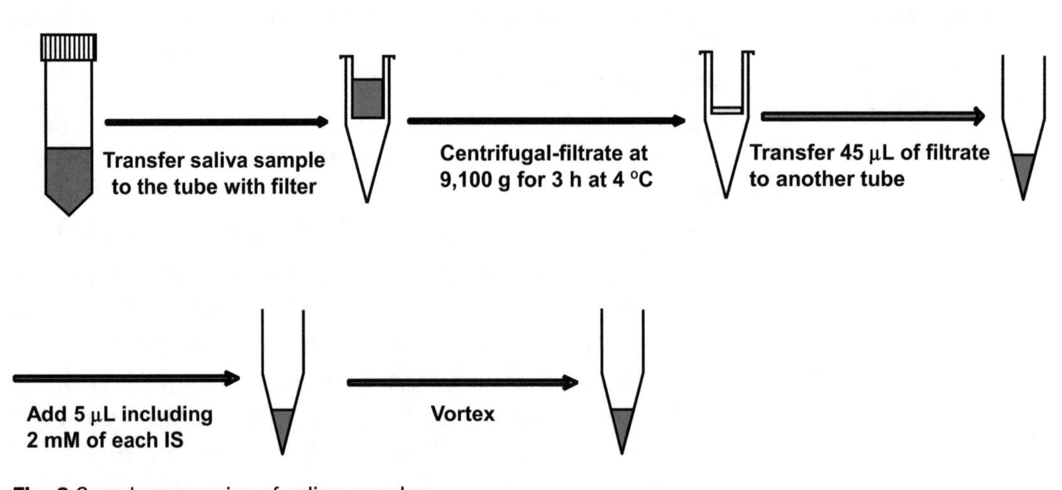

Fig. 2 Sample processing of saliva samples

3. The dissolved samples are centrifuged through a 5-kDa cutoff filter (Millipore, Bedford, MA) at 9100 $\times$ *g* for at least 3 h at 4 °C.

4. Transfer 45 μL of each sample to a 1.5 mL Eppendorf tube. Add 5 μL MilliQ water including 2 mM of methionine sulfone, 2-[N-morpholino]-ethanesulfonic acid (MES), D-Camphol-10-sulfonic acid, sodium salt, 3-aminopyrrolidine, and trimesate (IS in Fig. 2).

3.4 CE-MS Experiments

The metabolite standards, CE-time-of-flight (TOF)-MS instrumentation, and measurement conditions for cationic and anionic metabolites are described previously.

3.4.1 Measurement Instruments

Cation analysis is performed using an Agilent CE capillary electrophoresis system (G1600AX), an Agilent G1969A LC/MSD TOF system, an Agilent 1100 series isocratic HPLC pump, a G3251A Agilent CEMS adapter kit, and a G1607A Agilent CE-ESI-MS sprayer kit (Agilent Technologies, Waldbronn, Germany). Anion analysis is performed using an Agilent CE capillary electrophoresis system (G1600AX), an Agilent G1969A LC/MSD TOF system, an Agilent 1100 series isocratic HPLC pump, a G3251A Agilent CE-MS adapter kit, and a G1607A Agilent CE-electrospray ionization (ESI) source-MS sprayer kit (Agilent Technologies). For the cation and anion analyses, the CE-MS adapter kit includes a capillary cassette that facilitates thermostatic control of the capillary. The CE-ESI-MS sprayer kit simplifies coupling of the CE system with the MS system and is equipped with an electrospray source. For system control and data acquisition, Agilent ChemStation software for CE (A10.02) and Agilent MassHunter software for TOF-MS (B.02.00) are used while the other versions are acceptable. The original Agilent SST316Ti stainless steel ESI needle is replaced with a passivated SST316Ti stainless steel and platinum needle (passivated with 1% formic acid and a 20% aqueous solution of isopropanol at 80 °C for 30 min) for anion analysis.

3.4.2 Parameters of Cation Analyses

For cationic metabolite analysis using CE-TOF-MS, sample separation is performed in fused silica capillaries (50 μm i.d. × 100 cm total length) filled with 1 mol/L formic acid as the running electrolyte. The capillary is flushed before each sample injection with formic acid (1 mol/L) for 20 min before the first use and for 4 min for tissue sample analyses, 5 min at ammonium formate, 5 min at Milli-Q and 5 min at run buffer for saliva sample analyses. Sample solutions (approximately 3 nL) are injected at 50 mbar for 5 s and a voltage of 30 kV is applied. The capillary temperature is maintained at 20 °C and the temperature of the sample tray is kept below 5 °C. The sheath liquid, composed of methanol/water (50% v/v) and 0.1 μmol/L hexakis(2,2-difluoroethoxy)phosphazene (Hexakis), is delivered at 10 μL/min. ESI-TOF-MS is conducted in the positive ion mode. The capillary voltage is set at 4 kV and the flow rate of nitrogen gas (heater temperature = 300 °C) is set at 7 psig. In TOF-MS, the fragmentor, skimmer, and OCT RF voltages are 75, 50, and 125 V, respectively. Automatic recalibration of each acquired spectrum is performed using reference standards ([13C isotopic ion of protonated methanol dimer (2MeOH + H)]+, *m/z* 66.0631) and ([protonated Hexakis (M + H)]+, *m/z* 622.0290). Mass spectra are acquired at a rate of 1.5 cycles/s over a *m/z* range of 50–1000.

3.4.3 Parameters of Anion Analysis

For anionic metabolite analysis using CE-TOF-MS, a commercially available Cosmo(+) capillary (50 μm × 105 cm, Nacalai Tesque, Kyoto, Japan), chemically coated with a cationic polymer, is used for separation. Ammonium acetate solution (50 mmol/L; pH 8.5) is used as the electrolyte for separation. Before the first use, the new capillary is flushed successively with the running electrolyte (pH 8.5), 50 mmol/L acetic acid (pH 3.4), and then the electrolyte again for 10 min each. Before each injection, the capillary is equilibrated for 2 min by flushing with 50 mmol/L acetic acid (pH 3.4), and then with the running electrolyte for 5 min. A sample solution (approximately 30 nL) is injected at 50 mbar for 30 s, and a voltage of −30 kV was applied. The capillary temperature is maintained at 20 °C and the sample tray is cooled below 5 °C. An Agilent 1100 series pump equipped with a 1:100 splitter is used to deliver 10 μL/min of 5 mM ammonium acetate in 50% (v/v) methanol/water, containing 0.1 μM Hexakis, to the CE interface. Here, it is used as a sheath liquid surrounding the CE capillary to provide a stable electrical connection between the tip of the capillary and the grounded electrospray needle. ESI-TOF-MS is conducted in the negative ionization mode at a capillary voltage of 3.5 kV. For TOF-MS, the fragmentor, skimmer, and OCT RF voltages are set at 100, 50, and 200 V, respectively. The flow rate of the drying nitrogen gas (heater temperature = 300 °C) is maintained at 7 psig. Automatic recalibration of each acquired spectrum is performed using reference standards ([13C isotopic ion of deprotonated acetic acid dimer (2 CH3COOH–H)]-, *m/z* 120.03841), and ([Hexakis deprotonated acetic acid (M + CH3COOH–H)]-, *m/z* 680.03554). Exact mass data are acquired at a rate of 1.5 spectra/s over a *m/z* range of 50–1000.

3.4.4 Data Processing

Data processing can be conducted using Agilent MassHunter software when the target metabolites to be analyzed are already defined. However, this software cannot be used for non-targeted analysis (e.g., comprehensive analyses). For non-targeted analyses, we use our in-house software named MasterHands [17]. The vendor-supplied raw data is converted to a more general format (e.g., mzXML and netCDF). The data are sliced within 0.02 *m/z* range to generate extracted electropherograms (EIC). Noise-filtering, baseline-correction, and peak-detection to integrate the peak area are conducted on each EIC. Accurate *m/z* values are calculated based on the background-subtracted mass spectrum. Redundant features such as fragment and adduct ions were eliminated.

After the data analysis of individual data, alignment and peak matching across multiple datasets are conducted. Migration times are corrected by migration times of internal standards or more sophisticated way using dynamic programming [18]. All compounds are identified by matching the *m/z* value and the corrected migration time to corresponding metabolites in a standard library. For all peak identifications, we confirm the similarity of the isotope distributions between subjects and candidates. For unknown peaks, estimation using artificial intelligence is one possible solution [19].

Concentrations are calculated using external standards based on relative area: the area divided by the area of the internal standards. Before the sample measurements, we measure a mixture of our standard library including internal standards (methionine sulfone and camphor 10-sulfonic acid for cation and anion, respectively) and concentrations in the samples are calculated by a single-point calibration method. To compare the metabolite concentration with our data, *see* **Note 3** for details.

4 Notes

1. Calculation of absolute concentration using internal standards and metabolite a mixture is enough to eliminate batch effect while the measurement of quality control samples in each batch is preferable to eliminating unexpected bias [20].
2. The effect of storage temperature on salivary metabolite profiles are analyzed and, as expected, lower temperature (on ice) is required to obtain reproducible profiles. The profiles of saliva for 1 week, the storage temperature of −18 °C instead of −80 °C is enough to obtain reproducible profiles [21].
3. The database of CE-MS-based profile is available at MMMDB [22].

Acknowledgments

We acknowledge the grants from Yamagata Prefectural Government and Tsuruoka City, Japan.

References

1. Reo NV (2002) NMR-based metabolomics. Drug Chem Toxicol 25(4):375–382
2. Aharoni A et al (2002) Nontargeted metabolome analysis by use of Fourier transform ion cyclotron mass spectrometry. OMICS 6 (3):217–234
3. Castrillo JI et al (2003) An optimized protocol for metabolome analysis in yeast using direct infusion electrospray mass spectrometry. Phytochemistry 62(6):929–937
4. Fiehn O et al (2000) Identification of uncommon plant metabolites based on calculation of elemental compositions using gas chromatography and quadrupole mass spectrometry. Anal Chem 72(15):3573–3580
5. Plumb R et al (2003) Metabonomic analysis of mouse urine by liquid-chromatography-time of flight mass spectrometry (LC-TOFMS): detection of strain, diurnal and gender differences. Analyst 128(7):819–823
6. Soga T et al (2003) Quantitative metabolome analysis using capillary electrophoresis mass spectrometry. J Proteome Res 2(5):488–494

7. Monton MR, Soga T (2007) Metabolome analysis by capillary electrophoresis-mass spectrometry. J Chromatogr A 1168 (1–2):237–246. discussion 236
8. Itoi T et al (2017) Serum metabolomic profiles for human pancreatic cancer discrimination. Int J Mol Sci 18(4):767
9. Soga T et al (2011) Serum metabolomics reveals gamma-glutamyl dipeptides as biomarkers for discrimination among different forms of liver disease. J Hepatol 55(4):896–905
10. Hirayama A et al (2012) Metabolic profiling reveals new serum biomarkers for differentiating diabetic nephropathy. Anal Bioanal Chem 404(10):3101–3109
11. Nakajima T et al (2018) Urinary polyamine biomarker panels with machine-learning differentiated colorectal cancers, benign disease, and healthy controls. Int J Mol Sci 19(3):756
12. Ishikawa S et al (2016) Identification of salivary metabolomic biomarkers for oral cancer screening. Sci Rep 6:31520
13. Asai Y et al (2018) Elevated polyamines in saliva of pancreatic cancer. Cancers (Basel) 10 (2):43
14. Sugimoto M et al (2010) Capillary electrophoresis mass spectrometry-based saliva metabolomics identified oral, breast and pancreatic cancer-specific profiles. Metabolomics 6 (1):78–95
15. Tsuruoka M et al (2013) Capillary electrophoresis-mass spectrometry-based metabolome analysis of serum and saliva from neurodegenerative dementia patients. Electrophoresis 34(19):2865–2872
16. Hirayama A et al (2015) Effects of processing and storage conditions on charged metabolomic profiles in blood. Electrophoresis 36 (18):2148–2155
17. Sugimoto M et al (2012) Bioinformatics tools for mass spectroscopy-based metabolomic data processing and analysis. Curr Bioinforma 7 (1):96–108
18. Sugimoto M et al (2010) Differential metabolomics software for capillary electrophoresis-mass spectrometry data analysis. Metabolomics 6(1):27–41
19. Sugimoto M et al (2010) Prediction of metabolite identity from accurate mass, migration time prediction and isotopic pattern information in CE-TOFMS data. Electrophoresis 31 (14):2311–2318
20. Harada S et al (2018) Reliability of plasma polar metabolite concentrations in a large-scale cohort study using capillary electrophoresis-mass spectrometry. PLoS One 13(1):e0191230
21. Tomita A et al (2018) Effect of storage conditions on salivary polyamines quantified via liquid chromatography-mass spectrometry. Sci Rep 8(1):12075
22. Sugimoto M et al (2012) MMMDB: mouse multiple tissue metabolome database. Nucleic Acids Res 40(Database issue):D809–D814

Chapter 11

Lipidomic Analysis of Oxygenated Polyunsaturated Fatty Acid–Derived Inflammatory Mediators in Neurodegenerative Diseases

Mauricio Mastrogiovanni, Estefanía Ifrán, Andrés Trostchansky, and Homero Rubbo

Abstract

The lack of current biomarkers to follow neurodegeneration highlights the need of a comprehensive method to help to understand the biological mechanisms of disease. A consistent neuropathological feature of several neurodegenerative diseases is the extensive inflammation, called neuroinflammation. The final products of inflammatory processes may be detected as a screening tool in blood samples. Thus, identifying novel inflammatory biomarkers could improve the design of novel strategies for early diagnosis, disease progression, and/or novel therapeutic approaches. This chapter describes an HPLC-MS/MS based method which allows for detection and quantitation of several bioactive lipids related to inflammatory processes occurring in neurodegenerative diseases.

Key words Neurodegeneration, Inflammation, Mass spectrometry, Lipidomic

1 Introduction

Neuroinflammation in the central nervous system (CNS) is considered a hallmark of several neurodegenerative diseases including amyotrophic lateral sclerosis, Parkinson and Alzheimer diseases, among others [1, 2].

Metabolomic studies search for small molecules present in cells, tissues, or biological samples, whereas the observation of changes in addition to physiological modifications of signaling pathways may aid in elucidating where these changes are occurring. Blood biomarkers should be used as tools for monitoring the onset and progression of the disease, the appearance of clinical symptoms as well as the efficiency of potential treatments. A wide range of blood metabolites from <1000 to 1500 Da can be used as potential disease biomarkers, in particular those related to polyunsaturated

Paul L. Wood (ed.), *Metabolomics*, Neuromethods, vol. 159, https://doi.org/10.1007/978-1-0716-0864-7_11,

fatty acids present in brain, including arachidonic acid (AA), eicosapentaenoic acid (EPA), and docosahexaenoic acid (DHA) [3, 4].

AA can be metabolized by the prostaglandin endoperoxide H synthase (PGHS) or lipoxygenase (LOX) pathways being the precursor of a wide variety of anti- or proinflammatory compounds such as prostaglandins, leukotrienes, and hydroperoxyl (HpETE) or hydroxyl (HETE) derivatives which can be followed in blood and used as disease biomarkers [5, 6]. In the same way, EPA and DHA can also be metabolized by PGHS or LOX pathways leading to specific bioactive lipids within the group of specialized proresolving mediators (SPM). SPMs include lipoxins generated from AA, E-series resolvins from EPA as well as D-series resolvins, neuroprotectins, and maresins coming from DHA [7, 8]. SPMs actively turn off the inflammatory response by acting on distinct G protein–coupled receptors expressed on immune cells that activates dual anti-inflammatory and proresolution programs [7, 8]. Few reports have detected alterations in metabolism of SPMs in patients with neurodegenerative diseases [9, 10] or even promotion of resolution programs by exogenously administrated SPM [11]. Many of these compounds are able to cross the brain–blood barrier (BBB), thus being able to be detected by lipidomic analysis [12].

2 Materials

Prepare all solutions using ultrapure water (resistivity of 18 MΩ cm at 25 °C). Analytical grade reagents should be used for lipid extraction. HPLC grade reagents should be used for liquid chromatography. Prepare and store all reagents at room temperature.

2.1 Equipment

1. SPE vacuum manifold (Phenomenex, Torrance, CA).
2. SpeedVac Savant SPD1010 (ThermoScientific, NY).
3. Glass vials (2 ml) with 200 μl conical glass inserts and 8 mm screw caps with PTFE septa.
4. Agilent Infinity 1260 HPLC pump with refrigerated autosampler coupled to hybrid triple quadrupole/linear ion trap mass spectrometer QTRAP4500 (ABSciex, Framingham, MA).
5. Reversed phase HPLC column (Luna C18(2), 5 μm, 2.1 × 100 mm; Phenomenex, Torrance, CA).

2.2 Reagents and Supplies

2.2.1 Standard Solutions for Quantification

1. The following standards for the analysis are obtained from Cayman Chemicals (Ann Arbor, MI, USA):
   - Linoleic acid metabolites: 9-hydroxyoctadecadienoic acid (9-HODE) and 13-hydroxyoctadecadienoic acid (13-HODE).
   - Arachidonic acid metabolites: 6-keto-prostaglandin F1α (6-keto PGF1α), prostaglandin D_2 (PGD_2), prostaglandin

E_2 (PGE_2), prostaglandin $F_{2\alpha}$ ($PGF_{2\alpha}$), 8-*isoprotaglandin* $F_{2\alpha}$, thromboxane B_2(TxB_2), 5-hydroxyeicosatetraenoic acid (5-HETE), 12-hydroxyeicosatetraenoic acid (12-HETE), 15-hydroxyeicosatetraenoic acid (15-HETE).

– Eicosapentaenoic acid metabolites: 18-hydroxyeicosapentenoic acid (18-HEPE) and Resolvin E1 (RvE1).

– Docosahexaenoic acid metabolites: Resolvin D1 (RvD1), Resolvin D2 (RvD2), Resolvin D3 (RvD3), Resolvin D5 (RvD5), Maresin 1 (Mar1).

2.2.2 Internal Standard Cocktail

1. The following deuterated standards are obtained from Cayman Chemicals (Ann Arbor, MI, USA): 6-keto $PGF_{1\alpha}$-d4, PGD_2-d4, PGE_2-d4, $PGF_{2\alpha}$-d4, T_XB_2-d4, 5-HETE-d8, 12-HETE-d8, 15-HETE-d8, 9-HODE-d4, 13-HODE-d4, RvD1-d4, RvE1-d4, Mar1-d5, 5-iso prostaglandin $F_{2\alpha}$-VI-d11, 8-iso prostaglandin $F_{2\alpha}$-d4.

2.2.3 Biological Sample Preparation

1. 20% Methanol solution.
2. HCl (6 M).

2.2.4 Solid-Phase Extraction

1. Methanol (HPLC grade; Avantor, Center Valley, PA, USA).
2. Strata™-X SPE cartridges (Polymeric sorbent 33 μ, 60 mg, 3 ml; Phenomenex, Terrence, CA).

2.2.5 HPLC-MS/MS

1. Mobile phase A: water with 0.1% formic acid.
2. Mobile phase B: acetonitrile with 0.1% formic acid.
3. Needle wash: methanol–acetonitrile–water (20:20:60 v/v).

3 Methods

3.1 Standard Solutions

3.1.1 Stock Solutions

1. Lipid standards are provided at various concentrations in solution in an organic solvent (usually methyl acetate or ethanol).
2. Stock mix solutions of 500 pg/μl should be prepared from each lipid standard by adding the appropriate volume of stock solution to a 2 ml amber vial, and making the volume up to 1 ml with methanol. These will be used to construct the calibration mix and internal standard mix.
3. Stock solutions should be flushed with nitrogen and sealed with sealing film, before storage at −20 °C or −80 °C according to the manufacturer's indications. Mix solutions could be kept at −20 °C for 1 month without significant signal loss.

3.1.2 Calibration Curves for Quantification

1. The calibration curves are prepared by series dilution to give concentrations ranging from 1 pg/μl to 100 pg/μl (*see* **Note 1**).
2. All samples should contain an internal standard mix (25 pg/μl).

3.2 Collection and Storage of Biological Samples

3.2.1 Plasma (See Note 2)

1. Collect samples on ice, aliquot and freeze as soon as possible. Store at −80 °C, avoiding freeze–thaw cycles. If transport is necessary, this should be done on dry ice.
2. Plasma samples (prepared from EDTA tubes) should be aliquoted (500 μl) in cryogenic vials, and stored at −80 °C. This avoids freeze–thaw cycles.

3.3 Biological Sample Preparation

Solvent extraction should be performed on ice, keeping the samples out of direct light.

3.3.1 Plasma Samples

1. Thaw samples on ice.
2. Transfer the sample to an Eppendorf tube by pipette, measuring and taking note of the sample volume.
3. Add internal standard mix (2.5 μl of the prepared 500 pg/μ l mix).
4. Dilute the sample with equal volume of 20% methanol in acidified water (pH 3).
5. Mix the solution gently.
6. Centrifuge (30 min, 4 °C, 14,000 × *g*).
7. Transfer the clear supernatant directly onto the preconditioned SPE cartridge.

3.4 Solid-Phase Extraction

3.4.1 SPE Cartridge Conditioning

1. Label and attach the required number of SPE cartridges to the SPE vacuum manifold, making sure any unused ports are closed and there is a waste bucket underneath the cartridges for waste solvent collection.
2. Activate the cartridges with 3 ml 100% methanol, followed by 2 × 2 ml 10% methanol in acidified water, under a gentle vacuum (approximately 3 In Hg).
3. Close the tap under each cartridge as soon as the last of the water has flowed through.

3.4.2 SPE of Lipid Extracts

1. Transfer the supernatant samples onto the conditioned cartridges.
2. Let the sample flow through the SPE cartridges in a drop wise manner with the vacuum switched off.
3. Take care not to allow any air to pass through the cartridge sorbent and close the tap under each cartridge as soon as the last of the sample has flowed through.

4. Wash all the cartridges sequentially with 3 ml of 10% methanol in acidified water. If necessary this could be performed under low vacuum, letting the eluate run into the waste and closing the tap under each cartridge as the last of the solvent runs through in each stage. Finally all cartridges should be submitted to high vacuum (11 In Hg) for no more than 5 min.
5. Turn off the vacuum and remove the waste bucket from the vacuum manifold, replacing it with clean, round-bottomed 2 ml Eppendorf tubes positioned under each SPE cartridge in a rack.
6. Elute the lipids with methanol (1 ml).
7. Remove the tubes from the vacuum manifold and transfer to SpeedVac previously warmed up at 45 °C. Evaporation to dryness might require almost 2 h.

3.5 Sample Reconstitution and Preparation for HPLC-MS/MS Analysis

1. Reconstitute dried samples in 50 μl of methanol–water (1:1), vortex to ensure the entire inner surface is coated, and centrifuge briefly (10 s) with a benchtop minifuge.
2. Transfer the lipid extract to a 200 μl conical glass insert inside a 2 ml HPLC vial. Cap with an 8 mm screw cap with a PTFE septum.

3.6 Chromatography

1. Condition the column by running HPLC solvents A and B at a ratio of 70:30 and at a flow rate of 0.3 ml/min for at least 20 column volumes. At the same time, put the MS into operate mode. Check that there is a stable back pressure on the HPLC and clean baseline on the MS.
2. Column compartment should be set at 30 °C.
3. Set the autosampler to 5 °C and add sample vials to the autosampler plates. Injection volume should be set to 40 μl for each sample.
4. Elution settings are as follows: linear gradient of solvent B 30–95% from 0.1 to 11 min, maintain at 95% for 4 min and reequilibrate with initial conditions for 5 min. Flow rate is set at 0.3 ml/min during all chromatography steps.

3.7 Mass Spectrometry Data Acquisition

1. Data was collected on a QTRAP4500 hybrid triple quadrupole-linear ion trap. This spectrometer is equipped with a Turbo V ESI source operated in negative mode with following settings: curtain gas (CUR): 50, ion spray voltage (IS): -4500 V, temperature (TEM): 400 °C, ion source gas 1 (GS1): 30 psi (N2), GS2: 30 psi (N2), CEP: −10 V, and CXP: −13 V. Other parameters related to specific analytes are detailed below.

Table 1
Optimized sMRM pairs and parameters for bioactive lipids. Product ions in bold are set as primary sMRM pairs

Precursor	Pathway	Common name	Abbreviation	Precursor ion	Product Ions	Retention time (min)	Window (s)	DP (V)	EP (V)	CE (V)
AA	COX	20-hydroxy-$PGF_{2\alpha}$	20oh $PGF_{2\alpha}$	369,5	**165**; 351; 325; 193	1,5	60	−40	−10	−39
AA	COX	9-hydroxy-$PGF_{2\alpha}$	9oh $PGF_{2\alpha}$	369,5	**192**; 351; 325; 193	1,5	60	−60	−10	−35
AA	COX	19-hydroxy-PGE_2	19oh PGE_2	367,5	**243**; 349; 331; 287	1,5	60	−20	−10	−31
AA	COX	20-hydroxy-PGE_2	20oh PGE_2	367,5	**175**; 349; 331; 287	1,5	60	−30	−10	−27
AA	COX	Tetranor-prostaglandin F metabolite	Tetranor-PGFM	329,4	**247**; 311; 287; 267	1,5	60	−40	−10	−25
AA	COX	Tetranor-prostaglandin E metabolite	Tetranor-PGEM	327,4	**291**; 309; 273; 245	1,5	60	−30	−10	−23
EPA	COX	Δ17-6-keto-prostaglandin $F_{1\alpha}$	Δ17 6 k $PGF_{1\alpha}$	367,5	**163**; 349; 331; 243	1,5	60	−60	−10	−34
AA	COX	Tetranor-prostaglandin D metabolite	Tetranor-PGDM	327,4	**247**; 309; 245; 223	1,5	60	−20	−10	−20
		Internal standard	*(d4) 6 k PGF1α*	*373,5*	*167; 293; 249; 211*	*2,3*	*30*	*−60*	*−10*	*−34*
AA	COX	6-keto-prostaglandin $F_{1\alpha}$	6 k PGF1α	369,5	**245**; 351; 315; 163	2,3	30	−60	−10	−34
AA	Non-enz	2,3-dinor-8iso-prostaglandin $F_{2\alpha}$	2,3 dinor 8-iso $PGF_{2\alpha}$	325,4	**237**; 263; 219; 137	2,4	60	−30	−10	−19
		Internal standard	*(d4) Resolvin E1*	*353,3*	*197*	*2,8*	*30*	*−40*	*−10*	*−20*
EPA	LOX	Resolvin E1	Resolvin E1	349,3	**195**; 161; 205	2,8	30	−40	−10	−25

EPA	COX	Thromboxane B_3	T_XB_3	367,5	**169**; 195; 177; 125	2,8	60	−40	−10	−27
EPA	Non-enz	8-iso prostaglandin $F_{3\alpha}$	8-iso $PGF_{3\alpha}$	351,5	**307**; 333; 253; 245	2,9	60	−30	−10	−28
AA	COX	6-keto-prostaglandin E_1	6 k PGE_1	367,5	**331**; 349; 143; 125	3,0	60	−40	−10	−25
AA	COX	2,3-dinor-11-beta-prostaglandin $F_{2\alpha}$	2,3 dinor 11b $PGF_{2\alpha}$	325,4	**227**; 245; 163; 145	3,0	60	−30	−10	−22
AA	LOX	20-hydroxy-leukotriene B_4	20oh LTB_4	351,5	**195**; 333; 315; 177	3,0	60	−40	−10	−23
AA	LOX	20-carboxy-leukotriene B_4	20cooh LTB_4	365,5	**303**; 347; 200; 195	3,1	60	−40	−10	−26
AA	COX	2,3-dinor-thromboxane B_2	2,3 dinor T_XB_2	341,4	**137**; 171; 155; 123	3,1	60	−20	−10	−31
EPA	COX	Prostaglandin $F_{3\alpha}$	$PGF_{3\alpha}$	351,5	**193**; 333; 307; 245	4,0	60	−50	−10	−30
DgLA	COX	Thromboxane B_1	TXB1	371,5	**171**; 327; 291; 197	4,4	30	−30	−10	−27
AA	Non-enz	Isoprostane $F_{2\alpha}$-III	8-iso $PGF_{2\alpha}$III	353,5	**193**; 309; 291; 247	4,7	30	−40	−10	−35
		Internal standard	*(d4) TXB2*	*373,2*	*173; 293; 209; 199*	*4,6*	*30*	*−50*	*−10*	*−22*
AA	COX	Thromboxane B_2	TxB_2	369,2	**169**; 325; 195; 177	4,6	30	−50	−10	−22
		Internal standard	*(d4) 8-iso PGF2α VI*	*357,3*	*197; 313; 295; 251*	*4,7*	*30*	*−40*	*−10*	*−35*
AA	COX	6,15-diketo-, 13,14-dihydro-prostaglandin $F_{1\alpha}$	6,15 dk-,dh-$PGF_{1\alpha}$	369,5	**267**; 351; 223; 123	4,8	60	−40	−10	−37

(continued)

Table 1 (continued)

Precursor	Pathway	Common name	Abbreviation	Precursor ion	Product Ions	Retention time (min)	Window (s)	DP (V)	EP (V)	CE (V)
EPA	COX	Prostaglandin E_3	PGE_3	349,5	**269**; 313; 189	5,6	60	−30	−10	−24
AA	COX	11-beta-prostaglandin $F_{2\alpha}$	11b $PGF_{2\alpha}$	353,5	**335**; 309; 299; 247	5,0	30	−50	−10	−35
AA	Non-enz	Isoprostane $F_{2\alpha}$-IV	5-iso $PGF_{2\alpha}$VI	353,5	**115**; 335; 317; 309	5,4	30	−60	−10	−35
		Internal standard	*(d11) 5-iso PGF2αVI*	*364,4*	*115; 320; 346*	*5,4*	*30*	*−60*	*−10*	*−35*
AA	LOX	14,15-leukotriene D4	14,15 LTD4	495,7	**177**	5,5	60	−60	−10	−25
AA	COX	Prostaglandin $F_{2\alpha}$	$PGF_{2\alpha}$	353,5	**193**; 309; 291	5,6	30	−50	−10	−35
		Internal standard	*(d4) PGF2α*	*357,3*	*197; 313; 295*	*5,6*	*30*	*−50*	*−10*	*−35*
EPA	COX	Prostaglandin D_3	PGD_3	349,5	**269**; 233; 189;	6,0	60	−30	−10	−24
DgLA	COX	Prostaglandin $F_{1\alpha}$	$PGF_{1\alpha}$	355,5	**293**; 337; 319; 311	5,8	60	−60	−10	−33
AA	Non-enz	8-iso-15-keto prostaglandin $F_{2\beta}$	8-iso-15 k $PGF_{2\beta}$	351,5	**219**; 315; 289; 271	6,0	60	−50	−10	−22
AA	COX	Prostaglandin E_2	PGE_2	351,5	**271**; 315; 235; 189	6,2	30	−50	−10	−23
		Internal standard	*(d4) PGE_2*	*355,5*	*275; 255; 193*	*6,2*	*30*	*−50*	*−10*	*−23*
AA	COX	11-beta-prostaglandin E_2	11b PGE_2	351,5	**271**; 333; 315; 189	6,4	60	−40	−10	−23

AA	COX	11-dehydro-thromboxane B_2	11d-T_XB_2	367,5	**305**; 349; 243; 161	6,4	60	−20	−10	−26
DgLA	COX	Prostaglandin E_1	PGE_1	353,5	**235**; 317; 273; 191	6,4	60	−40	−10	−29
AA	COX	15-keto-prostaglandin $F_{2\alpha}$	15 k $PGF_{2\alpha}$	351,5	**219**; 333; 315; 289	6,4	60	−50	−10	−32
AA	COX	13,14-dihydro-prostaglandin $F_{2\alpha}$	dh $PGF_{2\alpha}$	355,5	**311**; 337; 283; 275	6,4	60	−60	−10	−29
AA	LOX	Lipoxin B_4	LXB_4	351,5	**221**; 333; 315; 233	6,4	60	−50	−10	−21
DHA	LOX	Resolvin D3	RvD3	375,1	**147**; 137; 181;	6,5	30	−50	−10	−25
AA	COX	Prostaglandin D_2	PGD_2	351,5	**271**; 315; 233; 189	6,6	30	−50	−10	−23
		Internal standard	*(d4) PGD_2*	*355,5*	*275; 239; 193*	*6,6*	*30*	*−50*	*−10*	*−23*
AA	LOX	Leukotriene D_4	LTD_4	495,7	**177**; 477; 413; 142	6,6	60	−50	−10	−29
AA	COX	11beta-13,14-dihydro-15-keto-prostaglandin $F_{2\alpha}$	11b dhk $PGF_{2\alpha}$	353,5	**221**; 317; 291; 219	6,6	60	−60	−10	−37
DgLA	COX	Prostaglandin D_1	PGD_1	353,5	**235**; 317; 273; 123	6,7	30	−40	−10	−29
DHA	LOX	Resolvin D2	RvD2	375,1	**175**; 141; 215	6,8	30	−50	−10	−30
AA	COX	15-keto-prostaglandin E_2	15 k PGE_2	349,5	**235**; 331; 287; 173	6,8	60	−30	−10	−26
DgLA	COX	15-keto-prostaglandin $F_{1\alpha}$	15 k $PGF_{1\alpha}$	353,5	**221**; 317; 291; 193	6,8	60	−50	−10	−38

(continued)

Table 1 (continued)

Precursor	Pathway	Common name	Abbreviation	Precursor ion	Product Ions	Retention time (min)	Window (s)	DP (V)	EP (V)	CE (V)
AA	LOX	14,15-leukotriene D_4	14,15 LTE_4	438,6	**333**; 420; 351; 317	6,9	60	−40	−10	−22
AA	COX	15-keto-prostaglandin D_2	15 k PGD_2	349,5	**235**; 331; 287; 173	7,0	60	−30	−10	−40
		Internal standard	*(d5) RvD1*	*380,5*	***141**; 121*	*7,3*	*30*	*−20*	*−10*	*−25*
DHA	LOX	Resolvin D1	RvD1	375,1	**141**; 215; 121; 233	7,3	30	−20	−10	−25
AA	COX	13,14-dihydro-15-keto prostaglandin $F_{2\alpha}$	Dhk $PGF_{2\alpha}$	353,5	**291**; 335; 191	7,0	30	−80	−10	−28
AA	LOX	11-trans-leukotriene D_4	11 t LTD_4	495,7	**177**; 477; 413; 143	7,2	60	−50	−10	−29
AA	LOX	Leukotriene C_4	LTC_4	624,8	**272**; 606; 254; 179	7,3	60	−50	−10	−22
DHA	LOX	Protectin D1	PD1	359,1	**153**	7,3	60	−20	−10	−21
AA	LOX	5(S),6(R)-Lipoxin A_4	6R-LXA_4	351,5	**167**; 235; 217; 115	7,3	60	−20	−10	−21
AA	COX	13,14-dihydro-15-keto prostaglandin E_2	dhk PGE_2	351,5	**207**; 333; 315; 235	7,3	60	−40	−10	−26
AA	LOX	5(S),15(R)-Lipoxin A_4	15R-LXA_4	351,5	**165**; 217; 115; 235	7,3	60	−20	−10	−23
ADA	COX	Dihomo prostaglandin $F_{2\alpha}$	Dihomo $PGF_{2\alpha}$	381,5	**337**; 301; 319; 217	7,4	60	−40	−10	−37
AA	LOX	5(S),6(S)-Lipoxin A_4	6S-LXA_4	351,5	**217**; 333; 235; 115	7,5	60	−20	−10	−18

AA	COX	Prostaglandin K_2	PGK_2	349,5	**249**; 331; 287; 205	7,8	60	−40	−10	−31
gLA	COX	Prostaglandin K_1	PGK_1	351,5	**251**; 333; 289; 233	7,8	60	−40	−10	−26
AA	COX	13,14-dihydro-15-keto prostaglandin D_2	dhk PGD_2	351,5	**207**; 333; 315; 175	8,0	60	−40	−10	−26
AA	LOX	Leukotriene E_4	LTE_4	438,6	**333**; 420; 351; 317	8,1	60	−30	−10	−25
AA/ADA	COX	Dihomo prostaglandin E_2	Dihomo PGE_2	379,5	**299**	8,1	60	−40	−10	−30
AA	COX	Bicyclo prostaglandin E_2	Bicyclo PGE_2	333,5	**175**; 235; 271; 315	8,1	60	−40	−10	−30
AA	COX	Prostaglandin J_2	PGJ_2	333,5	**189**; 251; 271; 315	8,4	30	−40	−10	−22
ADA	COX	Dihomo prostaglandin D_2	Dihomo PGD_2	379,5	**299**	8,5	60	−50	−10	−34
AA	LOX	11-trans-leukotriene C_4	11 t LTC_4	624,8	**272**; 606; 254; 179	8,5	60	−50	−10	−34
AA	COX	Prostaglandin A_2	PGA_2	333,5	**271**; 315; 235; 189	8,5	60	−20	−10	−20
AA	LOX	11-trans-leukotriene E_4	11t LTE_4	438,6	**333**; 420; 351; 235	8,6	60	−50	−10	−33
AA	LOX	8S,15S-dihydroxy-eicosatetraenoic acid	8,15-diHETE	335,5	**235**; 317; 273; 155	9,0	60	−40	−10	−26
AA	LOX	5S,15S-dihydroxy-eicosatetraenoic acid	5,15-diHETE	335,5	**201**; 317; 255; 173	9,1	60	−40	−10	−26
DHA	LOX	15-trans Neuroprotectin D_1	15 t-PD1	359,1	**206**; 153	9,0	60	−20	−10	−27
		Internal standard	*(d5) MaR1*	*364,5*	*177; 157*	*9,1*	*30*	*−50*	*−10*	*−20*
DHA	LOX	Maresin 1	7(R) MaR1	359,1	**177**; 341; 297; 250	9,1	30	−30	−10	−22

(continued)

Table 1 (continued)

Precursor	Pathway	Common name	Abbreviation	Precursor ion	Product Ions	Retention time (min)	Window (s)	DP (V)	EP (V)	CE (V)
DHA	LOX	Resolvin D5	RvD5	359,1	**199**; 141; 261	9,2	30	−50	−10	−20
DHA	LOX	10S,17S-dihydroxy Docosahexaenoic acid	10S-PD1	359,1	**153**; 315; 297; 123	9,2	60	−20	−10	−21
AA	LOX	Leukotriene B_4	LTB_4	335,5	**195**; 317; 273; 151	9,5	30	−45	−10	−23
AA	LOX	6-trans-,12-epi-Leukotnene B_4	6 t,12epi LTB_4	335,5	**195**; 317; 273; 151	9,5	30	−45	−10	−22
AA	Synthetic	15-deoxy-Δ12,14-PGD_2	15d PGD_2	333,5	**271**; 333; 315; 203	9,5	60	−30	−10	−22
AA	LOX	12-epi-leukotriene B_4	12epi LTB_4	335,5	**195**; 317; 273; 203	9,5	30	−45	−10	−22
AA	LOX	12-oxo-leukotriene B_4	12oxo LTB_4	333,4	**253**; 317; 273; 195	9,5	60	−50	−10	−22
AA	LOX	6-trans-leukotriene B_4	6 t LTB_4	335,5	**195**; 317; 273; 203	9,5	30	−45	−10	−22
ADA	COX	Dihomo prostaglandin J_2	Dihomo PGJ_2	361,5	**299**	9,6	60	−50	−10	−29
LA	CYP	12,13-dihydroxy-octadecenoic acid	12,13 diHOME	313,5	**183**; 295; 277; 195	9,6	60	−50	−10	−29
LA	CYP	9,10-dihydroxy-octadecenoic acid	9,10 diHOME	313,5	**201**; 295; 277; 171	9,8	60	−50	−10	−29
AA	LOX	Tetranor 12-hydroxy-eicosatetraenoic acid	Tetranor 12-HETE	265,4	**109**; 221; 165; 139	9,8	60	−20	−10	−18

AA	COX	Prostaglandin B_2	PGB_2	333,2	**271**; 315; 235; 175	9,8	60	−40 −10	−25
AA	CYP	14,15-dihydroxy-eicosatrienoic acid	14,15-diHETrE	337,5	**207**; 319; 163; 129	10,0	60	−30 −10	−24
DHA	CYP	19,20-dihydroxy-docosapentaenoic acid	19,20 DiHDPA	361,5	**229**; 343; 281; 273	10,0	60	−40 −10	−22
AA	COX	12S-hydroxy-heptadecatrienoic acid	12-HHTrE	279,4	**217**; 261; 179; 135	10,1	60	−30 −10	−21
EPA	LOX	Lipoxin A_5	LXA_5	349,5	**215**; 251; 233; 115	10,1	60	−30 −10	−25
AA	LOX	14,15-leukotriene C_4	14,15 LTC_4	624,8	**272**; 606; 254; 179	10,1	60	−30 −10	−32
AdA	COX	Dihomo 15-deoxy-PGD_2	Dihomo 15d PGD_2	361,5	**299**	10,3	60	−40 −10	−25
AA	CYP	11,12-dihydroxy-eicosatrienoic acid	11,12-diHETrE	337,5	**167**; 319; 277; 255	10,3	60	−40 −10	−25
AA	CYP	8,9-dihydroxy-eicosatrienoic acid	8,9-diHETrE	337,5	**127**; 319; 185; 111	10,6	60	−30 −10	−27
AA	LOX	Hepoxilin B3	HXB3	335,5	**183**; 317; 291; 163	10,6	60	−40 −10	−21
LA	LOX	9-hydroxy-octadecatrienoic acid	9-HOTrE	293,4	**171**; 275; 231	10,7	60	−40 −10	−22
AA	LOX	Hepoxilin A3	HXA3	335,5	**195**; 317; 273; 161	10,6	60	−60 −10	−26
AA	COX	15-deoxy-Δ12,14-PGJ_2	15d PGJ_2	315,4	**203**; 297; 271; 217	10,8	60	−30 −10	−20
EPA	CYP	18-hydroxy-eicosapentaenoic acid	18-HEPE	317,3	**215**; 299; 259; 255	10,8	30	−50 −10	−20

(continued)

Table 1
(continued)

Precursor	Pathway	Common name	Abbreviation	Precursor ion	Product Ions	Retention time (min)	Window (s)	DP (V)	EP (V)	CE (V)
AA	CYP	19-hydroxy-eicosatetraenoic acid	19-HETE	319,2	**231**; 257; 275; 301	11,0	60	−40	−10	−23
EPA	LOX	11-hydroxy-eicosapentaenoic acid	11-HEPE	317,5	**215**; 299; 259; 255	11,0	60	−40	−10	−20
EPA	LOX	15-hydroxy-eicosapentaenoic acid	15-HEPE	317,5	**219**; 299; 256; 175	11,0	60	−40	−10	−18
AA	CYP	5,6-dihydroxy-eicosatrienoic acid	5,6-diHETrE	337,5	**145**; 319; 255; 191	11,0	60	−40	−10	−22
gLA	LOX	13-hydroxy-g-octadecatrienoic acid	13-HOTrE(y)	293,4	**193**; 275; 231; 149	10,9	60	−40	−10	−19
AA	Sintético	15-deoxy-prostaglandin A_2	15d PGA_2	315,4	**255**; 297; 271; 233	11,0	60	−40	−10	−20
EPA	CYP	5,6 dihydroxy-eicosatetraenoic acid	5,6-diHETE	335,5	**115**; 317; 291; 273	11,0	60	−50	−10	−29
AA	CYP	20-hydroxy-eicosatetraenoic acid	20-HETE	319,2	**245**; 257; 275; 301	11,0	60	−50	−10	−24
aLA	LOX	13-hydroxy-octadecatrienoic acid	13-HOTrE	293,4	**195**; 275; 223; 205	11,0	60	−40	−10	−28
AA	CYP	20-carboxy arachidonic acid	20cooh AA	333,5	**271**; 289; 297; 315	11,0	60	−60	−10	−23
EPA	LOX	12-hydroxy-eicosapentaenoic acid	12-HEPE	317,5	**179**; 299; 255; 135	11,0	60	−30	−10	−19
AA	CYP	18-hydroxy-eicosatetraenoic acid	18-HETE	319,2	**261**; 257; 217; 301	11,2	30	−60	−10	−20

EPA	Non-enz	9-hydroxy-eicosapentaenoic acid	9-HEPE	317,5	**149**; 299; 273; 255	11,3	60	−40	−10	−20
AA	CYP	17-hydroxy-eicosatetraenoic acid	17-HETE	319,2	**247**; 203; 257; 301	11,3	30	−50	−10	−20
AA	CYP	16-hydroxy-eicosatetraenoic acid	16-HETE	319,2	**189**; 233; 257; 301	11,4	60	−30	−10	−21
EPA	LOX	8-hydroxy-eicosapentaenoic acid	8-HEPE	317,5	**155**; 299; 255; 161	11,4	60	−50	−10	−29
EPA	LOX	5-hydroxy-eicosapentaenoic acid	5-HEPE	317,5	**115**; 299; 255; 235	11,4	60	−30	−10	−22
EPA	CYP	17,18-epoxy eicosatetraenoic acid	17(18) EpETE	317,5	**259**; 299; 255; 215	11,4	60	−40	−10	−18
DHA	Non-enz	20-hydroxy-docosahexaenoic acid	20 HDoHE	343,5	**241**; 325; 299; 281	11,6	60	−30	−10	−18
LA	LOX	9-hydroxy-octadecadienoic acid	9-HODE	295,5	**171**; 277; 233; 213	11,5	30	−60	−10	−23
		Internal standard	*(d4) 9-HODE*	*299,5*	*172; 281; 253; 217*	*11,5*	*30*	*−60*	*−10*	*−23*
		Internal standard	*(d4) 13-HODE*	*299,5*	*198; 281*	*11,4*	*30*	*−60*	*−10*	*−23*
LA	LOX	13-hydroxy-octadecadienoic acid	13-HODE	295,5	**195**; 277	11,5	30	−60	−10	−23
AA	LOX	15-hydroxy-eicosatetraenoic acid	15-HETE	319,2	**175**; 219; 257; 301	11,7	30	−40	−10	−19
LA	LOX	13-oxo-octadecatrienoic acid	13-oxoODE	293,4	**167**; 249; 195; 113	11,7	60	−50	−10	−29
		Internal standard	*(d8) 15-HETE*	*327,3*	*226; 309; 265; 182*	*11,7*	*30*	*−40*	*−10*	*−19*
DHA	Non-enz	16-hydroxy-docosahexaenoic acid	16 HDoHE	343,5	**233**; 325; 281; 189	11,8	60	−50	−10	−19

(continued)

Table 1 (continued)

Precursor	Pathway	Common name	Abbreviation	Precursor ion	Product Ions	Retention time (min)	Window (s)	DP (V)	EP (V)	CE (V)
DHA	Aspirin COX / LOX	17-hydroxy docosahexaenoic acid	17 HDoHE	343,5	**229**; 281; 201; 325	11,9	60	−20	−10	−19
AA	LOX	15-oxo-eicosatetraenoic acid	15-oxoETE	317,5	**113**; 299; 273; 139	11,9	60	−20	−10	−25
LA	LOX	9-oxo-octadecadienoic acid	9-oxoODE	293,4	**185**; 249; 151; 113	11,9	60	−50	−10	−28
AA	LOX/non-enz	13-hydroxy-docosahexaenoic acid	13 HDoHE	343,5	**221**; 325; 281; 233	11,9	60	−30	−10	−19
EPA	CYP	14,15-epoxy eicosatetraenoic acid	14(15) EpETE	317,5	**207**; 299; 255; 219	11,5	30	−30	−10	−19
AA	Non-enz	11-hydroxy-eicosatetraenoic acid	11-HETE	319,2	**167**; 257; 275; 301	11,9	60	−40	−10	−23
DHA	LOX/non-enz	14-hydroxy-docosahexaenoic acid	14 HDoHE	343,5	**205**; 325; 281; 161	12,0	60	−30	−10	−18
AA	LOX	12-hydroxy-eicosatetraenoic acid	12-HETE	319,2	**135**; 179; 257; 301	12,1	30	−50	−10	−19
DHA	LOX/non-enz	11-hydroxy-docosahexaenoic acid	11 HDoHE	343,5	**149**; 325; 281; 165	12,1	60	−20	−10	−19
		Internal standard	*(d8) 12-HETE*	*327,3*	***184**; 309; 283; 265*	*12,1*	*30*	*−50*	*−10*	*−19*
AA	Non-enz	8-hydroxy-eicosatetraenoic acid	8-HETE	319,2	**155**; 163; 257; 301	12,2	60	−40	−10	−19
DHA	LOX/non-enz	10-hydroxy docosahexaenoic acid	10 HDoHE	343,5	**153**; 325; 299; 281	12,0	60	−50	−10	−19

AA	LOX	12-oxo-eicosatetraenoic acid	12-oxoETE	317,5	**153**; 299; 273; 235	12,2	60	−50	−10	−23
DHA	LOX/non-enz	7-hydroxy-docosahexaenoic acid	7 HDoHE	343,5	**141**; 325; 281; 201	12,2	60	−40	−10	−19
AA	Non-enz	9-hydroxy-eicosatetraenoic acid	9-HETE	319,2	**123**; 179; 257; 301	12,2	60	−40	−10	−20
DHA	Non-enz	8-hydroxy-docosahexaenoic acid	8 HDoHE	343,5	**109**; 299; 281; 189	12,2	60	−40	−10	−20
DgLA	LOX	15-hydroxy-eicosatrienoic acid	15-HETrE	321,5	**221**; 259	12,2	60	−30	−10	−21
AA	LOX	5-hydroxy-eicosatetraenoic acid	5-HETE	319,2	**115**; 203; 257; 301	12,4	30	−40	−10	−20
DHA	CYP	19,20-epoxy docosapentaenoic acid	19(20) EpDPE	343,5	**241**; 325; 299; 281	12,4	60	−50	−10	−18
		Internal standard	*(d8) 5-HETE*	*327,3*	*116; 309; 265; 210*	*12,4*	*30*	*−40*	*−10*	*−20*
DgLA	LOX	8-hydroxy-eicosatrienoic acid	8-HETrE	321,5	**157**; 303; 259; 163	12,5	60	−20	−10	−22
DPA	LOX	17-keto docosapentaenoic acid	17 k DPA	343,5	**247**; 325; 299; 231	12,5	60	−40	−10	−23
AA	CYP	14(15)-epoxy-eicosatrienoic acid	14,15-EET	319,2	**175**; 219; 257; 301	12,5	60	−30	−10	−17
LA	CYP	12(13)-epoxy-octadecenoic acid	12,13 EpOME	295,5	**195**; 277	12,5	60	−50	−10	−23
LA	CYP	9(10)-epoxy-octadecenoic acid	9,10 EpOME	295,5	**171**; 277; 233; 125	12,5	60	−60	−10	−21
DHA	CYP	16,17-epoxy docosapentaenoic acid	16(17) EpDPE	343,5	**193**; 299; 281; 161	12,5	60	−40	−10	−19
DHA	LOX/non-enz	4-hydroxy-docosahexaenoic acid	4 HDoHE	343,5	**101**; 325; 299; 281	12,5	60	−60	−10	−18

(continued)

Table 1 (continued)

Precursor	Pathway	Common name	Abbreviation	Precursor ion	Product Ions	Retention time (min)	Window (s)	DP (V)	EP (V)	CE (V)
AA	LOX	5-oxo-eicosatetraenoic acid	5-oxoETE	317,5	**203**; 299; 273; 245	12,6	60	−60	−10	−22
AA	COX	2,3-dinor-6-keto prostaglandin $F_{1\alpha}$	2,3 dinor-6 k $PGF_{1\alpha}$	363,5	**281**; 327; 301; 199	12,7	60	−30	−10	−23
AA	CYP	11(12)-epoxy-eicosatrienoic acid	11,12-EET	319,2	**167**; 179; 257; 301	12,9	60	−30	−10	−17
AA	CYP	8(9)-epoxy-eicosatrienoic acid	8,9-EET	319,2	**155**; 257; 275; 301	13,0	60	−30	−10	−18
AA	CYP	5(6)-epoxy-eicosatrienoic acid	5,6-EET	319,2	**191**; 257; 275; 301	13,1	60	−30	−10	−17
EDA	LOX	15-oxo-eicosadienoic acid	15 oxoEDE	321,5	**223**; 303; 195; 113	13,3	60	−80	−10	−32
MA	LOX	5-hydroxy-eicosatrienoic acid	5-HETrE	321,5	**205**; 303; 259; 115	13,5	60	−30	−10	−19
–	–	Eicosapentaenoic acid	EPA	301,3	**257**; 283; 223; 203	14,1	30	−40	−10	−16
–	–	Adrenic acid	Arachidonic acid	303,3	**259**; 205	14,9	30	−55	−10	−20
		Internal standard	*(d8) Arachiconic acid*	*311,3*	*267; 183*	*14,9*	*30*	*−55*	*−10*	*−20*
–	–	Docosahexaenoic acid	DHA	327,3	**283**; 310; 249; 229	14,6	30	−40	−10	−19
–	–	Arachidonic acid	Adrenic acid	331,3	**287**; 313; 233	14,8	60	−70	−10	−20

2. The mass spectrometer is operated in *Scheduled Multiple Reaction Monitoring* (sMRM) mode. *Multiple Reaction Monitoring* (MRM) is an MS approach that monitors the transition of a precursor ion to a specific product ion fragment. These defined precursor and product ions are known as an MRM pair. A sMRM methodology takes into account the association of an MRM pair with the liquid chromatography retention time. sMRM is an improvement over MRM allowing for better data collection and more analytes to be monitored in a single analysis. The MS operating software (Analyst 1.6.2, ABSciex) takes into account the total number of sMRM pairs and their respective retention times, and optimizes how the mass spectrometer scans in for maximum Dwell time for each transition. In these lipid mediator method a total cycle time of 2 s is set for better acquisition and a 30–60 s retention time window for each sMRM pair is sufficient to account for metabolite peaks and slight shifts in their retention times [13]. Moreover, we utilize *sMRM Algorithm Pro*, an algorithm contained in Analyst 1.6.2 which potentiates the acquisitions since it can trigger secondary MRM transitions from primary transitions for improved dwell times and cycle times by only acquiring secondary transitions when the compound is eluting and above a threshold. This is an improvement in the method since secondary sMRM helps identity confirmation of molecules detected. In summary, with this current methodology we analyze a total of 160 metabolites complies more than 600 sMRM pairs with only one injection per sample in the same chromatographic analysis (Table 1).

3.8 Data Analysis

1. Analyst 1.6.2 software can be used to construct calibration curves of each compound from the calibration standards that were run. It normalizes the peak area of each analyte against the peak area of the appropriate internal standard (internal standard with the nearest retention time) to provide an area ratio value. The area ratio is then plotted against the known concentration of each calibration standard and calibration curves are calculated by the least squares regression method.
2. The samples are processed in the same way, generating an area ratio for each compound in each sample. The concentration corresponding to this ratio is then calculated and the software provides a table of the concentration of each compound in each lipid extract (as pg/μl, *see* **Note 3**).

This method allows for analyzing major metabolites derived from AA, EPA, DHA by enzymatic and nonenzymatic pathways in plasma samples from controls as well as neurodegenerative diseases patients.

4 Notes

1. Limit of detection varies for different metabolites within 0.1–1.0 pg/μl (4–40 pg) range.
2. The described method is specifically developed for plasma. This lipidomic approach can also be applied to CSF samples. However, it should be noted that many of these metabolites are present in lower concentration than in plasma.
3. Dilution factor of 10 should be considered when processing data, since 500 μl of plasma are eventually reconstituted in 50 μl for injection to HPLC-MS/MS analysis.

Acknowledgments

This work was supported by grants from CSIC-Grupos N° 536 to HR, Comisión Sectorial de Investigación Científica-I + D to AT, and -Iniciación ID:416 to MM. MM was supported by a postgraduate fellowship by Universidad de la República—Uruguay.

References

1. Wood P (ed) (2003) Neuroinflammation: mechanisms and management, 2nd edn. Humana Press, Totowa, NJ. https://doi.org/10.1007/978-1-59259-297-5
2. Hooten KG, Beers DR, Zhao W, Appel SH (2015) Protective and toxic neuroinflammation in amyotrophic lateral sclerosis. Neurotherapeutics 12(2):364–375. https://doi.org/10.1007/s13311-014-0329-3
3. Rozen S, Cudkowicz ME, Bogdanov M, Matson WR, Kristal BS, Beecher C, Harrison S, Vouros P, Flarakos J, Vigneau-Callahan K, Matson TD, Newhall KM, Beal MF, Brown RH, Kaddurah-Daouk R (2005) Metabolomic analysis and signatures in motor neuron disease. Metabolomics 1(2):101–108. https://doi.org/10.1007/s11306-005-4810-1
4. Kendall AC, Nicolaou A (2017) Lipidomics analyses of oxygenated metabolites of polyunsaturated fatty acids. In: Wood P (ed) Neuromethods. Humana Press, New York, pp 211–228. https://doi.org/10.1007/978-1-4939-6946-3_15
5. Rouzer CA, Marnett LJ (2005) Structural and functional differences between cyclooxygenases: fatty acid oxygenases with a critical role in cell signaling. Biochem Biophys Res Commun 338:34–44
6. Brash AR (1999) Lipoxygenases: occurrence, functions, catalysis, and acquisition of substrate. J Biol Chem 274:23679–23682
7. Dalli J, Serhan CN (2018) Immunoresolvents signaling molecules at intersection between the brain and immune system. Curr Opin Immunol 50:48–54
8. Serhan CN, Levy BD (2018) Resolvins in inflammation: emergence of the pro-resolving superfamily of mediators protection versus uncontrolled inflammation: first responders and resolution. J Clin Invest 128:2657–2669
9. Prüss H, Rosche B, Sullivan AB, Brommer B, Wengert O, Gronert K, Schwab JM (2013) Proresolution lipid mediators in multiple sclerosis - differential, disease severity-dependent synthesis - a clinical pilot trial. PLoS One 8:4–8
10. Wang X, Zhu M, Hjorth E, Cortés-Toro V, Eyjolfsdottir H, Graff C, Nennesmo I, Palmblad J, Eriksdotter M, Sambamurti K, Fitzgerald JM, Serhan CN, Granholm A-C, Schultzberg M (2015) Resolution of inflammation is altered in Alzheimer's disease. Alzheimers Dement 40:40–50
11. Francos-Quijorna I, Santos-Nogueira E, Gronert K, Sullivan AB, Kopp MA, Brommer B, David S, Schwab JM, López-Vales R (2017) Maresin 1 promotes inflammatory resolution, neuroprotection, and

functional neurological recovery after spinal cord injury. J Neurosci 37:11731–11743

12. Wenk MR (2010) Lipidomics: new tools and applications. Cell 143:888–895
13. Dumlao DS, Buczynski MW, Norris PC, Harkewicz R, Dennis EA (2011) High-throughput lipidomic analysis of fatty acid derived eicosanoids and N-acylethanolamines. Biochim Biophys Acta Mol Cell Biol Lipids 1811:724–736

Chapter 12

Characterization and Quantification of the Fatty Acid Amidome

Kristen A. Jeffries, Emma K. Farrell, Ryan L. Anderson, Gabriela Suarez, Amanda J. Goyette Osborne, Mc Kenzi K. Heide, and David J. Merkler

Abstract

Fatty acid amides are a diverse family of underappreciated, biologically occurring lipids. Herein, we detail the methods we have used to identify, characterize, and quantify a set of fatty acid amides, the fatty acid amidome, from cultured mammalian cells and insects.

Key words Fatty acid amides, Amidome, Drosophila

1 Introduction

Fatty acid amides, R-CO-NH-R′, represent an intriguing set of biologically occurring lipid-derived molecules. The structure of fatty acid amides is relatively simple, the R-CO- moiety being derived from a fatty acid and the -NH-R′ moiety being derived from a biogenic amine (Table 1). Greater than 70 different fatty acid amides have been identified from living systems [1]. The best known and most understood members of the fatty acid amide family are *N*-arachidonoylethanolamine (anandamide), *N*-(17-hydroxylinolenoyl)-L-glutamine (volicitin), and oleamide. Anandamide and oleamide are produced by mammals while volicitin is produced in insects. Anandamide is the endogenous ligand to cannabinoid receptor 1 (CB_1) [2], oleamide regulates sleep [3], and volicitin is an elicitor of plant volatiles [4]. Much less is known about the functions of the other ~70 known fatty acid amides, but it seems likely that these lipid amides are biologically significant given the importance of anandamide, volicitin, and oleamide. Of course, questions about fatty acid amide function are intimately linked to questions about their endogenous receptors. Unknown fatty acid

Kristen A. Jeffries, Emma K. Farrell, and Ryan L. Anderson contributed equally to this work.

Paul L. Wood (ed.), *Metabolomics*, Neuromethods, vol. 159, https://doi.org/10.1007/978-1-0716-0864-7_12,

Table 1
The biologically-occurring fatty acid amides[a]

Fatty acid amide	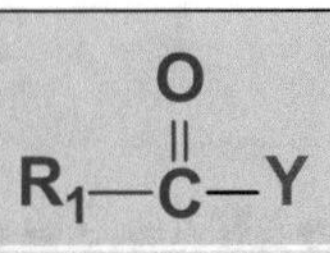 $R_1-C(=O)-Y$
Primary fatty acid amide	Y = NH_2
N-acylethanolamine	Y = NH-CH_2-CH_2-OH
N-acylglycine	Y = NH-CH_2-COOH
N-acyltaurine	Y = HN-CH_2-CH_2-SO_3H
N-acyl-γ-aminobutyric acid	Y = NH-CH_2-CH_2-CH_2-COOH
N-acylamino acid	Y = NH-CH(R_2)-COOH
N-acyldopamine	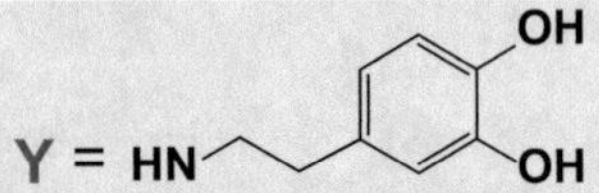 Y = HN–CH_2CH_2–(3,4-dihydroxyphenyl) [structure: OH, OH]
N-acylserotonin	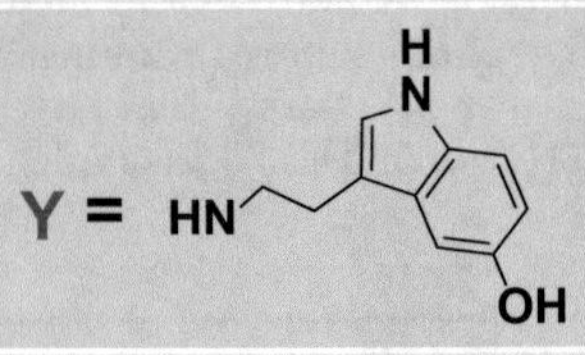 Y = HN–CH_2CH_2–(5-hydroxyindol-3-yl) [structure: H N, OH]
N-acyltyramine	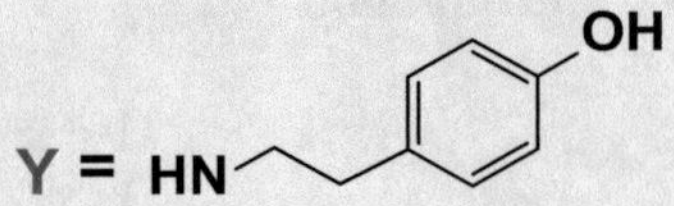 Y = HN–CH_2CH_2–(4-hydroxyphenyl) [structure: OH]

[a]**R_1-CO-** is the fatty acyl moiety, **R_2** represents the diversity of structures found in canonical 20 amino acids found in proteins other than glycine, and **Y** represents the biogenic amine

amide receptors point to other unknown fatty acid amide binding proteins or enzymes involved in fatty acid amide biosynthesis, degradation, metabolism, and transport [5].

One major challenge in measuring the fatty acid amidome is the low abundance of fatty acid amides. Work in cultured mouse neuroblastoma $N_{18}TG_2$ cells has demonstrated that oleamide represents ~0.1% of the total oleoyl-derived metabolites in these cells [6, 7]. Combine their low abundance with the lack of a convenient chromophore, the structural similarity between the members of the fatty acid amide family, and the inherent variability in living organisms, and it is easy to understand the difficulty in measuring the fatty acid amidome. Nonetheless, methods have been developed to identify and quantify the fatty acid amidome, and these data have proven critical to our current understanding of fatty acid amides [8–15]. The review by Piscitelli and Bradshaw [15] is highly recommended for a detailed summary of the methods used since the early 1990s to identify and quantify fatty acid amides. Such methods will be useful in moving forward to resolve some of the unanswered questions about fatty acid amides. For example, identifying the tissues and/or organelles that possess relatively high levels of a specific fatty acid amide may help address the function question. In our laboratory, the evaluation of the fatty acid amidome after decreasing the expression of a targeted enzyme has provided valuable insights into the pathways of fatty acid amide biosynthesis [7, 16, 17]. Herein, we detail the method we have used to evaluate the fatty acid amidome from cultured mammalian cells and from a number of insects [7, 14, 16–18].

2 Materials

2.1 Biological Sources for Fatty Acid Amides

We have identified fatty acid amides from different biological sources: cultured mammalian cells (mouse neuroblastoma $N_{18}TG_2$ cells and sheep choroid plexus cells) and a number of insects (*Drosophila melanogaster*, *Bombyx mori*, *Tribolium castaneum*, *Rhynchophorus cruentatus*, and *Oncopeltus fasciatus*). $N_{18}TG_2$ cells are purchased from the Deutsche Sammlung von Mikroorganism und Zellkuturn GmBH, and sheep choroid plexus (SCP) cells are purchased from the American Type Culture Collection. *Drosophila melanogaster* UAS stocks (CG9486) are purchased from Vienna *Drosophila* Resource Center. *D. melanogaster* (Oregon R), *Oncopeltus fasciatus* (milkweed bug), and *Bombyx mori* eggs are from Carolina Biological. *Tribolium castaneum* (red flour beetle) are a gift from Dr. Susan J. Brown, Kansas State University. *Rhynchophorus cruentatus* (palmetto weevil) are captured locally using a device provided by Dr. Robin M. Giblin-Davis [19], flash-frozen in liquid N_2 once brought to the laboratory, and stored at −80 °C until analyzed for their fatty acid amide content.

2.2 Fatty Acid Amide Standards

Synthetic standards for some of fatty acid amides are commercially available, including the *N*-acylglycines, *N*-acylethanolamines, *N*-acyldopamines, *N*-acylserotonins, and the primary fatty acid amides. Suppliers for fatty acid amides that we have used include Cayman Chemical Co., Bachem, and Sigma-Aldrich. When not available, we have synthesized specific *N*-acylglycines, *N*-acyltryptamines, and primary fatty amides using published methods [20–22].

3 Methods

3.1 Cell Culture

All cells are grown in 225 cm^2 culture dishes. $N_{18}TG_2$ cells are grown in Dulbecco's Modified Eagle Medium (DMEM) supplemented with 100 μM 6-thioguanine. SCP are grown in Eagle's Minimal Essential Medium (EMEM) with 100 μM sodium pyruvate. The media for both SCP and $N_{18}TG_2$ cells are supplemented with 10% fetal bovine serum (FBS), 100 I.U./mL penicillin, and 1.0 mg/mL streptomycin. Cells are incubated at 37 °C with 5% CO_2 according to supplier instructions. If fatty acid enrichment is desired, cultures are grown to 80–90% confluency, the culture medium removed, and then replaced with media containing 0.5% FBS or horse serum and a mixture of 2.5 mM fatty acid/ 0.25 mM bovine serum albumin (BSA). We found better incorporation of the fatty acyl moiety into the fatty acid amide from an exogenously added fatty acid by mixing together the fatty acid with BSA. After an incubation period of 12–48 h, the medium is collected, the cells are washed with PBS, and trypsin is added to detach them from the growing surface. The cells are collected by centrifugation and stored at −80 °C until analyzed for their fatty acid amide content. The spent media is centrifuged again to remove any residual cells and the supernatant stored at −80 °C until analyzed for their fatty acid amide content.

3.2 Care and Maintenance of the Insects

1. *Drosophila melanogaster* are maintained in capped plastic tubes containing 4–24 Instant Drosophila Medium at room temperature. After 5 days, *D. melanogaster* are collected by immobilizing them with ice, flash-frozen in liquid N_2, and the frozen flies shaken vigorously to detach the head from the thorax-abdomen. The heads are separated from thorax-abdomens by sifting these through a wire mesh. The head and thorax-abdomen samples are stored separated at −80 °C until analyzed for their fatty acid amide content.
2. *Bombyx mori* eggs are obtained from Carolina Biological, immediately placed into a petri dish, and fed Silkworm Artificial Dry Diet after hatching. Silkworms are maintained at room temperature and are allowed to progress through their life cycle from pupae to the moth stage [23]. *B. mori* are collected from first instar to the fifth instar, based on the number of molts. After each collection, the *B. mori* are flash-frozen in liquid N_2, and stored at −80 °C until analyzed for their fatty acid amide content.
3. *Tribolium castaneum* are hardy insects and are straightforward to maintain. *T. castaneum* are reared in presifted 100% organic bread flour (obtained locally). The stocks are maintained in an incubator at 30 °C in petri dishes before being transferred to larger plastic containers capped with lids possessing holes of a sufficient size to allow free atmosphere exchange but prevent the escape of the beetles. Sieves of varying mesh gauges are used to separate adults from larvae. They are then flash-frozen in liquid N_2, and stored at −80 °C until analyzed for their fatty acid amide content.

4. *Oncopeltus fasciatus* are maintained in a 4 qt. screw top Tupperware container, which contained 20–30 insects at room temperature with a source of water and food (sunflower kernels, purchased locally). *O. fasciatus* are collected by immobilizing them with ice, flash-frozen in liquid N_2, and stored at −80 °C until analyzed for their fatty acid amide content.

3.3 Lipid Extraction/Preparation of the Fatty Acid Amidome

3.3.1 Lipid Extraction from the Cultured Mammalian Cells

Lipids, including the long-chain fatty acid amides, are extracted from the cultured cells ($N_{18}TG_2$ or SCP cells) using a procedure similar to that described by Sultana and Johnson [11]. Methanol (4 mL) is added to cell pellets and samples are sonicated for 15 min. on ice (*see* **Notes 1** and **2**). Samples are centrifuged at 4,500 × *g* for 10 min, and the supernatant is separated from the pellet, and is dried under N_2 in a warm water bath at 40 °C. The pellet is reextracted with 4 mL 1:1:0.1 (v/v/v) chloroform–methanol–water, sonicated for 10 min, vortexed for 2 min, and centrifuged 10 min at 4,500 × *g*. The supernatant from this step is added to the dried supernatant from the first extraction and the combined supernatants are dried under N_2 at 40 °C. The pellet is reextracted for a third time with 4.8 mL of chloroform–methanol 2:1 (v/v) and 0.8 mL of 0.5 M KCl/0.08 M H_3PO_4, sonicated 2 min., vortexed for 2 min, and centrifuged for 10 min at 4,500 × *g*. The lower lipid phase from the centrifugation step is combined with the dried extraction mixture from the first and second supernatants, and the resulting combination is taken to dryness under N_2 at 40 °C.

3.3.2 Lipid Extraction from D. melanogaster and Other Insects

1. The long-chain fatty acid amides are extracted from insects using a method slightly modified from that described in Subheading 3.3.1. *D. melanogaster* heads or thorax-abdomen in 1.0 g batches are ground in a chilled mortar with 30 mL of methanol and the resulting paste is sonicated for 15 min on ice. Cellular debris is removed by centrifugation and the supernatant is dried under N_2 at 40 °C.

   The pellet is reextracted with 30 mL of 1:1:0.1 (v/v/v) chloroform–methanol–water followed by sonication for 10 min on ice. The supernatant is collected by centrifugation and added to the dried supernatant from the first extraction. The resulting combination is then dried under N_2 at 40 °C. The pellet is reextracted for a third time with a 41 mL solution prepared by mixing together 36 mL of 2:1 (v/v) chloroform–methanol and 5 mL of 0.5 M KCl/0.8 M H_3PO_4, followed by sonication for 2 min on ice. After vigorously mixing the pellet with this solution for 2 min using a vortex, the mixture is centrifuged to create a phase separation. The lower lipid phase is removed, and then added to the dried mixture of first and second extractions. The combination of the 3 extractions is dried under N_2 at 40 °C and stored at −20 °C until analyzed for their fatty acid amide content.

This protocol is used in our initial studies of fatty acid amides in *D. melanogaster* [14] and has since been applied to identification of fatty acid amides in *B. mori*, *T. castaneum*, *R. cruentatus*, and *O. fasciatus*. We have used the protocol to identify fatty acid amides from the whole insect, from insect segments or body components, and from the different life cycle stages in *B. mori* [17]. The protocol has proven useful in samples of different amounts (0.4–3.0 g). The only change is to linearly adjust the volumes to match the mass of the batch size.

3.3.3 Solid-Phase Lipid Enrichment of Fatty Acid Amides

1. The dried lipid extract either from the cultured mammalian cells (*see* Subheading 3.3.1) or from the insect homogenates (*see* Subheading 3.3.2) is subjected to a solid-phase extraction step to enrich the lipid metabolites with fatty acid amides [11]. Silica columns (top layer of 0.5 g of sand and bottom layer of 0.5 g of silica) are washed with *n*-hexane (*see* **Note 3**). The dried lipid extracts are redissolved in 0.1 mL of *n*-hexane and applied to a washed silica column. The silica column is developed as follows: 4 mL of *n*-hexane, 1.0 mL of 99:1 (v/v) *n*-hexane–acetic acid, 1 mL of 90:10 (v/v) *n*-hexane–ethyl acetate, 1 mL of 80:20 (v/v) *n*-hexane–ethyl acetate, 1 mL of 70:30 (v/v) *n*-hexane–ethyl acetate, 1.5 mL of 2:1 (v/v) chloroform–isopropanol, and, lastly, 0.5 mL of methanol. The eluant from the final three organic wash steps, the 70:30 (v/v)*n*-hexane–ethyl acetate, chloroform–isopropanol, and the methanol washes, contained fatty acid amides. These are combined, taken to dryness under N_2 at 40–45 °C, and stored at −20 °C until analyzed for their fatty acid amide content.
2. To test the effectiveness of the extraction protocol, a nonnatural fatty acid or fatty acid amide is added to cell pellet or insect homogenate. Percent recoveries of the nonnatural fatty acid or fatty acid amide ranged from 80% to 100%.

3.3.4 Liquid Chromatography–Quadrupole Time-of-Flight Mass Spectrometry

1. The dried extract from the solid-phase extraction step are reconstituted in HPLC-grade methanol and spiked with an internal standard, 10 pmoles of *N*-arachidonoylglycine-d_8 (a final volume of 100 μL) (*see* **Notes 4** and **5**). Other fatty acid amide standards can be used, as long as there is no overlap between the *m/z* for the standard as those of an endogenous analyte. An Agilent 1260 liquid chromatography system with a Kinetex™ 2.6 μm C_{18} 100 Å (50 × 2.1 mm) reverse phase column is connected to an Agilent 6540 quadrupole time-of-flight mass spectrometer equipped with a Dual Agilent Jet-stream ESI source in positive ion mode. The LC/QTOF-MS methods that we employed are described in Jeffries et al. [14]. Mobile phase A consists of water with 0.1% formic acid, and mobile phase B consists of acetonitrile with 0.1% formic

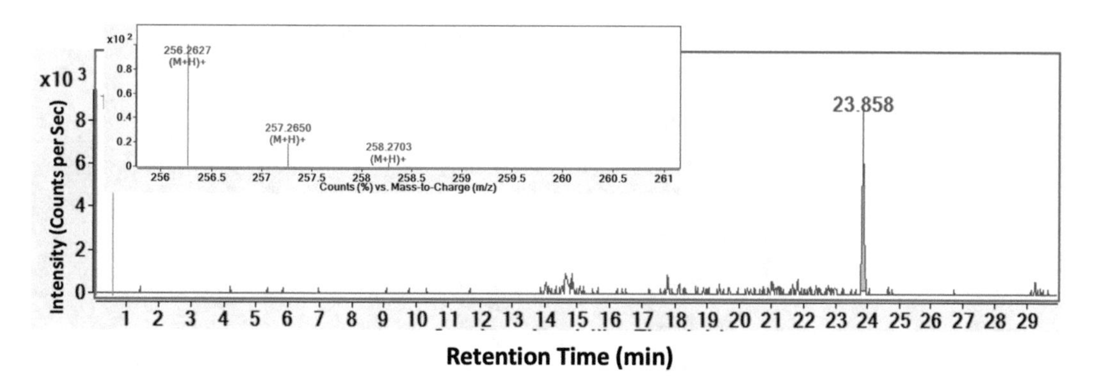

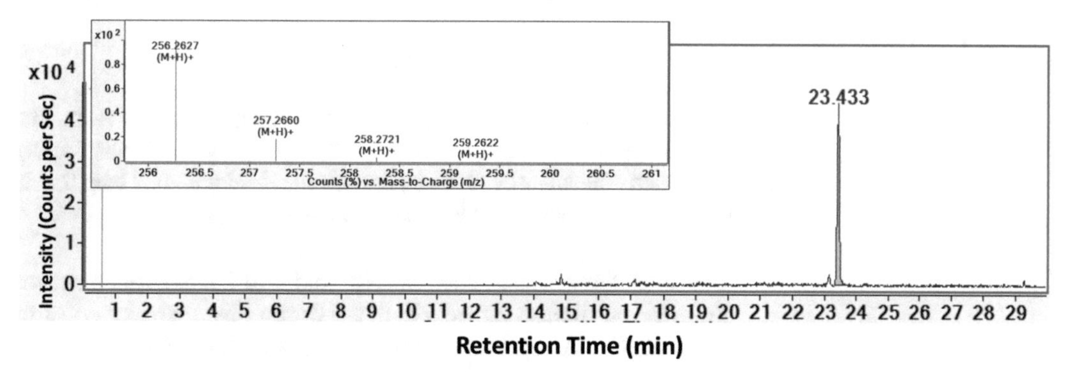

Fig. 1 Identification of Palmitamide from *O. fasciatus* by LQ/QTOF-MS. The mass spectrum and EIC peak of the palmitamide standard (upper panel) matched those of palmitamide (lower panel) identified from 0.5 g homogenates *O. fasciatus* (8–10 insects). These data are representative of data we have collected for each fatty acid amide identified from cultured mammalian cells and insect homogenates

acid. The gradient starts at 10% B and is linearly increased to 100% B over the course of 5 min followed by a hold of 3 min at 100% B. Equilibrium is achieved by holding at 10% B for 8 min. The flow rate is 0.6 mL/min and each injection is 10–25 μL. Data are collected at an acquisition rate of 2 spectra/s in in the extended dynamic range mode (2 GHz). The capillary voltage is set to 3.5 kV; the drying gas temperature is set to 300 °C with a gas flow rate of 8 L/min; the sheath gas temperature is set to 350 °C with a gas flow rate of 11 L/min; and the nebulizer pressure is set to 35 psig. Extracted ion chromatograms (EICs) are obtained from the total ion chromatograms (TICs) for each of the long-chain fatty acid amides using Agilent MassHunter Qualitative Analysis B.04.00. The fatty acid amides in each extraction are identified by comparison to synthetic standards by molecular ion *m/z* and retention time (Fig. 1).

2. Quantification of the identified fatty acid amides is performed by integrating the area under the chromatographic peak and comparing that value against standard curves constructed using

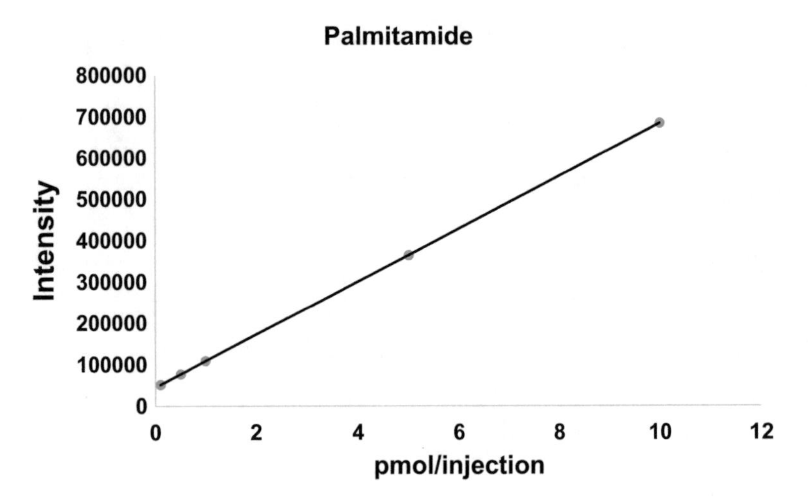

Fig. 2 Standard Curve for Palmitamide Generated using the LQ/QTOF-MS. The standard samples contained 0.1, 0.5, 1.0, 5.0, or 10 pmol of palmitamide per 20 μL injection. The intensity of the injections ranged from 80,000 to 650,000. The data fit well to a linear line with an R^2 value of 0.9993

the same fatty acid amide (*see* **Notes 6** and 7). Standard curves are in the linear range of 0.1–10 pmoles on the column ($r^2 > 0.99$). A representative curve is shown in Fig. 2.

N-Arachidonoylglycine-d_8, 1 pmole per 10 μL injection, is spiked into each extraction to measure instrument performance and for data normalization. Solvent and slip additive blanks are run to evaluate background levels for specific fatty acid amides, particularly important for the quantification of oleamide, erucamide, and palmitamide [24, 25]. Background levels of the fatty acid amide are subtracted from analyte levels and are <10% of the endogenous levels found in the cultured cells or insect homogenates. The levels of the *N*-acyldopamines represent the sum total of the amount of the *N*-acyldopamine and the amount of the *N*-acyldopamine quinone. An outline of the workflow for the identification and quantification of the fatty acid amidome is shown in Fig. 3.

4 Notes

1. We always carry out the lipid extractions in triplicate from three individual samples.
2. Slip additive contamination occurred frequently even when plastic tubes, pipettes, and columns are avoided during extractions. Blank samples are run in parallel during extractions and enrichment steps and are subtracted from analyte amounts.

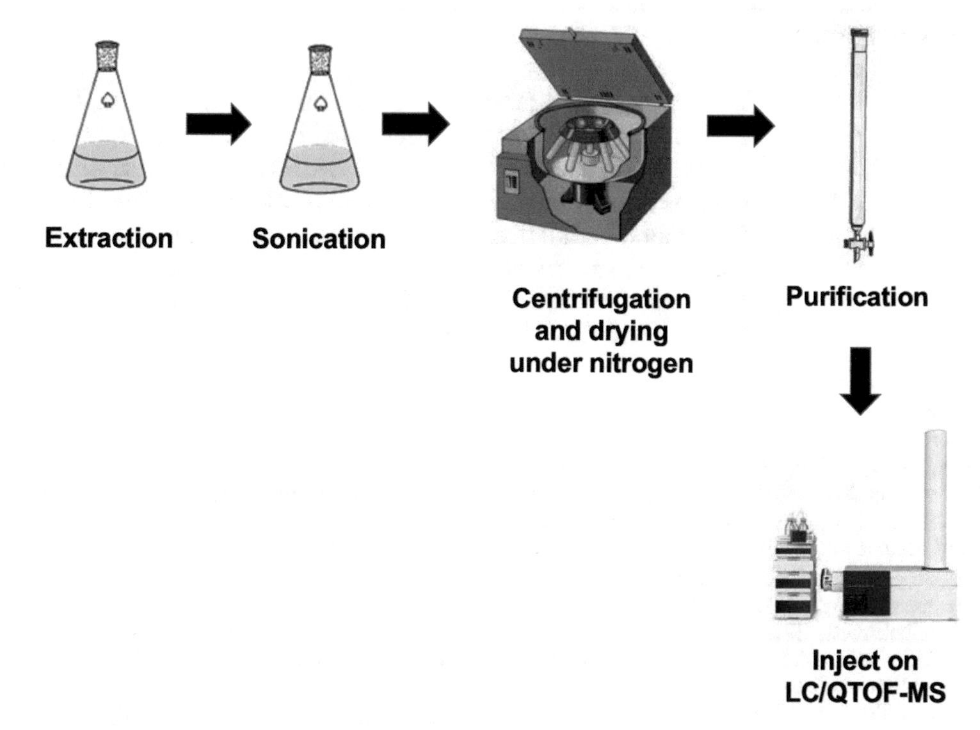

Fig. 3 Work flow for the identification and quantification of the fatty acid amidome

3. The columns we use for solid-phase extraction are constructed with a lower layer of 0.5 g of silica resin and a top layer of 0.5 g of laboratory-grade sand. This is done to protect the silica from the direct application of solvents to the column. We have found that the direct application of solvents to the silica column can cause channels in the resin, leading to inconsistent elution of fatty acid amides.
4. Due to the low quantities of fatty acid amides in biological samples, HPLC vials are not reused. Rather, 100 μL inserts with springs are used and discarded after analysis.
5. Thermo Scientific Pierce C_{18} Zip Tips are used to clean up the extractions from insects according to the manufacturer's protocol. We found that this step is not necessary for samples from cultured cells.
6. The fatty acid amide standards are stored dry at −20 °C or dissolved in HPLC-grade methanol and stored at −80 °C. Standard curves are made fresh from stock solutions to avoid concentration variance due to solvent evaporation and are rerun after instrument calibration is performed.
7. We found that it is best to prepare the individual and separate fatty acid amide standards and not work from a solution that contains a mixture of fatty acid amide standards.

Acknowledgments

This work is dedicated to the memory of the late Dr. Mitchell E. Johnson and has been supported, in part, by grants from the University of South Florida (a Creative Scholarship Grant from the College of Arts and Sciences), the Shirley W. and William L. Griffin Charitable Foundation, the National Institute of Drug Abuse at the National Institutes of Health (R03-DA034323), and the National Institute of General Medical Science of the National Institutes of Health (R15-GM107864) to D.J.M.

References

1. Bradshaw HB, Lee SH, McHugh D (2009) Orphan endogenous lipids and orphan GPCRs: a good match. Prostaglandins Other Lipid Mediat 89:131–134
2. Devane WA, Hanuš L, Breuer A, Pertwee RG, Stevenson LA, Griffin G, Gibson D, Mandelbaum A, Etinger A, Mechoulam R (1992) Isolation and structure of a brain constituent that binds to the cannabinoid receptor. Science 258:1946–1949
3. Cravatt BF, Prospero-Garcia O, Siuzdak G, Gilula NB, Henriksen SJ, Boger DL, Lerner RA (1992) Chemical characterization of a family of brain lipids that induce sleep. Science 268 (1995):1506–1509
4. Alborn HT, Turlings TCJ, Jones TH, Stenhagen G, Loughrin JH, Tumlinson JH (1997) An elicitor of plant volatiles from beet armyworm oral secretion. Science 276:945–949
5. Merkler DJ, Leahy JW (2018) Binding-based proteomic profiling and the fatty acid amides. Trends Res 1. https://doi.org/10.15761/TR.1000120
6. Bisogno T, Sepe N, De Petrocellis L, Mechoulam R, Di Marzo V (1997) The sleep inducing factor oleamide is produced by mouse neuroblastoma cells. Biochem Biophys Res Commun 239:473–479
7. Merkler DJ, Chew GH, Gee AJ, Merkler KA, Sorondo JP-O, Johnson ME (2004) Oleic acid derived metabolites in mouse neuroblastoma $N_{18}TG_2$ cells. Biochemistry 43:12667–12674
8. Fontana A, Di Marzo V, Cadas H, Piomelli D (1995) Analysis of anandamide, an endogenous cannabinoid substance, and of other natural *N*-acylethanolamines. Prostaglandins Leukot Essent Fatty Acids 53:301–308
9. Schmid PC, Kresbach RJ, Perry SR, Dettmer TM, Maasson JL, Schmid HHO (1995) Occurrence and postmortem generation of anandamide and other long-chain *N*-acylethanolamines in mammalian brain. FEBS Lett 375:117–120
10. Tan B, Bradshaw HB, Rimmerman N, Srinivasan H, Yu YW, Krey JF, Monn MF, Chen JS-C, Hu SS-J, Pickens SR, Walker JM (2006) Targeted lipidomics: discovery of new fatty acyl amides. AAPS J 8:E461–E465
11. Sultana T, Johnson ME (2006) Sample preparation and gas chromatography of primary fatty acid amides. J Chromatogr A 1101:278–285
12. Tan B, O'Dell DK, Yu YW, Monn MF, Hughes HV, Burstein S, Walker JM (2010) Identification of endogenous acyl amino acids based on a targeted lipidomics approach. J Lipid Res 51:112–119
13. Tortoriello G, Rhodes BP, Takacs SM, Stuart JM, Basnet A, Raboune S, Widlanski TS, Doherty P, Harkany T, Bradshaw HB (2013) Targeted lipidomics in *Drosophila melanogaster* identifies novel 2-monoacylglycerols and *N*-acyl amides. PLoS One 8:e67865
14. Jeffries KA, Dempsey DR, Behari AL, Anderson RL, Merkler DJ (2014) *Drosophila melanogaster* as a model system to study long-chain fatty acid amide metabolism. FEBS Lett 588:1596–1602
15. Piscitelli F, Bradshaw HB (2017) Endocannabinoid analytical methodologies: techniques that discoveries that drive techniques. Adv Pharmacol 80:1–30
16. Jeffries KA, Dempsey DR, Farrell EK, Anderson RL, Garbade GJ, Gurina TS, Gruhonjic I, Gunderson CA, Merkler DJ (2016) Glycine *N*-acyltransfearse-like 3 is responsible for long-chain *N*-acylglycine formation in $N_{18}TG_2$ cells. J Lipid Res 57:781–790
17. Anderson RL (2018) Fatty acid amides and their biosynthetic enzymes found in insect model systems. Ph.D. Thesis, University of South Florida, Tampa, FL

18. Anderson RL, Battistini MR, Wallis DJ, Shoji C, O'Flynn BG, Dillashaw JE, Merkler DJ (2018) *Bm*-iAANAT and its potential role in fatty acid amide biosynthesis in *Bombyx mori*. Prostaglandins Leukot Essent Fatty Acids 135:10–17
19. Vanderbilt CF, Giblin-Davis RM, Weissling TJ (1998) Mating behavior and sexual response to aggregation phemomone of *Rhynchophorus cruentatus* (Coleptera: Curculionidae). Fla Entomol 81:351–360
20. Philbrook GE (1954) Synthesis of the lower aliphatic amides. J Org Chem 19:623–625
21. Wilcox BJ, Ritenour-Rodgers KJ, Asser AS, Baumgart LE, Baumgart MA, Boger DL, DeBlassio JL, deLong MA, Glufke U, Henz ME, King L III, Merkler KA, Patterson JE, Robleski JJ, Vederas JC, Merkler DJ (1999) *N*-Acylglycine amidation: iImplications for the biosynthesis of the fatty acid primary amides. Biochemistry 38:3235–3245
22. Grundmann F, Dill V, Dowling A, Thanwisai A, Bode E, Chantratita N, ffrench-Constant R, Bode HB (2012) Identification and isolation of insecticidal oxazoles from *Pseudomonas* spp. Beilstein J Org Chem 8:749–752
23. Soumya M, Harnatha Reddy A, Nageswari G, Venkatappa B (2017) Silkworm (*Bombyx mori*) and its constituents: a fascinating insect in science and research. J Entomol Zool Stud 5:1701–1705
24. Nielson RC (1991) Extraction and quantification of polyolefin additives. J Liq Chromatogr 14:503–519
25. McDonald GR, Hudson AL, Dunn SMJ, You H, Baker GB, Whittal RM, Martin JW, Jha A, Edmondson DE, Holt A (2008) Bioactive contaminants leach from disposable laboratory plasticware. Science 322:917

Chapter 13

Metabolomics Analysis of Complex Biological Specimens Using Nuclear Magnetic Resonance Spectroscopy

Khushboo Gulati, Sharanya Sarkar, and Krishna Mohan Poluri

Abstract

Metabolomics is an emerging technology that promises the extensive analyses of small molecules in the biological specimen. Metabolomics is recognized as an imperative tool in several distinct areas such as infectious disease research, mapping cellular metabolic processes, investigating tumor metabolism, and environmental and nutritional research such as in phytochemistry, foodomics, and toxicology. Tools such as mass spectroscopy and nuclear magnetic resonance (NMR) spectroscopy are widely utilized for studying metabolomics. However, NMR has been proven to be more robust owing to its high-throughput and sensitive assessment of metabolic profiles. NMR-based metabolomics involve a sequence of steps starting from sample preparation to data analysis in order to unravel information with respect to both quality and quantity of metabolites. The current chapter elucidates the basic protocol used to perform both the qualitative and quantitative analysis of metabolites in complex biological samples using NMR spectroscopy.

Key words Nuclear magnetic resonance spectroscopy, Metabolomics, Biological specimen, Tumor metabolism, Molecular markers

1 Introduction

Metabolomics is the comprehensive qualitative and quantitative analysis of small molecules known as metabolites (less than 1500 Da) in living systems. These metabolites are classified as primary and secondary metabolites. Primary metabolites are the molecules that are involved in common functions (like respiration) among different species. In contrast, secondary metabolites are species specific which play roles by interacting with external environment or with other molecules/organisms [1]. Together these metabolites are known as metabolome. The metabolome of an organism may alter due to several factors including sex, age, environmental conditions, nutrition, diseased state, genetic alterations, physiological conditions, and/or therapeutic interventions. [2]. Studying the metabolome of an organism provides a potential to determine both the physiological and pathological state of an

Paul L. Wood (ed.), *Metabolomics*, Neuromethods, vol. 159, https://doi.org/10.1007/978-1-0716-0864-7_13,

organism. The major goal of metabolomics is the identification and quantification of metabolic state of an organism at a particular time point. The combination of metabolomics with other omic studies such as genomics, transcriptomics, and proteomics provides a route for the comprehensive understanding of the biological system which can be further used to determine the healthy or altered/ diseased state of an organism [3]. Of all the omic studies, metabolomics offers more detailed information by the identification of set of complex molecular markers associated with the process [4]. High sensitivity and specificity of metabolic profiles also make metabolomics studies imperative among all the omic studies. Moreover, metabolic profiling is a simple, easy, and cost-effective way to discern the complete biological process [5].

Metabolomics is not only confined to assessing the cellular metabolic processes related to humans/animals, but also extensively used to study plant chemistry (phytochemistry), toxicology, foodomics, and so on [6–10]

Metabolomics field has evolved as a result of advancement in various high-throughput analytical techniques for the precise identification of thousands of metabolites in a relatively short time.

MS (mass spectrometry) and NMR (nuclear magnetic resonance spectroscopy) are the two major techniques that are used to study metabolomics. Of these two techniques, NMR has taken lead over MS despite higher sensitivity of MS. This is due to the lower reproducibility and ambiguities/uncertainties associated with spectral assignments in MS. In contrast, NMR provides high reproducibility and aids in identification of compounds with identical molecular weights. NMR is very useful for the detection of compounds that are hard to ionize and demands derivatization in MS. NMR also aids in exploring the structures of unknown compounds [1]. In addition, NMR offers an advantage of probing the mechanisms and dynamics related to various metabolic pathways via stable isotopic labeling techniques [11]. Site-specific NMR imaging approach has also been efficiently applied to study metabolic activities in living organisms [12]. NMR requires little or no sample preparation. It is a nondestructive method as the NMR sample once prepared can be used for the multiple NMR experiments and can also be reused for the other analytical techniques. NMR is an unbiased technique as all the protonated compounds will be detected in the sample despite their differential physical properties. All these attributes make NMR the method of choice to study metabolomics. Advancements in NMR technology such as high field strength superconducting magnets, cryogenically cooled probes, microcoil probes, and high-performance radio frequency coils have further enhanced the sensitivity of NMR by several folds [13, 14]. Moreover, high-resolution magic angle spinning (HRMAS) allows studying metabolomics in intact tissues [15]. Hence, all these developments make NMR an ideal solution

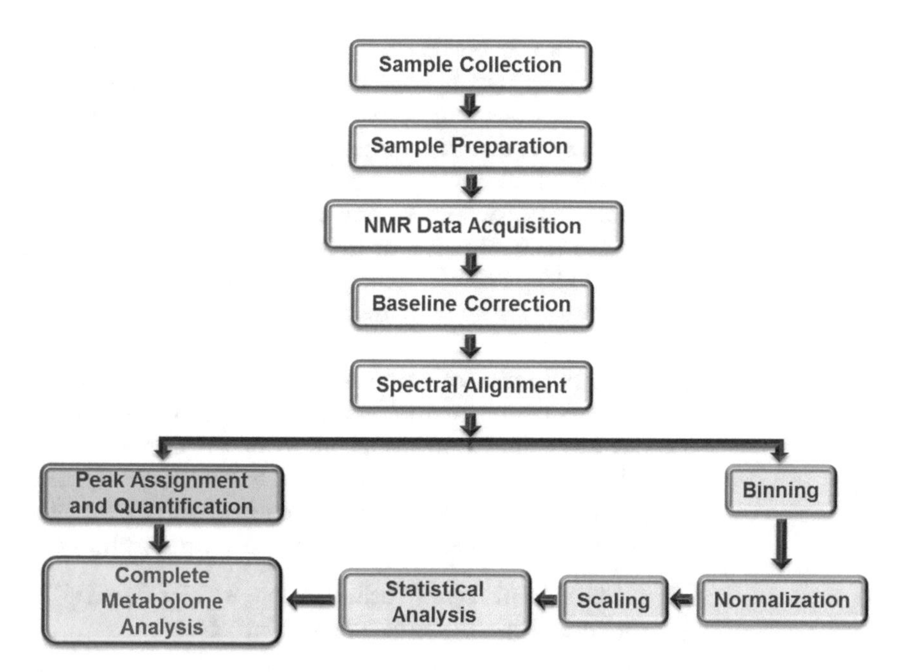

Fig. 1 Flowchart showing the essential steps involved in NMR based metabolomics studies

to accomplish multitude of complex metabolomic studies/tasks. The present chapter will discuss the basic work flow to carry out NMR-based metabolomics which includes sample preparation methods for common biofluids, basic NMR approaches for data collection followed by statistical data analysis (Fig. 1).

2 Materials

2.1 Requirements

Studying metabolomics using NMR is a relatively simple process that involves several steps, comprising sample preparation, data acquisition, data preprocessing, and statistical analysis. To fulfill these steps, some basic equipment, reagents, and software packages are required which are as follows:

Equipment: NMR tubes, microcentrifuge tubes, centrifuge, homogenizer/mortar and pestle, lyophilizer, weighing machine, speed vacuum concentrator, pH meter, pipette(s), vortex machine, liquid nitrogen, ice, −80 °C, −20 °C, and 4 °C refrigerators.

Reagents: D_2O, organic solvents. Sodium azide, buffers, internal standards such as sodium trimethylsilyl [2,2,3,3-$2H_4$] propionate (TSP), 4,4-dimethyl-4-silapentane-1-sulfonic acid (DSS), and tetramethylsilane (TMS).

Software packages: Bruker TopSpin, Amix, Chenomx and MetaboAnalyst etc. for NMR data processing and analysis.

2.2 Sample Preparation

In order to study NMR based metabolomics for different samples, same methodology is used for data collection and data analysis. However, distinct sample types (urine, plasma, other body fluids and tumors) require differential preparation methods. All the methods employed usually aim to prepare samples with minimal metabolite decay and devoid of any artifacts, contaminants, or factors giving rise to qualitative or quantitative variations in metabolites and so on [16]. Sample preparation protocols for some representative biological samples are discussed below.

2.2.1 Urine

The metabolomics of urine has gained much importance in the recent years owing to the fact that it (1) gives information about toxicity levels and nutritional status of the concerned individual, (2) helps in disease diagnosis, and (3) helps in development of drug [17]. The basic protocol to prepare urine sample for NMR based metabolomics studies has been adapted from Bernini et al. [18].

1. Collect "midstream" urine sample in 15 mL propylene tubes that are properly sterilized (*see* **Note 1**).
2. Freeze within 4 h of collection and store at −80 °C (*see* **Note 2**).
3. Thaw the frozen samples at room temperature. To ensure homogeneity, make sure to vortex them before use.
4. Make aliquots of each of the collected samples.
5. Add 630 μL of the urine sample to 70 μL of sodium phosphate buffer (0.2 M Na_2HPO_4 and 0.2 M NaH_2PO_4 in 100% ^2H_2O, pH 7.0 containing 5 mM TSP and 0.2% sodium azide.
6. Centrifuge samples at 14,000 × *g* for 5 min to remove any leftover debris.
7. Transfer 600 μL of the supernatant into 5 mm NMR tube (*see* **Note 3**).

2.2.2 Plasma

NMR-based metabolomics of blood aids in estimation of small molecules, required for clinical testing. Hence, this tool is a potential candidate for disease diagnosis, owing to its sensitivity for the detection of markers present in the blood. The basic protocol for blood collection and processing has been adapted from Bernini et al. [19].

1. Withdraw blood from donors in BD Vacutainer® K2E 5.4 mg for plasma isolation. While sampling blood to derive plasma, the presence of anticoagulants like EDTA is essential. K2EDTA is thus spray dried onto the walls of these Vacutainers.
2. Centrifuge at around 800–1000 RCF (relative centrifugal force) for 10 min at 4 °C. For blood processing, do not exceed 2 h after blood collection for sample preparation. It is advisable to keep the samples at 4 °C during the entire procedure (*see* **Note 4**).

3. Collect the supernatant after centrifugation. To 300 μL of this, add 300 μL of phosphate sodium buffer (100 mM Na_2HPO_4, 0.2% NaN_3, and 5 mM TSP, pH 7.4). Then transfer 600 μL of the sample into a 5 mm NMR tube.

2.2.3 Tissue

NMR-based metabolomics is currently used to determine the metabolic profiles of tissues associated with cancer as well as lesions in the premalignant stage. It is also utilized as an indicator of cancer aggressiveness. Different research groups are employing differential methodologies for metabolite extraction from tissues and some of them are discussed below.

1. Freeze the dissected tissue with help of liquid nitrogen immediately postcollection (*see* **Note 5**).
2. Sample preparation protocols for polar and lipophilic metabolites using different solvents are elucidated below:
   (a) Polar metabolite extraction with acetonitrile: [20, 21].
   - Weigh the frozen tissue in intact form.
   - Homogenize in a solution containing acetonitrile and water in 1:1 ratio by volume (5 mL/g tissue).
   - Centrifuge at 12,000 × *g* for 10 min at 4 °C.
   - After collecting the supernatant, lyophilize it.
   - Until NMR spectrum is acquired, store samples at −80 °C.
   - Resuspend samples in 600 μL NMR buffer (100 mM sodium phosphate buffer, pH 7.4, in D_2O, containing 5 mM TSP and 0.2% NaN_3). Alternatively, D_2O containing TSP can be used.
   - Centrifuge the samples at 12,000 × *g* for 5 min after vortexing them.
   - Transfer 600 μL of the supernatant into an NMR tube.

   (b) Polar metabolite extraction with perchloric acid [22, 23].
   - Weigh the tissue in the frozen condition itself.
   - Cool a mortar with liquid nitrogen. After grinding the frozen tissue in liq. N_2, add 5 mL/g of chilled perchloric acid (6%). Electric homogenizer may also be used as an alternative to manual grinding.
   - One may continue grinding further to ensure that the acid and the tissue components are completely mixed. Then, thaw the sample inside the mortar.
   - As an alternative, homogenize in the chilled solution right away after weighing the frozen sample.
   - Place on ice for 10 min after vortexing the sample.

- Centrifuge for 10 min at 12,000 × *g* at 4 °C.
- Add 2 M K_2CO_3 to neutralize the acidic pH and bring the final value to 7.4.
- To make the potassium perchlorate salts settle down, leave the sample on ice for about 30 min. Ensure to add K_2CO_3 at a very slow rate to prevent sample depletion owing to effervescence.
- Centrifuge the sample and freeze-dry the supernatant. Store the samples at −80 °C during the time period prior to NMR spectra acquisition.
- The rest of the procedure (NMR sample preparation) is exactly the same as the protocol for acetonitrile-based extraction.

(c) Simultaneous extraction of polar and lipophilic metabolites with methanol/chloroform/water [24–27].
- Prepare ice cold solvents, namely methanol, chloroform and water.
- Transfer weighed frozen tissue into a glass container in the intact form.
- Add methanol (4 mL/g of tissue) and water (0.85 mL/g) for homogenizing the sample.
- Add chloroform (2 mL/g) after vortexing the sample. Vortex one more time after chloroform addition.
- Leave sample for 15 min on ice. Centrifuge at 1000 × *g* for 15 min at 4 °C. The solutions are expected to divide into two layers. The upper layer constituting methanol/water would contain the polar metabolites whereas the lower phase constituting methanol would consist of the lipophilic components. They are isolated by the protein content and the debris of the cells. In case there is no clear separation, centrifuge again.
- Transfer both layers of the sample into two different glass tubes one by one.
- Use a speed vacuum concentrator to eliminate solvents from the sample. Alternatively, use stream of nitrogen gas.
- Keep samples extracted from the aqueous phase at −80 °C.
- Store the lipophilic samples in deuterated organic solvents in the refrigerator. In order to block oxidation effects, the storage of the lipophilic extracts from tissue samples preferably should not exceed 2–3 days.

- Add 600 μL deuterated NMR solvent (2:1 mixture of chloroform-d ($CDCl_3$) containing 0.03 vol/vol TMS, and CD_3OD) to the extract and vortex gently.
- Centrifuge at 1000 × *g* for 5 min.
- Transfer 600 μL of the supernatant into an NMR tube.

3 Methods

3.1 Data Acquisition

Once the sample has been prepared, the next step in NMR based metabolomics is to acquire the data (spectra). Data acquisition comprises few basic steps. The process of data acquisition is outlined in a step by step manner.

1. Set the temperature to 300 K or else required.
2. Load the sample into the probe.
3. Allow time for equilibration.
4. Tune the probe to its most sensitive frequency. Match the probe so that the greatest amount of power at the base of the probe is transferred toward its coil.
5. Fix the value of the RF carrier frequency offset to that of water resonance.
6. Ascertain the power of water saturation.
7. Adjust the power for the shaped pulse for the purpose of excitation sculpting.
8. For a given power level, control the 90° pulse length.
9. (Optional) Readjust the frequency offset for suppressing water signal peaks (*see* **Note 6**).
10. Convey all the above settings to the required experiment.
11. Select suitable pulse sequence according to the need of the experiment (*see* **Notes 7–9**).
12. Apply a line broadening of 0.3–1 Hz and zero-fill by twofolds to process the 1D spectrum.
13. Repeat the procedure on a representative sample as a measure for quality control.

3.2 Data Preprocessing

The acquired NMR data needs to be transformed into a format that is appropriate for multivariate data analysis. The preprocessing stage of NMR data comprises the following steps: (a) Baseline correction, (b) Alignment, (c) Binning, (d) Normalization, and (e) Scaling [28].

3.2.1 Baseline Correction

Few initial data points in the FID are misrepresented that constitute low frequency modulations in the FT (Fourier-Transformed) spectrum, which is the major resultant of baseline distortions in 1D

NMR spectra. These distortions countervail the intensity values, consequently making peak assignment and quantification error-prone. The study of metabolomics requires correcting such distortions so that small but important peaks corresponding to metabolites or other biomarkers can be accurately identified. Automatic baseline distortion correction methods like frequency domain methods are usually preferred for NMR spectra related to metabolomics [29]. Although the preferred method for baseline correction varies from lab to lab [29, 30] several software programs including TopSpin, NMRPipe, Bayesil, NMRProcFlow, and Automics are used to do baseline correction. Among them, Topspin is the most commonly employed software.

3.2.2 Alignment

To understand the differential pattern across NMR dataset, it is important to amend the peak shifts that are commonly seen in case of biological samples [31]. This requires aligning the spectra to an internal standard. TSP at 0.0 ppm is used as an internal standard in majority of studies. Apart from fixing global shifts, one has to align the spectra at the fine level as well. A number of algorithms have cropped up for this purpose. For example, TopSpin, icoshift, NMRProcFlow, MVAPACK, Automics, and mQTL are some of the commonly employed tools for the alignment of spectra (*see* **Note 10**).

3.2.3 Binning

This step is required in the preprocessing of the data to further perform downstream statistical analysis. Here, the spectrum is divided into small segments known as bins or buckets. Resulting bins in the spectra become a new variable in the spectrum. For example, binning a 8.6 ppm spectrum with a bin size of 0.02 ppm gives rise to 430 new features in the spectrum (*see* **Note 11**). Algorithms like Amix, NMRProcFlow, KIMBLE, MetaboLab, and dataChord Spectrum miner are frequently utilized for binning.

3.2.4 Normalization

NMR-based metabolomics requires the normalization of the data in order to provide information about the overall concentration of a particular metabolite in the sample accurately. Fundamentally, it is a mathematical operation (table row operation) that takes into consideration the metabolite dilution to make profiles comparable. For example, the concentration of some metabolites in a urine sample may apparently appear less if the concerned individual has a greater water intake than normal. Some of the normalization approaches are area normalization, probabilistic quotient normalization (PQN), range normalization, and normalization to a reference metabolite. Normalization of a spectrum to its total intensity is the most widely used approach. Common tools that are used to normalize the NMR spectra include MetaboAnalyst, mQTL, KIMBLE, MVAPACK, and NMRProcFlow.

3.2.5 Scaling

Metabolites in a single sample can vary in their concentration or abundance, ranging in orders of magnitude. This naturally creates a problem as it allows the highest peaks in the spectrum to exert the greatest influence in the statistical data models. Preprocessing thus requires scaling, a table column operation that attempts to balance signal intensity variances. Some of the known scaling methods include autoscaling, pareto scaling, mean centering, range scaling, and vast scaling (*see* **Note 12**). Amix, Mnova NMR, Pathomx, and Chenomx are some of the software programs used for scaling.

3.3 Statistical Analysis

After NMR data preprocessing, the next step is to carry out the multivariate analysis, or in simpler terms, the statistical analysis. The chemometric tools that are related to such statistical analysis can be broadly classified into unsupervised and supervised. Unsupervised analysis aims to overview the data and mark any implicit patterns or trends in the data, like a group's intrinsic variation or patterns associated with clustering. Principal components analysis (PCA) and hierarchical clustering analysis (HCA) are some of the well-known unsupervised methods [2]. Supervised analysis, on the other hand, utilizes prior knowledge of the class membership and identifies signals that are unique between groups. It analyzes features to infer clustering behavior between the various classes. Partial least squares discriminant analysis (PLS-DA) and orthogonal partial least squares discriminant analysis (OPLS-DA) are the most widely used supervised analysis techniques. Being a regression extension of PCA, PLS-DA augments the covariance to the class variable and enhances the separation between groups of observations to the maximum limit. One can also utilize the index values of the Variable Importance in Projection (VIP) in a PLS-DA to identify the hallmark metabolites that can distinguish the two groups [2] (*see* **Note 13**). Figure 2a represents the multivariate analysis of urine samples from colorectal cancer patients (CRC) and healthy controls (HCs). Score plots obtained using unsupervised pattern recogini-tion analysis via PCA was unable to distinguish the metabolic diffiences between CRC patients and HCs. In contrast, supervised pattern recoginition analysis via OPLS-DA resulted in score plots that clearly mark the seperation between the two groups (Fig. 2b) [32].

Here, we provide a detailed step-by-step protocol to normalize, scale, and perform the statistical analysis of the NMR-acquired data using MetaboAnalyst, which is a set of online tools for interpreting and assessing the metabolomics data (https://www.metaboanalyst.ca/).

1. Type https://www.metaboanalyst.ca/ in the browser search bar.
2. Click on “Click here” to start Statistical Analysis.
3. In the “Upload Data” section, submit your data either in the tab-delimited format or as a zip file.

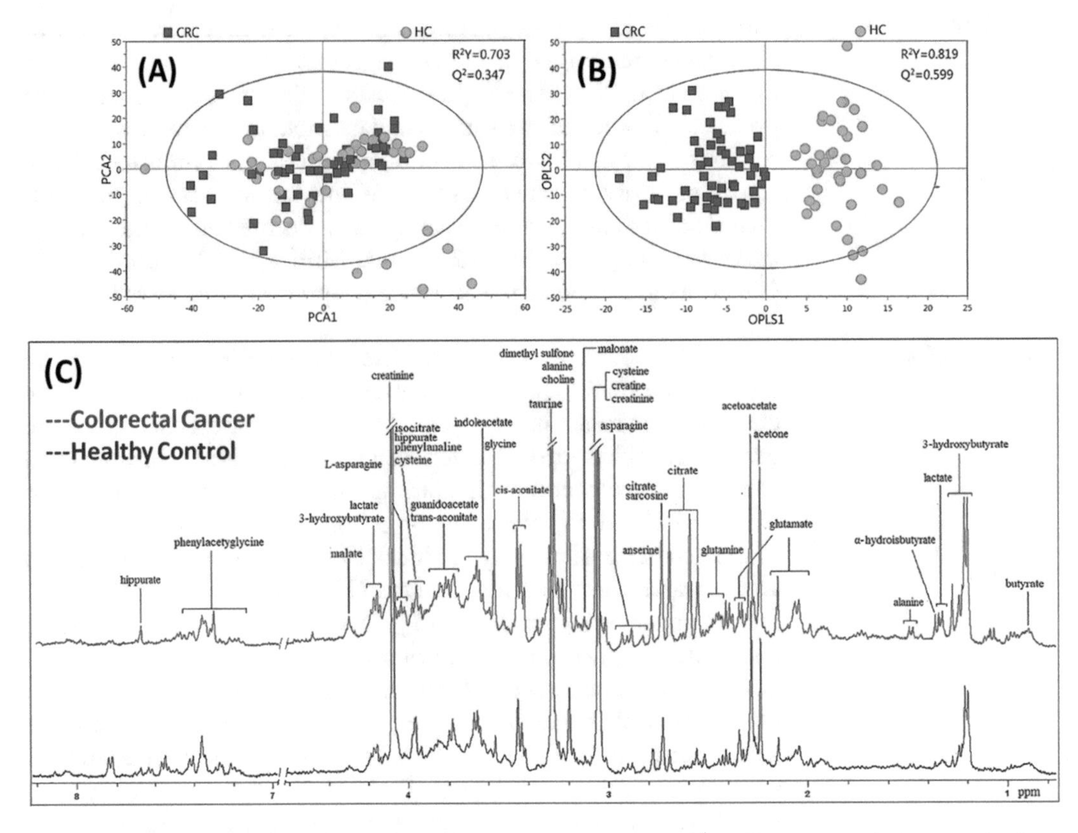

Fig. 2 Comparative (**a**) PCA, (**b**) OPLS-DA scatter plots, and (**c**) Assigned ^{1}H NMR spectra of urine samples from colorectal cancer (CRC) patients and healthy control (HC) samples. (The figure has been adapted from the research study published by Wang et al. [29])

4. Check "Data Integrity" to ensure that the uploaded data is in the exact required format. "Skip" if your data is clean.
5. "Filter" the data to remove unwanted variables and baseline noises.
6. Proceed to "Normalization Overview." This section contains various options under Sample Normalization, Data Transformation, and Data Scaling. Choose the options according to your experimental design.
7. Click on "Normalize" > "View Result." This displays the normalized data whereby the skewed data is converted to a bell-shaped Gaussian curve.
8. Proceed to Analysis page that displays five options for statistical analysis, namely, Univariate Analysis, Chemometrics Analysis, Feature Identification, Cluster Analysis, and Classification & Feature Selection. PCA and PLS-DA under Chemometrics Analysis are the most commonly used analysis methods.

9. Click on the various plot option tabs under "PCA" and "PLS-DA" like "Screen plot," "2D scores plot," "3D scores plot," "Loadings plot," and "Biplot." You may also extract other necessary representative information like Dendrogram and Heatmap.
10. Finally click on "Download" > "Generate Report" to get the detailed analysis report of the NMR data in PDF format.

3.4 NMR Peak Assignment and Quantification

Detection of metabolites: The important aspect of NMR based metabolomics is the identification of metabolites in a sample. 2D ^{1}H homonuclear correlation experiments that are based on spin–spin couplings of the nuclear spins are commonly employed for this purpose (*see* **Note 14**). The two-dimensional ^{1}H–^{1}H TOCSY experiment is also a popular choice in this respect as it maps the total spin system by furnishing correlations for distant spins, in addition to the directly connected ones (*see* **Note 15**) [33]. 2D selective heteronuclear J-resolved experiments can be utilized for measuring long-range coupling constants between proton and carbon with high degree of accuracy. ^{1}H–^{13}C heteronuclear single quantum coherence spectroscopy (HSQC) that involves greater dispersion of ^{13}C chemical shifts, is utilized when greater resolution is required for identifying metabolites unambiguously. Correlations of heteronuclear pairs of atoms that are isolated by multiple bonds can also be provided by heteronuclear multiple bond correlation (HMBC). Unknown metabolites can also be detected by a process known as spiking that requires adding the purified metabolite in small amounts to the biofluid (sample) to be tested, followed by NMR spectrum acquisition (*see* **Notes 16** and **17**).

Metabolite identification requires two simple steps: deconvolution of the metabolite mixture into individual components and metabolite identification by querying one or more databanks. Some notable NMR metabolomics databanks that are available publicly are the Biological Magnetic Resonance Data Bank (BMRB) and the Human Metabolome Database (HMDB) [34, 35]. They contain experimental data of pure metabolites that act as standards (*see* **Note 18**). Figure 2c provides an overview of all the metabolites in ^{1}H-NMR spectra of urine samples of CRC patients and HC as reported by Wang et al. [32].

Quantification of metabolites: The second objective of metabolomics is to quantify or measure the concentrations of the detected metabolites. The most convenient method that gives accurate quantification of metabolites is 1D NMR. This method employs an internal standard such as DSS/TSP with known concentration. Before carrying out the experiment, this reference compound is simply added to the sample [6]. The volumes of the peaks corresponding to the metabolites with unknown concentrations are measured. They are then related to that of DSS to know the absolute concentrations of the metabolites. Quantification by 1D NMR works well only when the peaks are isolated. Overlaps in peaks in 1D NMR can be resolved by 2D NMR, but quantification by the latter is more demanding, thus making it an active field of research currently. Homonuclear J-resolved experiments can analyze complex samples accurately with the aid of line-shape fitting [36, 37]. One can also quantitate metabolites by spiking the sample repeatedly. This is done by adding a mixture of metabolites that have predetermined concentrations, followed by recording a 2D spectrum for each of the spiking levels (*see* **Note 19**).

An extremely popular, patented software known as Chenomx is used for both identification and quantification of metabolites. It provides the output in the form of tables that consist of identified compounds and their corresponding concentrations.

4 Notes

1. Pass a little quantity of urine before collecting into the sterilized container. This "midstream" collection will help to avoid urethral bacteria. Moreover, multiple collections from an individual are preferred over a time period to minimize variations as much as possible.
2. Store urine samples at −25 °C immediately after postcollection to ensure sample stability. In case it is not feasible, store at 4 °C immediately by mixing with sodium azide, followed by its transfer to −80 °C or − 25 °C after no more than 24 h. This would help ameliorate the problem of acetate contamination from bacterial growth.
3. NMR tubes with small diameters and shaped tubes (3 mm tubes and S-tubes) can be chosen to alleviate the effects of high salt concentrations in samples on the tuning and matching of probes.
4. Process the blood at 4 °C within 2 h from collection as it would help in the authentic evaluation of glucose and its downstream products like lactate and pyruvate by preventing glucose degradation.
5. To minimize the degradation of the metabolites and get insight into the in vivo metabolism, the tissue requires freezing in liquid nitrogen or on dry ice immediately postcollection. This is due to the metabolites like phosphates and intermediates of the glycolytic cycle that can change on a timescale of milliseconds.
6. Strong signals from water can damp radiation, especially in cryoprobes. Therefore, employ water suppression via presaturation to bring about the attenuation of water signal of

biofluids, better resolving the peaks corresponding to the metabolites of our interest. Alternatively, utilize advanced water suppression schemes like WATERGATE (water suppression by gradient-tailored excitation) to eliminate water signals, while allowing minimal impact on the exchangeable protons.

7. Certain pulse sequence schemes have been implemented in standardized protocols for metabolomics that are same for both solution-state NMR as well as solid-state NMR of tissues utilizing HR-MAS NMR. The 1D ^{1}H NOESY, accompanied by water suppression, is simple and requires minimal hardware and parameter optimization. This pulse sequence is especially important for urine samples owing to the superior quality of water suppression and consistency in the spectra.

   CPMG (Carr–Purcell–Meiboom–Gill), on the other hand, is the most agreed upon choice for samples like serum and tissue extracts that consist of high MW (molecular weight) components like proteins having restricted rotational times and shorter T_2, resulting in broader resonances. The T_2 filter in CPMG helps to thwart such signals thereby resulting in a spectrum with high metabolite resolution. Another pulse sequence known as diffusion-edited NMR helps in studying macromolecules like proteins and lipoproteins that have limited translational mobility. Therefore, one can obtain high-resolution profile of small metabolites by subtracting the diffusion-edited spectrum from a standard 1D NMR spectrum.

8. HR-MAS NMR employs water presaturation and T_2 editing pulse sequences to block the "noise" signals arising from lipids and other macromolecules. Tissue metabolism has been known to be affected by temperature and rapid spinning during NMR data acquisition. Therefore, acquire spectra at around 1–5 °C ensuring that the sample does not freeze. In addition, keep spin rates within 2–5000 Hz in order to block spinning sidebands in the spectra.

9. Conventional multidimensional NMR involves linear data acquisition. It has been proposed to be replaced by nonlinear or nonuniform sampling approaches that would give rise to sparse time-domain data sets. Such data can be processed utilizing the use of suitable processing methods, such as forward maximum entropy reconstruction or maximum entropy algorithms. This would ultimately reduce the time for recording the spectra as compared to a linearly sampled one. In order to speed up metabolomics studies, Hadamard-encoded or projection spectroscopy approaches can be employed to efficiently detect metabolites from biological samples in an unambiguous manner. Ultrahigh-field magnets and novel techniques like dynamic nuclear polarization may pave the way toward submicromolar NMR metabolomics.

10. Users must be aware while aligning the spectra as it has the potential to introduce artifacts and affect quantification of metabolites.
11. Earlier, bins were of 0.04 ppm, but they often resulted in low resolution and made certain peaks split between two bins. Therefore, narrower bins of 0.02 ppm or bins with variable size are more preferred. In case of targeted studies on known or expected metabolites, binning can be skipped altogether and one can move directly to metabolite quantification after peak alignment.
12. "Autoscaling," a commonly applied method, scales all data to unit variance but may induce the noisy variables to be predominant. "Pareto scaling" does not utilize the standard deviation as its scaling factor, but rather uses its square root. It is the most frequently used approach for NMR spectra. The scaling factor in "range scaling" is the difference between the greatest and least values for each variable, while in "meancentering" all values are fine-tuned to vary around zero instead of the mean. "Vast scaling" is developed to emend the measurement deviation enlargement.
13. The fingerprint metabolites need to be rigorously validated using univariate analysis (like student t-test, ROC (receiver operating characteristic) curve analysis) since PLS-DA model has the tendency to over fit the data The cross-validation would therefore ensure that the separation is statistically significant, and not generated due to random noise.
14. The spin-echo modulation of 2D ^{1}H homonuclear J-resolved experiment furnishes the accurate measurement of the J-coupling constants and simplifies the spectra in the direct dimension of acquisition. The ^{1}H NMR experiment for metabolite identification with the greatest resolution is the "J-res" experiment. The projection of the 2D spectrum onto the horizontal axis is equivalent to a fully "decoupled" proton spectrum.
15. To attain coherence transfer in the TOCSY experiment, a period of isotropic mixing or "spin-lock" has to be provided. Spin lock can be achieved simply by a continuous wave irradiation of high strength or more advanced schemes like MLEV-17 or DIPSI-2. Isotropic mixing usually requires 50–100 ms. In case longer mixing times are intended in some experiments, precaution should be taken as it can damage apparatus like cryogenically cooled probes owing to excessive heating.
16. In case the metabolite of interest is present in the sample, peaks corresponding to the spiked specimen overlap exactly with those assigned originally to the metabolite in the biofluid. The quantity of the spiked metabolite should be between 25% and 50% of that in the biofluid. Less spiking can lead to

ambiguity regarding the increase in peak intensity. High spiking can inundate the original peaks and thus provide uncertain results regarding signal overlap.

17. 2D NMR can also be employed to discover biomarkers by comparing samples from distinct groups, followed by "bucketing." For this, the spectral regions need to be relatively integrated. Statistical analysis then allows the user to determine the spectral regions that are relevant for a particular group. Hierarchical alignment of two-dimensional spectra-pattern recognition (HATS-PR) can be used to identify such biomarkers by the principle of pattern recognition. Alternatively, image processing protocols can be employed to compare the spectra and determine the presence of unique signals present in one population. This is done utilizing differential analysis by 2D NMR Spectroscopy (DANS).
18. Identification with the aid of databases requires matching of many NMR parameters (like chemical shifts, J-couplings) of the concerned metabolite to those of the pure compound. However, when a subgroup of these independent parameters is used for identification, the risk of false identification is increased. Problem also arises when signals are not found in the database, thereby posing a challenge in the identification of the unknown compound. This is owing to the fact that a very small fraction of the total number of possible metabolites has their NMR spectra stored in the databanks, ranging in the order of 1000–2000 in number. Such uncataloged compounds need to be purified with the help of existing separation techniques and then identified with other spectroscopic tools.
19. For NMR-based quantification, one should ensure that the spins relax completely in between scans in order to avert effects owing to differential saturation.

Acknowledgments

The authors thank Dr. Dinesh Kumar, CBMR-Lucknow, for the technical inputs. This work is supported by the following research grants to KMP: CRG/2018/001329 grant from DST-SERB, GKC-01/2016-17/212/NMCG-Research from NMCG-MoWR, and BEST-18-KMP/IITR/109 from BEST Pvt. Ltd.

References

1. Markley JL, Brüschweiler R, Edison AS, Eghbalnia HR, Powers R, Raftery D et al (2017) The future of NMR-based metabolomics. Curr Opin Biotechnol 43:34–40
2. Cho H-W, Kim SB, Jeong MK, Park Y, Gletsu N, Ziegler TR et al (2008) Discovery of metabolite features for the modelling and analysis of high-resolution NMR spectra. Int J Data Min Bioinform 2:176
3. Weckwerth W (2003) Metabolomics in systems biology. Annu Rev Plant Biol 54:669–689

4. Guleria A, Kumar A, Kumar U, Raj R, Kumar D (2018) NMR based metabolomics: an exquisite and facile method for evaluating therapeutic efficacy and screening drug toxicity. Curr Top Med Chem 18:1827–1849
5. Keun HC (2018) NMR-based metabolomics, vol 14. Royal Society of Chemistry, Cambridge
6. Kim HK, Choi YH, Verpoorte R (2010) NMR-based metabolomic analysis of plants. Nat Protoc 5:536
7. Ramirez T, Daneshian M, Kamp H, Bois FY, Clench MR, Coen M et al (2013) Metabolomics in toxicology and preclinical research. ALTEX 30:209
8. Arora N, Kumari P, Kumar A, Gangwar R, Gulati K, Pruthi PA et al (2019) Delineating the molecular responses of a halotolerant microalga using integrated omics approach to identify genetic engineering targets for enhanced TAG production. Biotechnol Biofuels 12:2
9. Trimigno A, Marincola FC, Dellarosa N, Picone G, Laghi L (2015) Definition of food quality by NMR-based foodomics. Curr Opin Food Sci 4:99–104
10. Arora N, Dubey D, Sharma M, Patel A, Guleria A, Pruthi PA et al (2018) NMR-based metabolomic approach to elucidate the differential cellular responses during mitigation of arsenic (III, V) in a green microalga. ACS Omega 3:11847–11856
11. Gowda GN, Raftery D (2015) Can NMR solve some significant challenges in metabolomics? J Magn Reson 260:144–160
12. Fan TW-M, Lane AN (2016) Applications of NMR spectroscopy to systems biochemistry. Prog Nucl Magn Reson Spectrosc 92:18–53
13. Rolin D (2012) Metabolomics coming of age with its technological diversity, vol 67. Academic, Oxford
14. Putri SP, Yamamoto S, Tsugawa H, Fukusaki E (2013) Current metabolomics: technological advances. J Biosci Bioeng 116:9–16
15. Beckonert O, Coen M, Keun HC, Wang Y, Ebbels TM, Holmes E et al (2010) High-resolution magic-angle-spinning NMR spectroscopy for metabolic profiling of intact tissues. Nat Protoc 5:1019
16. Beckonert O, Keun HC, Ebbels TM, Bundy J, Holmes E, Lindon JC et al (2007) Metabolic profiling, metabolomic and metabonomic procedures for NMR spectroscopy of urine, plasma, serum and tissue extracts. Nat Protoc 2:2692
17. Slaff B, Sengupta A, Weljie A (2018) NMR spectroscopy of urine. NMR-based metabolomics. Royal Society of Chemistry, Cambridge, pp 39–84
18. Bernini P, Bertini I, Luchinat C, Nepi S, Saccenti E, Schäfer H et al (2009) Individual human phenotypes in metabolic space and time. J Proteome Res 8:4264–4271
19. Bernini P, Bertini I, Luchinat C, Nincheri P, Staderini S, Turano P (2011) Standard operating procedures for pre-analytical handling of blood and urine for metabolomic studies and biobanks. J Biomol NMR 49:231–243
20. Waters NJ, Holmes E, Waterfield CJ, Farrant RD, Nicholson JK (2002) NMR and pattern recognition studies on liver extracts and intact livers from rats treated with α-naphthylisothiocyanate. Biochem Pharmacol 64:67–77
21. Coen M, Lenz EM, Nicholson JK, Wilson ID, Pognan F, Lindon JC (2003) An integrated metabonomic investigation of acetaminophen toxicity in the mouse using NMR spectroscopy. Chem Res Toxicol 16:295–303
22. Sonnewald U, Isern E, Gribbestad I, Unsgård G (1994) UDP-N-acetylhexosamines and hypotaurine in human glioblastoma, normal brain tissue and cell cultures: 1H/NMR spectroscopy study. Anticancer Res 14:793–798
23. Henke J, Willker W, Engelmann J, Leibfritz D (1996) Combined extraction techniques of tumour cells and lipid/phospholipid assignment by two dimensional NMR spectroscopy. Anticancer Res 16:1417–1427
24. Lin CY, Wu H, Tjeerdema RS, Viant MR (2007) Evaluation of metabolite extraction strategies from tissue samples using NMR metabolomics. Metabolomics 3:55–67
25. Bligh EG, Dyer WJ (1959) A rapid method of total lipid extraction and purification. Can J Biochem Physiol 37:911–917
26. Tyagi RK, Azrad A, Degani H, Salomon Y (1996) Simultaneous extraction of cellular lipids and water-soluble metabolites: evaluation by NMR spectroscopy. Magn Reson Med 35:194–200
27. Beckonert O, Monnerjahn J, Bonk U, Leibfritz D (2003) Visualizing metabolic changes in breast-cancer tissue using 1H-NMR spectroscopy and self-organizing maps. NMR Biomed 16:1–11
28. Brennan L (2014) NMR-based metabolomics: from sample preparation to applications in nutrition research. Prog Nucl Magn Reson Spectrosc 83:42–49
29. Xi Y, Rocke DM (2008) Baseline correction for NMR spectroscopic metabolomics data analysis. BMC Bioinformatics 9:324

30. Chang D, Banack CD, Shah SL (2007) Robust baseline correction algorithm for signal dense NMR spectra. J Magn Reson 187:288–292
31. Defernez M, Colquhoun IJ (2003) Factors affecting the robustness of metabolite fingerprinting using 1H NMR spectra. Phytochemistry 62:1009–1017
32. Wang Z, Lin Y, Liang J, Huang Y, Ma C, Liu X et al (2017) Nmr-based metabolomic techniques identify potential urinary biomarkers for early colorectal cancer detection. Oncotarget 8:105819
33. Braunschweiler L, Ernst R (1983) Coherence transfer by isotropic mixing: application to proton correlation spectroscopy. J Magn Reson 53:521–528
34. Ulrich EL, Akutsu H, Doreleijers JF, Harano Y, Ioannidis YE, Lin J et al (2007) BioMagResBank. Nucleic Acids Res 36:402–408
35. Wishart DS, Knox C, Guo AC, Eisner R, Young N, Gautam B et al (2008) HMDB: a knowledgebase for the human metabolome. Nucleic Acids Res 37:603–610
36. Ludwig C, Viant MR (2010) Two-dimensional J-resolved NMR spectroscopy: review of a key methodology in the metabolomics toolbox. Phytochem Anal 21:22–32
37. Parsons HM, Ludwig C, Viant MR (2009) Line-shape analysis of J-resolved NMR spectra: application to metabolomics and quantification of intensity errors from signal processing and high signal congestion. Magn Reson Chem 47:86–95

Chapter 14

Untargeted Metabolomics Methods to Analyze Blood-Derived Samples

Danuta Dudzik and Antonia García

Abstract

Untargeted metabolomics is applied to a wide range of preclinical and clinical studies. The metabolome is the furthest downstream product of the genome and is considered to provide a direct measure of the phenotype at the molecular level. Therefore, metabolomics seeks for the comprehensive detection and offers the broadest possible metabolome coverage in type of molecules and level of concentration. Nevertheless, the large number of molecules with physicochemical and structural diversity constituting the metabolome possess a bottleneck and is pointed as one of the greatest analytical challenges. Due to its high sensitivity and throughput, a commonly used mass spectrometry (MS) platform is often considered as a "gold standard" that plays a crucial role in untargeted metabolomics methods. Furthermore, the combination of separation techniques (LC, liquid chromatography; GC, gas chromatography; or CE, capillary electrophoresis) with mass spectrometry tremendously expands the capability of the analysis of highly complex biological samples with the highest metabolite coverage. This chapter exemplifies the strategies used for untargeted metabolomics analysis of blood derived samples. It gives experimental details on basic steps like blood plasma/serum withdrawal, metabolite extraction and analysis for LC-MS, GC-MS, and CE-MS untargeted metabolomic studies.

Key words Nontargeted metabolomics, Metabolic phenotyping, Liquid chromatography, Capillary electrophoresis, Gas chromatography, Mass spectrometry, Fingerprinting

1 Introduction

Untargeted metabolomics is a nonbiased, global approach that aims to measure all the metabolites present in a given biological sample and to use this information to understand health and disease and for diagnostic or drug monitoring purposes. It provides a snapshot of the physiological status of the organism at the molecular level at the specific time that sample was collected [1, 2]. However, no single analytical technique can cover the whole range of metabolites present in a complex biological samples [3]. Currently, among the analytical platforms employed in the analysis of metabolome, nuclear magnetic resonance (NMR) and mass spectrometry (MS) are the most widely used analytical techniques

Paul L. Wood (ed.), *Metabolomics*, Neuromethods, vol. 159, https://doi.org/10.1007/978-1-0716-0864-7_14,

[3]. However, despite high analytical reproducibility and simplicity of sample preparation, major drawbacks in NMR for untargeted metabolomics are the poor sensitivity and spectral complexity [4]. Whereas while mass spectrometry is much more sensitive and higher throughput, it requires a previous separation step for better functionalities. Therefore, the key separation techniques including liquid chromatography (LC), gas chromatography (GC) and capillary electrophoresis (CE) are used to separate the metabolome based on its chemical and physical complexity. Those systems coupled to mass spectrometry provides the comprehensive view of the metabolome. Liquid chromatography in combination with mass spectrometry (LC-MS) provides the broadest metabolite coverage in comparisons to other platforms and is considered as the technique of choice in many metabolomics studies that enable the detection of thousands of compounds, including low-abundant metabolites during a single analytical run. An undeniable advantage of LC-MS based metabolomics is its high sensitivity and high throughput [5, 6]. Herein the metabolite separation and detection are strongly influenced by the metabolite extraction procedure, the chemical nature of the mobile and the stationary phases used [3, 5]. To date, reverse phase liquid chromatography (RPLC), mainly suited for the detection of low polar compounds is the most commonly used chromatographic technique due to its good reproducibility and wide applicability [4, 6]. The type and the number of metabolites detected depends also on the extraction procedure and the type of the ionization (positive and negative), used during the analysis. For the LC-MS sample preparation, proteins are removed and the metabolites are dissolved in the appropriate solvent without metabolite degradation [7]. Nevertheless, RPLC is not suitable for all metabolites like highly polar and charged metabolites that are difficult to be retained and coelute with the void volume. Hence, analyses of those compounds usually rely on other complementary separation techniques such as GC and CE [4]. Gas chromatography together with mass spectrometry (GC-MS) displays good sensitivity, peak resolution, reproducibility, and robustness. Nonetheless, it requires very extensive sample preparation due to the chemical derivatization that limits its applicability to the analysis of volatile and thermally stable compounds. This approach allows the analysis of distinct classes of rather low-polar volatile metabolites of fats and esters, and high polar metabolites, such as amino acids, biogenic amines, small saccharides and carboxylic acids in a single chromatographic run and single ionization mode [4, 6]. Capillary electrophoresis coupled to mass spectrometry (CE-MS) is considered as a complementary analytical technique for the analysis of ionic classes of metabolites or degradation products from primary metabolism, as well as secondary metabolites that have poor separation in LC-MS and GC-MS platform. Typical sample preparation for CE-MS analysis usually

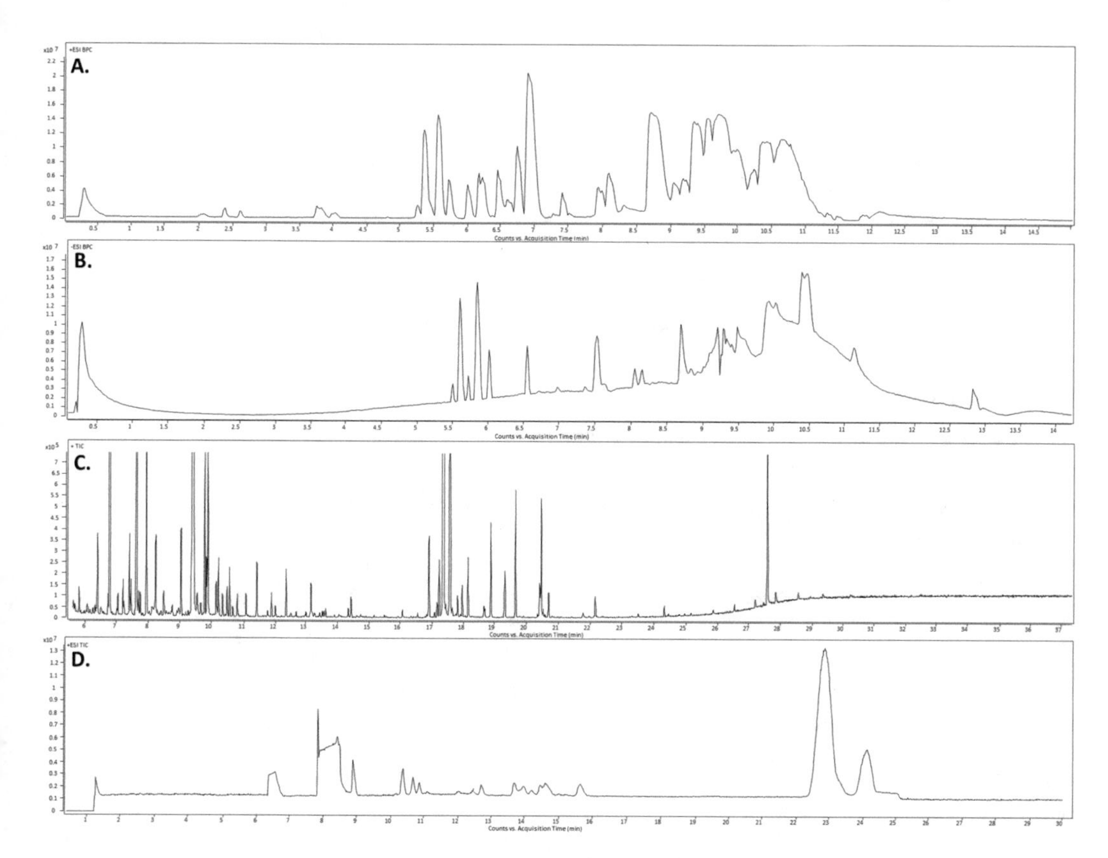

Fig. 1 Representative chromatograms obtained from the analysis of a plasma sample from a healthy volunteer. Data obtained from (**a**) LC-MS positive ionization mode; (**b**) LC-MS negative ionization mode; (**c**) GC-MS and (**d**) CE-MS for untargeted metabolomics analysis

requires minimal treatment, that is, sample dilution and/or deproteinization [4]. A representative LC-MS, GC-MS, and CE-MS chromatograms of plasma sample from a healthy volunteer are shown in Fig. 1. Nevertheless, not only does the success of metabolomics analysis depend on the analytical platform used, but it requires careful considerations in terms of the whole metabolomics workflow (Fig. 2). The quality of these steps is a key factor that determine the final outcomes of the metabolomics study. Among the most crucial steps, experimental design, sample collection, and metabolite extraction are of particular importance. Following the quality control and quality assurance procedures and protocols, reviewed elsewhere [8, 9], ensure that the analyzed sample is representative and reflects the current state of studied system.

This chapter focuses on the description of the analytical protocols for plasma and serum blood sample harvesting and metabolite extraction for multiplatform untargeted metabolomics analysis based on LC-MS, GC-MS, and CE-MS analytical systems.

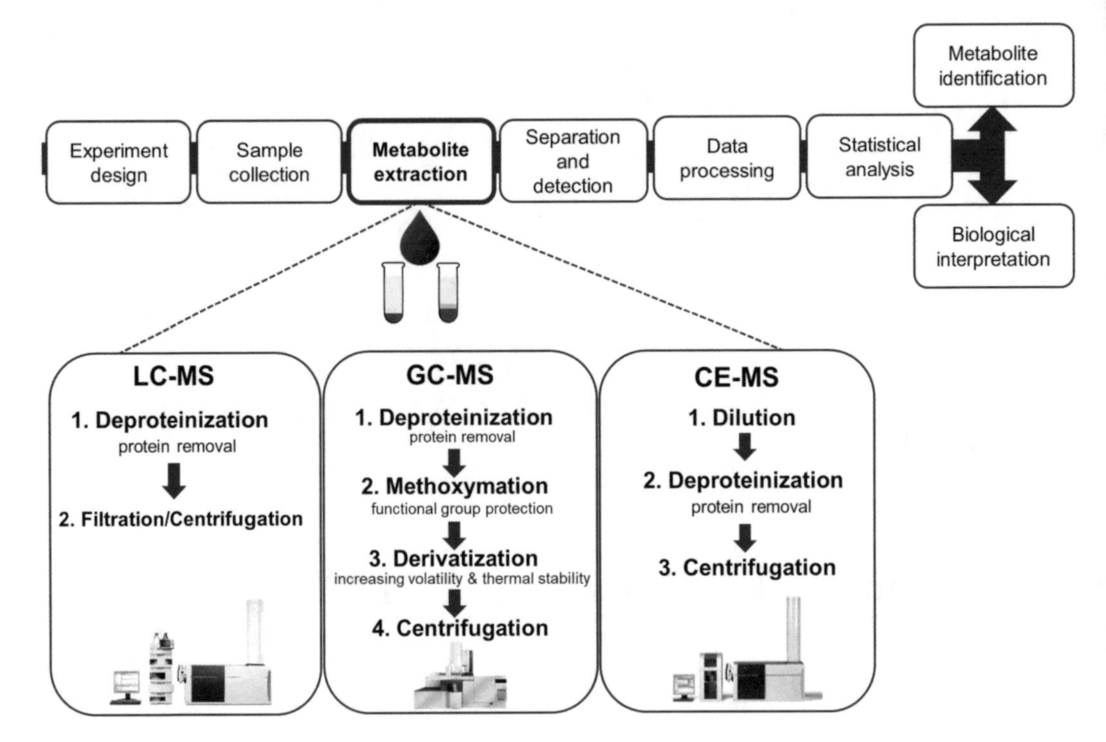

Fig. 2 The workflow with the simplified procedure for plasma/serum metabolite extraction in MS-based untargeted metabolomics study

2 Materials

2.1 Samples

The procedures for sample collection and handling have to strictly follow guidelines and specific SOPs developed to ensure high sample quality [8, 9]. The patient clinical metadata should be taken and kept. The sample processing should be carried out at controlled temperature (4 °C) and samples stored at −80 °C. To avoid bias and unwanted variation, all the samples of the study should be randomized and treated in the same way throughout the thaw, metabolite extraction, and analysis.

2.1.1 Plasma Preparation

1. Collect whole blood after an overnight fasting of at least 8 h, from a peripheral vein directly into tubes containing Na_2EDTA as anticoagulant (*see* **Notes 1** and **2**).
2. Mix the tube well after collection in the Vacutainer.
3. Separate cells from plasma by centrifugation ~2000 × *g*, 4 °C, 20 min.
4. The clear, upper layer is plasma (*see* **Note 3**). Remove supernatant, not touching the buffy layer.
5. Aliquot the plasma in desired proportions into the labeled precooled sample storage tubes without aspirating blood cells (*see* **Note 4**).

2.1.2 Serum Preparation

1. Collect blood samples from a peripheral vein into the serum collection tube with clotting activator (Silicate).
2. Shake the vial—gently but thoroughly.
3. Store the vial at room temperature in an upright position to allow coagulation. Clot formation should be completed after 20–30 min in most cases. Time at room temperature until centrifugation should not exceed 40 min.
4. Centrifuge to separate the serum from the blood clot at ~2000 × *g*, 4 °C, 10 min.
5. Aliquot the sample in desired proportions into the labeled precooled sample storage tubes (*see* **Note 4**).

2.2 Reagents (See Note 5)

Ultrapure water (Milli-Q Plus 185 system Millipore, Billerica, MA, USA) is used in preparation of all buffers, standard solutions, and dilutions.

2.2.1 LC-MS

1. MS grade acetonitrile (Sigma-Aldrich).
2. Analytical grade ethanol (Sigma-Aldrich).
3. MS grade methanol (Sigma-Aldrich).
4. Analytical grade formic acid 99% (Sigma-Aldrich).

2.2.2 GC-MS

1. MS grade acetonitrile (Sigma-Aldrich).
2. Analytical grade heptane (Sigma-Aldrich).
3. Analytical grade 2-propanol (Sigma-Aldrich).
4. Standard containing grain fatty acid methyl ester (FAME) mixture (C8:0–C22:1n9) at a concentration of 10 mg/mL in dichloromethane (CRM47801 Sigma-Aldrich).
5. *N,O*-bis(trimethylsilyl)trifluoroacetamide (BSTFA) with 1% trimethylchlorosilane (TMCS; Pierce Chemical Co, Rockford, IL, USA) (*see* **Note 6**).
6. *O*-methoxyamine hydrochloride (Sigma-Aldrich).
7. Silylation grade pyridine (Sigma-Aldrich).
8. Tricosane (Sigma-Aldrich) as internal standard.
9. Helium (99.9999%).

2.2.3 CE-MS

1. MS grade formic acid 99% (Sigma-Aldrich).
2. Sodium hydroxide (NaOH) 97% (Panreac, Barcelona, Spain).
3. Methionine sulfone (Sigma-Aldrich) as internal standard.
4. L-valinol, glycyl-leucine, ethionine standards (Sigma-Aldrich).
5. 5 mM purine, 2.5 mM HP922 (Agilent Technologies, USA).

2.3 Solvents and Buffer Solutions (See Notes 7 and 8)

2.3.1 LC-MS

1. Mobile phase A: ultrapure water with 0.1% formic acid.
2. Mobile phase B: acetonitrile with 0.1% formic acid.

2.3.2 GC-MS

1. *O*-methoxyamine hydrochloride 15 mg/mL in pyridine.

   Weigh 375 mg *O*-methoxyamine hydrochloride and transfer it to a 25 mL flask with pyridine, fill up to the mark with the same solvent. Add a small magnet and stir with magnetic stirring until the solution is completely clear. Transfer the solution to a hermetic tube. Store in the freezer at −20 °C up less than 1 month.
2. Stock solution 1000 mg/L (ppm) of tricosane in heptane. Stored at −80 °C. Before the analysis the solution is thawed and diluted up to 20 ppm.

2.3.3 CE-MS

1. Background electrolyte (BGE): 1.0 M formic acid diluted from concentrated formic acid in 10% methanol.
2. Sheath liquid used for detection: 50 mL methanol, 50 mL ultrapure water, add the reference mass standards (10 μL 5 mM purine and 30 μL 2.5 mM HP922).
3. 0.2 M formic acid with 5% acetonitrile and 0.4 mM methionine sulfone.
4. The internal standard, methionine sulfone is prepared in ultrapure water at a concentration of 25 mM and diluted adequately before adding to the samples.
5. Individual stock solutions 1000 mg/L (ppm) of the reference compounds (L-Valinol, Glycyl-leucine, Ethionine) are prepared in ultrapure water and stored at −80 °C. Before the analysis, the solutions are thawed and diluted with ultrapure water to 100 ppm each.
6. Sodium hydroxide (NaOH) 97% is used as obtained.

2.4 Other Supplies/Consumables/Equipment (See Note 9)

- Ultracentrifuge (5415 R Eppendorf).
- Refrigerated centrifuge (Heraeus Megafuge 1.0R, Heraeus Instruments).
- Solvent evaporator (Speedvac concentrator Thermo Fisher Scientific, Waltham, MA).
- Laboratory vortex mixer (TopMix FB15024, Fisher Scientific).
- Laboratory ultrasonic bath (3000513, Selecta P).
- Crushed ice and container.
- Cryotubes to store samples in −80 °C temp.

- LC-MS, GC-MS, CE-MS vials (e.g., Chromacol vials, [Chromacol Ltd., UK] with 0.3 mL fixed insert, Waters vials, or others) (*see* **Note 10**).
- Centrifree ® ultracentrifugation device (Millipore Ireland Ltd., Ireland) 30 kDa protein cutoff.
- Eppendorf® tubes.
- Calibrated pipettes.
- Nylon syringe filter 0.2 μm (Millipore, Billerica, MA, USA).

2.4.1 LC-MS

- UHPLC column RP Zorbax Extend C18 column (2.1 × 50 mm, 1.8 μm; Agilent Technologies).
- UHPLC system (Agilent 1290 Infinity LC System), equipped with a degasser, two binary pumps, and a thermostated autosampler coupled with Q-TOF LC-MS (6550) system (Agilent).

2.4.2 GC-MS

- Digitheat 80 L stove (Selecta, Barcelona, Spain).
- GC column (DB-5MS (30 m length, 0.25 mm ID, 0.25 μm film thickness; 95% dimethyl–5% diphenylpolysiloxane); Agilent Technologies, cat. no. 122-5532G, coupled to a precolumn (10 m DuraGuard; 0.25 mm ID × 10 m, Agilent Technologies).
- Ultrainert low pressure drop liner with wool GC liner (Agilent Technologies, cat. no. 5190–3165).
- GC system (Agilent Technologies 7890A), equipped with an autosampler (Agilent 7693) and interfaced to an inert mass spectrometer with triple-Axis detector (5975C, Agilent).

2.4.3 CE-MS

- Fused silica capillary, length 96 cm, ID: 50 μm (Agilent Technologies).
- Agilent 7100 capillary electrophoresis system coupled to an Agilent 6224 TOF mass spectrometer.

3 Methods

3.1 Metabolite Fingerprinting by Liquid Chromatography Coupled to Mass Spectrometry (LC-MS)

1. Thaw samples on ice (*see* **Note 11**).
2. Vortex-mix for 2 min (*see* **Note 12**).
3. On ice, pipet 100 μL of plasma to an Eppendorf® tube.
4. Add 300 μL of precooled at −20 °C MetOH–EtOH (1:1) (*see* **Note 13**).
5. Vortex-mix samples for 1 min.
6. Incubate the samples on ice for 20 min (*see* **Note 14**).

7. Vortex-mix for a few seconds.
8. Centrifuge the sample at 15,700 × *g* for 20 min at 4 °C (*see* **Note 15**).
9. Transfer 100 μL of the supernatant to LC vial equipped with a 200 μL insert (*see* **Note 16**).

3.2 Metabolite Fingerprinting by Gas Chromatography Coupled to Mass Spectrometry (GC-MS)

1. Thaw sample on ice (*see* **Note 11**).
2. Vortex samples for 2 min (*see* **Note 12**).
3. On ice, pipet 50 μL of plasma to an Eppendorf® tube.
4. Add 150 μL of precooled at −20 °C acetonitrile (*see* **Note 17**).
5. Vortex-mix for 2 min.
6. Incubate the samples on ice for 20 min (*see* **Note 14**).
7. Vortex-mix for a few seconds.
8. Centrifuge the sample at 15,400 × *g* for 10 min at 4 °C (*see* **Note 15**).
9. Transfer 100 μL of the supernatant to a GC vial equipped with a 200 μL insert (*see* **Note 18**).
10. Dry the extracts in a speed vacuum concentrator at 30 °C to complete dryness (*see* **Note 19**).

Derivatization (*See* **Notes 20** and **21**)

1. Methoximation
   (a) Add 10 μL of *O*-methoxyamine hydrochloride (15 mg/mL) in pyridine to each dried extract, and immediately close tubes afterward (*see* **Note 22**).
   (b) Vortex vigorously for 1 min.
   (c) Ultrasonicate the samples for 2 min and vortex vigorously for 2 min—repeat the process 3 times.
   (d) Cover the GC vials and incubate in the dark at room temperature for 16 h (*see* **Note 23**).
2. Silylation
   (a) Add 10 μL of BSTFA with 1% TMCS to each sample and immediately close the vial.
   (b) Vortex vigorously for 5 min.
   (c) Place the GC vials in an oven for 1 h at 70 °C for silylation.
   (d) Allow the sample to cool down to room temperature in a dark place for at least 15 min.
   (e) Add 100 μL of heptane containing 20 ppm of tricosane (IS) to each GC vial. Immediately close the vial.
   (f) Vortex vigorously for 2 min.
   (g) Before analysis, centrifuge the samples at 1000 × *g* for 15 min at 18 °C (*see* **Notes 24** and **25**).

3.3 Metabolite Fingerprinting by Capillary Electrophoresis Coupled to Mass Spectrometry (CE-MS)

1. Thaw samples on ice (*see* **Note 11**).
2. Vortex-mix for 2 min (*see* **Note 12**).
3. Take 100 μL of plasma in the Eppendorf® tube.
4. Add 100 μL of 0.2 M formic acid (with 5% acetonitrile and 0.4 mM methionine sulfone, IS).
5. Vortex-mix for approximately 1 min.
6. Transfer the mixed samples to a Centrifree Millipore (30 kDa) filter (*see* **Note 26**).
7. Centrifuge at 2000 × *g* for 70 min at 4 °C (*see* **Notes 15** and **27**).
8. Transfer the filtrate directly to the CE vial for analysis.

3.4 Quality Assurance

3.4.1 Quality Control Samples

Prepare Quality control (QC) samples (*see* **Note 28**) together with the experimental samples by following the same sample treatment. Calculate the exact number of QC samples for each analytical platform (*see* **Note 29**).

3.4.2 Blank Sample

Consider the use of extraction blank, prepared following exactly the same sample preparation procedure as experimental samples but without the biological sample. Use MilliQ water instead of the sample of interest. Mobile phase blank, contains only the mobile phase used for analytical run. Cleaning blank, prepared with strong solvent (e.g., isopropanol). If possible, collection blank is desirable. Such blank is collected from the same source as the samples, but will be free of the analyte. All the conditions related to the storage, pretreatment, extraction, concentration, and analysis have to be the same as for the experimental samples (*see* **Note 30**).

3.4.3 Sequence

Samples must be randomized, both for sample preparation and analysis, in order to ensure that there is no bias introduced and no correlation between biological factors and acquisition order [9, 10].

General Worklist Format

1. IS (GC-MS) or standard mix (CE-MS) (*see* **Note 31**).
2. Blank sample (*see* **Note 32**).
3. Standard containing grain fatty acid methyl ester (FAME) mixture (GC-MS) (*see* **Note 33**).
4. QC samples (*see* **Note 34**).
5. Experimental samples.
6. QC sample every 5–10 samples (*see* **Note 34**).
7. Blank sample (*see* **Note 35**).

3.5 Equipment Setup

The analytical conditions and equipment setup are presented as previously described [11–14].

3.5.1 LC-QTOF-MS

UHPLC system (Agilent 1290 Infinity LC System) composed of a degasser, two binary pumps, and a thermostated autosampler (maintained at 4 °C) coupled with Q-TOF LC-MS (6550 iFunnel) system (Agilent Technologies). In total, 0.5 μL of extracted plasma samples is injected onto a reverse-phase Zorbax Extend C18 column (2.1 × 50 mm, 1.8 μm; Agilent Technologies) thermostated at 60 °C. The flow rate of mobile phase (A: water with 0.1% formic acid and B: acetonitrile with 0.1% formic acid) is 0.6 mL/min, running in gradient method from 5% to 100% B phase. During all analyses, two reference masses are used: *m/z* 121.0509 ($C_5H_4N_4$) and *m/z* 922.0098 ($C_{18}H_{18}O_6N_3P_3F_{24}$) for positive ionization mode (ESI+) and *m/z* 112.9856 [$C_2O_2F_3(NH_4)$] and *m/z* 1033.9881 ($C_{18}H_{18}O_6N_3P_3F_{24}$) for negative ionization mode (ESI−), continuously infused into the system for constant mass correction. The system is operated in full scan mode from *m/z* 50 to 1000 for positive and *m/z* 50 to 1100 for negative mode. Capillary voltage is set to 3 kV for both positive and negative ionization modes; the drying gas flow rate is 12 L/min at 250 °C and gas nebulizer 52 psi, fragmentor voltage 175 V for positive and 250 V for negative ionization modes, skimmer 65 V, and octopole radio frequency voltage (OCT RF Vpp) 750 V. Data are collected in the centroid mode at a scan rate of 1.0 spectrum per second.

3.5.2 GC–EI–Q–MS

GC system (Agilent Technologies 7890A) consist of an autosampler (Agilent Technologies 7693) interfaced to inert mass spectrometer with triple-Axis detector (5975C, Agilent Technologies). In total 2 μL of the derivatized sample is injected through a GC-Column DB5-MS (30 m length, 0.25 mm ID, 0.25 μm film; 95% dimethyl–5% diphenylpolysiloxane) with a precolumn (10 m J&W integrated with Agilent 122-5532G). The injector port is held at 250 °C, and the split ratio is 1:10 with helium carrier gas flow rate 1.0 mL/min. The temperature gradient is programmed as follows: The initial oven temperature is set at 60 °C (held for 1 min), increased to 325 °C at 10 °C/min rate, and a cooldown period is applied for 10 min before the next injection. Retention time locking tool is used during the analysis to keep constant the retention time for IS at 21.18 min. The detector transfer line, the filament source and the quadrupole temperature are, respectively, set at 280, 230, and 150 °C. MS detection is performed with electron impact (EI) mode at−70 eV. The mass spectrometer is operated in scan mode over a mass range of 50–600 *m/z* at a rate of 2.7 scan/s.

3.5.3 CE–TOF–MS

Agilent 7100 Capillary Electrophoresis system coupled with a TOF mass spectrometer (6224 Agilent) with an electrospray ionization source. A 1200 series ISO Pump from Agilent Technologies is used to supply sheath liquid. The separation in a fused silica capillary (Agilent Technologies) (50 μm inner diameter and 96 cm length). New capillaries are preconditioned with flush of 1 M NaOH for 30 min followed by MilliQ® water for 30 min and background electrolyte, BGE (1.0 M formic acid in 10% methanol) for 30 min. Before each analysis, the capillary is flushed for 5 min (950 mbar pressure) with BGE. Samples are injected at 50 mbar pressure for 50 s. After each sample injection, the BGE is injected for 20 s at 100 mbar pressure to improve the reproducibility. The separation conditions included 25 mbar of pressure and + 30 kV of voltage. The current observed under these conditions is around 20 μA. Before each injection, the instrument automatically refill the BGE vial. The MS is operated in positive mode. Drying gas is set to 10 L/min, nebulizer 10 psi, voltage 3.5 kV, fragmentor 100 V, gas temperature 200 °C, and skimmer 65 V. The sheath liquid composition is methanol/water (1/1, v/v) with two reference standards: standards (10 μL of 5 mM $C_5H_4N_4$, *m/z* 121.0509 and 30 μL of 2.5 mM $C_{18}H_{18}O_6N_3P_3F_{24}$, *m/z* 922.0098), using a flow rate of 0.6 mL/min (split 1/100). For MS, the total separation time (35 min) is divided into two segments; the first segment 0–1 min without applying nebulization and the remaining 34 min with nebulization.

4 Notes

1. Concern of the type of wipes used for sterilizing the site of blood collection (e.g., 70% isopropyl alcohol).
2. Absolutely avoid including multiple anticoagulant sample types in the same experiment.
3. As a general guideline, 1.8 mL of whole blood will yield ~1.0 mL of plasma.
4. Exclude hemolytic samples.
5. Read Material Safety Data Sheet (SDS) for all reagent. Organic solvents and other reagents used in protocol are highly flammable, are harmful by inhalation, and can cause severe skin burns and eye damage, among others. Pyridine is extremely toxic and is a carcinogenic substance. Wear protective gloves, protective clothing, and eye protection. Wash face, hands, and any exposed skin thoroughly after handling. Handle the reagents in a fume hood.

6. N,O-bis-(trimethylsilyl)-trifluoroacetamide (BSTFA) with 1% (*v/v*) trimethylchlorosilane reagent is moisture sensitive and must be sealed to prevent deactivation.
7. Volumes are measured with calibrated pipettes.
8. Calculate the appropriate volume of each reactant solution taking into account the total number of experimental samples, quality control, and blank samples. Never prepare a new solution in the middle of the experiment.
9. Alternative supplies, consumables, and equipment can be used.
10. Choose correct vial, the septum, the closure, for your application. Septa made of PTFE are recommended for single injections and not appropriate for long-term sample storage; PTFE/silicone is recommended for multiple injections and sample storage. Vial closure—crimp cap squeezes the septum between the rim of the glass vial and the crimped aluminum cap preventing evaporation. The crimp cap vial requires crimping tools to carry out the sealing process. Screw caps are universal, and screwing the cap applies a mechanical force that squeezes the septum between the glass rim and the cap.
11. After thawing, the samples should be processed as quickly as possible to avoid changes in the sample composition.
12. Thawing is not a homogeneous process. Vortex the sample to avoid deviations of the metabolite concentration.
13. Maintain the ratio of MetOH–EtOH to sample volumes at 3:1 to guarantee the precipitation of the proteins.
14. Additional 20-min incubation step on ice enhance the metabolite yield.
15. It is recommended to precool the centrifuge before use.
16. If applicable samples can be filtered: nylon syringe filter 0.2 μm.
17. Maintain the ratio of acetonitrile to sample volumes at 3:1 to guarantee the precipitation of the proteins.
18. Use the GC vial suited for a crimped aluminum cap that prevents derivatized sample evaporation once the sample is ready for the analysis.
19. Check to assure the vial is completely dry. Complete sample drying minimize the variation between samples and protect the reaction of the aqueous media during methoximation and derivatization procedure.
20. Derivatization, accomplished by altering chemical functional groups including –COOH, –OH, –NH, and –SH, provides increased sample volatility and increases sample stability, enhances sensitivity and selectivity, and improves chromatographic efficiency. Methoximation is used to reduce the chromatographic complexity of the samples. *O*-methoxyamine

hydrochloride in pyridine to stabilize carbonyl moieties by suppressing keto–enol tautomerism and the formation of multiple acetal- or ketal- structures. It also helps to reduce the number of derivatives of reducing sugars and generates only two forms of the –N=C < derivative, syn/anti. Silylation is the most widely used derivatization procedure where, an active hydrogen is replaced by an alkylsilyl group, such as trimethylsilyl (TMS). According to the chemical nature of the acidic proton replaced, the reaction can be incomplete and lead to multiple derivatives due to single, double, or triple replacing.

21. In the derivatization procedure, condensation of reagents appears on the wall and lid of the reaction tubes. The samples should be vortexed vigorously in every step of the protocol.
22. Pyridine is extremely toxic. Wear protective gloves, clothes, and mask respirator. Handle the solvent in a fume hood.
23. Cover the samples with the aluminum foil to avoid pyridine photolysis.
24. The samples should be analyzed immediately after preparation. Do not use derivatized samples later than 72 h after derivatization. Do not store derivatized samples in the fridge or freezer and in contact with air.
25. Reanalysis of samples that have already been injected into the GC-MS does not result in reproducible data.
26. The device holds a sample volume of 1 mL with a hold-up volume of 10 μL.
27. Never surpass 2000 × g with these filters.
28. QC sample should be homogeneous and representative of the qualitative and quantitative composition of the study samples. The comprehensive overview of the type and the quality assurance strategy has been provided in detail elsewhere [9, 10, 15].
29. QC samples are analyzed at the beginning (for system equilibration and conditioning in order to achieve fully reproducible conditions) and at the end of an analytical run as well as at regular intervals through the analysis in order to monitor stability and reproducibility of the analytical process.
30. All the conditions related to the blank preparation (e.g., sample pretreatment, extraction, storage, and analysis) have to be the same as for experimental samples. Blank samples serve for identifying and removing artefacts from the data matrix, allow to identify features arising from derivatization process, solvents, cleanliness of glassware, equipment, contaminants, and other sample components originating from, for example, tubes, solvent impurities, sample additives, or preservatives [9].

31. Should be provided for the assessment of system suitability (examination of peak shape, retention or migration time, stability, repeatability, and signal intensity).
32. At least one blank should be included at the beginning and the end of the batch. It is not recommended to analyze blank samples in the middle of the analytical batch.
33. Fatty acid methyl ester (FAME) mixture for further data calibration.
34. The repeated injection of the QC, from every 5 up to 10 experimental samples, depend on the analytical platform and sample type.
35. Blank sample at the end of analytical run allows for examining the carryover effect.

Acknowledgments

The authors would like to express their gratitude to the financial support received from the Spanish Ministry of Science, Innovation and Universities (RTI2018-095166-B-I00) and Fondo Europeo de Desarrollo Regional (FEDER) and the Polish National Science Center (No 04-0225/09/529; DEC-2018/29/B/NZ7/02489).

References

1. Patti GJ, Yanes O, Siuzdak G (2012) Metabolomics: the apogee of the omics trilogy. Nat Rev Mol Cell Biol 13:263–269. https://doi.org/10.1038/nrm3314
2. Dunn WB, Broadhurst DI, Atherton HJ, Goodacre R, Griffin JL (2011) Systems level studies of mammalian metabolomes: the roles of mass spectrometry and nuclear magnetic resonance spectroscopy. Chem Soc Rev 40:387–426. https://doi.org/10.1039/b906712b
3. García-Cañaveras JC, López S, Castell JV, Donato MT, Lahoz A (2016) Extending metabolome coverage for untargeted metabolite profiling of adherent cultured hepatic cells. Anal Bioanal Chem 408:1217–1230. https://doi.org/10.1007/s00216-015-9227-8
4. Wang Y, Liu S, Hu Y, Wan J (2015) Current state of the art of mass spectrometry- based metabolomics studies – a review focusing on wide coverage, high throughput and easy. RSC Adv 5:78728–78737. https://doi.org/10.1039/C5RA14058G
5. Ortmayr K, Causon TJ, Hann S, Koellensperger G (2016) Increasing selectivity and coverage in LC-MS based metabolome analysis. TrAC Trends Anal Chem 82:358–366. https://doi.org/10.1016/J.TRAC.2016.06.011
6. Haggarty J, Burgess KE (2017) Recent advances in liquid and gas chromatography methodology for extending coverage of the metabolome. Curr Opin Biotechnol 43:77–85. https://doi.org/10.1016/j.copbio.2016.09.006
7. Yanes O, Tautenhahn R, Patti GJ, Siuzdak G (2011) Expanding coverage of the metabolome for global metabolite profiling. Anal Chem 83:2152–2161. https://doi.org/10.1021/ac102981k
8. Hernandes VV, Barbas C, Dudzik D (2017) A review of blood sample handling and pre-processing for metabolomics studies. Electrophoresis 38(18):2232–2241. https://doi.org/10.1002/elps.201700086
9. Dudzik D, Barbas-Bernardos C, García A, Barbas C (2018) Quality assurance procedures for mass spectrometry untargeted metabolomics. A review. J Pharm Biomed Anal

147:149–173. https://doi.org/10.1016/j.jpba.2017.07.044

10. Dunn WB, Wilson ID, Nicholls AW, Broadhurst D (2012) The importance of experimental design and QC samples in large-scale and MS-driven untargeted metabolomic studies of humans. Bioanalysis 4:2249–2264. https://doi.org/10.4155/bio.12.204
11. Dudzik D, Zorawski M, Skotnicki M, Zarzycki W, Kozlowska G, Bibik-Malinowska-K, Vallejo M, García A, Barbas C, Ramos MP (2014) Metabolic fingerprint of gestational diabetes mellitus. J Proteome 103:57–71. https://doi.org/10.1016/j.jprot.2014.03.025
12. Dudzik D, Zorawski M, Skotnicki M, Zarzycki W, García A, Angulo S, Lorenzo MP, Barbas C, Ramos MP (2017) GC–MS based gestational diabetes mellitus longitudinal study: identification of 2-and 3-hydroxybutyrate as potential prognostic biomarkers. J Pharm Biomed Anal 144:90–98. https://doi.org/10.1016/j.jpba.2017.02.056
13. Naz S, Garcia A, Rusak M, Barbas C (2013) Method development and validation for rat serum fingerprinting with CE-MS: application to ventilator-induced-lung-injury study. Anal Bioanal Chem 405:4849–4858. https://doi.org/10.1007/s00216-013-6882-5
14. Mastrangelo A, Ferrarini A, Rey-Stolle F, García A, Barbas C (2015) From sample treatment to biomarker discovery: a tutorial for untargeted metabolomics based on GC-(EI)-Q-MS. Anal Chim Acta 900:21–35. https://doi.org/10.1016/j.aca.2015.10.001
15. Sangster T, Major H, Plumb R, Wilson AJ, Wilson ID (2006) A pragmatic and readily implemented quality control strategy for HPLC-MS and GC-MS-based metabonomic analysis. Analyst 131:1075–1078. https://doi.org/10.1039/b604498k

Chapter 15

Multicompartmental High-Throughput Metabolomics Based on Mass Spectrometry

Raúl González-Domínguez, Álvaro González-Domínguez, Ana Sayago, and Ángeles Fernández-Recamales

Abstract

Multicompartmental metabolomics is a powerful strategy for studying organism metabolic homeostasis in a holistic manner. To this end, mass spectrometry–based platforms are very versatile tools for high-throughput metabolomics, of particular interest when dealing with large populations. This chapter details methods and tips for comprehensive analysis of multiple biological matrices, including serum, urine, brain, liver, kidney, spleen, and thymus, by applying a metabolomic multiplatform based on the combination of direct mass spectrometry analysis, ultrahigh-performance liquid chromatography and gas chromatography coupled to mass spectrometry.

Key words Metabolomics, Multicompartmental, High-throughput, Mass spectrometry

1 Introduction

1.1 High-Throughput Approaches in Metabolomics

Metabolomics is a fast-growing research field due to its great potential to holistically elucidate the characteristic metabolic perturbations associated to disease pathogenesis and progression, as well as in response to external stimuli (e.g., diet, lifestyle, medication, environmental pollution). Given the huge complexity of the metabolome, very powerful analytical approaches are needed in metabolomics, with wide coverage and high sensitivity, specificity, and reproducibility. Furthermore, increasing the throughput of these techniques is also becoming essential with the aim of facilitating the implementation of metabolomics in large-scale studies, as well as for reducing costs. For these purposes, mass spectrometry (MS) is the most versatile analytical platform, which can be applied in multiple and complementary instrumental configurations. Direct mass spectrometry (DMS) fingerprinting is the simplest MS-based approach for high-throughput metabolomics because of the lack of a preceding chromatographic step [1, 2]. Thereby, total analysis

Paul L. Wood (ed.), *Metabolomics*, Neuromethods, vol. 159, https://doi.org/10.1007/978-1-0716-0864-7_15,

time can be significantly minimized (less than 1 min), and common challenges associated with chromatography (e.g., retention time variability, peak broadening) are avoided, thus increasing technical stability. On the other hand, the mass spectrometer can also be coupled with various separation approaches to reduce the complexity of metabolomic fingerprints. Nowadays, liquid chromatography–mass spectrometry (LC-MS) is the main workhorse in metabolomics due to its versatility, enabling the measurement of a large number of metabolites with very diverse chemical nature by means of applying complementary retention mechanisms (e.g., reversed phase, RP; hydrophilic interaction liquid chromatography, HILIC) [3]. Although the use of traditional high-performance liquid chromatography (HPLC) has been extensively described in metabolomics, its low resolution considerably hinders the analysis of complex matrices. Consequently, very long chromatographic methods are usually employed, which limits its application when dealing with large sample populations. In this vein, ultrahigh-performance liquid chromatography (UHPLC) has emerged as a more appropriate strategy for high-throughput metabolomics, with higher sensitivity, reproducibility and reduced analysis time than conventional HPLC. Alternatively, the coupling of gas chromatography with mass spectrometry (GC-MS) also provides comparable resolution to that showed by UHPLC-MS, being a robust and sensitive tool for the analysis of low molecular weight metabolites [4]. This hyphenated approach requires the introduction of a time-consuming derivatization step, thus limiting its throughput. However, numerous efforts have been made to develop rapid GC methodologies, considered as the gold standard for metabolomic profiling of primary metabolites. Therefore, multiple MS-based approaches are currently available for high-throughput metabolomic screening, each one having its inherent analytical limitations. For this reason, the combination of complementary platforms is nowadays the most common strategy with the aim of increasing metabolomic coverage [5, 6].

1.2 Multicompartmental Metabolomics: Alzheimer's Disease as a Case Study

Multiple biological matrices can be employed in metabolomic research, including plasma, serum, urine, saliva, and other biofluids, as well as tissues and cells, each one having a characteristic metabolome. The study of tissue samples allows for the direct investigation of molecular mechanisms associated with disease pathologies or other external stimuli, but their availability is very limited in human cohorts. On the other hand, biological fluids are easily obtained in the clinical practice, thus simplifying the discovery of simple and cheap biomarkers.

Alzheimer's disease (AD) is a multifactorial neurodegenerative disorder, characterized by abnormal metabolism of the amyloid precursor protein and tau hyperphosphorylation, oxidative stress, mitochondrial dysfunction, neuroinflammatory processes, altered metal homeostasis and many other pathological processes [7–9]. Thus, numerous authors have previously described the application of metabolomics to elucidate the interrelated pathways implicated in AD pathogenesis and progression, as recently reviewed [6]. The analysis of brain samples enables the in situ investigation of neuropathological abnormalities associated with AD [10–12]. However, the use of post-mortem tissue precludes the study of pathogenesis at the first stages of disease, thus hindering the discovery of prognostic and early diagnostic biomarkers, and the development of efficient therapeutic approaches. Alternatively, cerebrospinal fluid (CSF) also shows a great potential to characterize metabolomic abnormalities associated with AD, given that its composition directly reflects the brain neurochemistry [13–16]. Finally, the use of peripheral and noninvasive biological fluids has also been described with the aim of identifying biomarkers easily translatable to the clinical practice, such as serum [17–20], plasma [21–23], urine [24], and saliva [25, 26]. Taking this into account, multicompartmental metabolomics is a very interesting strategy since allows obtaining a global picture of pathological mechanisms underlying AD. For this purpose, the use of transgenic animal models facilitates the analysis of paired samples to investigate the organism's homeostasis, including plasma and serum [27–29], urine [30, 31], brain tissue [10, 32–34], and other peripheral organs [35–38].

2 Materials

2.1 Chemicals

High purity methanol, ethanol, chloroform, dichloromethane, formic acid, ammonium formate, pyridine, methoxyamine hydrochloride, and N-methyl-N-trimethylsilyl-trifluoroacetamide (MSTFA) are used for sample treatment and chromatographic separation. Deionized water is obtained by using a Milli-Q Gradient system (Millipore, Watford, UK).

2.2 Animal Handling

Transgenic APP×PS1 mice, expressing the Swedish mutation of the amyloid precursor protein (APP) together with deleted presenilin 1 (PS1) in exon 9, as well as age-matched wild-type mice of the same genetic background (C57BL/6) are employed to carry out a multicompartmental metabolomic investigation of Alzheimer's disease. After reception, animals are acclimated in rooms with a 12 h light–dark cycle at 20–25 °C for 3 days, with water and food available ad libitum. Subsequently, mice are exsanguinated by cardiac puncture after isoflurane inhalation, and blood samples are immediately cooled for 30 min and then centrifuged at 3500 × *g* for 10 min at 4 °C to obtain serum samples (*see* **Note 1**). Brain, liver, kidneys, spleen, and thymus are also rapidly removed and rinsed with saline solution (0.9% NaCl, w/v). Brains are dissected into hippocampus, cortex, cerebellum, striatum, and olfactory bulbs. Urine is collected in sterile plastic containers and centrifuged at 2000 × *g* for 10 min at 4 °C. Finally, all these samples are snap-frozen in liquid nitrogen and stored at −80 °C until analysis. Animals must be handled according to the directive 2010/63/EU stipulated by the European Community.

2.3 Instrumentation

DMS and LC-MS methods described in this chapter have been previously optimized in a quadrupole time-of-flight (Q-TOF) mass spectrometer equipped with electrospray source, model QSTAR XL Hybrid system (Applied Biosystems, Foster City, CA, USA), using an integrated syringe pump and an Accela LC system (Thermo Fisher Scientific), respectively. For accurate mass measurements, the TOF analyzer is daily calibrated using renin and taurocholic acid in positive and negative ion modes, respectively. On the other hand, a Trace GC ULTRA coupled to ion trap mass spectrometry detector ITQ 900 (Thermo Fisher Scientific) is employed for GC/MS-based metabolomics. For sample extraction, cryogenic homogenizer model Freezer/Mill 6770 (SPEX Sample-Prep), pellet mixer (VWR International), and centrifuge (Eppendorf 5804R) are used.

3 Methods

3.1 Sample Treatment

A two-step extraction method is used to prepare serum samples [17].

3.1.1 Extraction of Serum Samples

1. Thaw serum samples on an ice bath.
2. Add 100 μL of serum into an Eppendorf tube placed on an ice bath, and mix with 400 μL of methanol–ethanol (1:1, v/v).
3. Shake during 5 min to precipitate proteins using a vortex mixed or an orbital shaker.
4. Centrifuge at 4000 × *g* for 10 min at 4 °C.
5. Transfer the supernatant to a new tube, dry under nitrogen stream (*see* **Note 2**), and reconstitute the residue with 100 μL of methanol–water (80:20, v/v) containing 0.1% (v/v) formic acid (polar extract) (*see* **Note 3**).
6. Add 400 μL of chloroform–methanol (1:1, v/v) to the protein precipitate.
7. Shake vigorously during 5 min using a vortex mixer or an orbital shaker.
8. Centrifuge at 10,000 × *g* for 10 min at 4 °C.
9. Dry the supernatant under nitrogen stream (*see* **Note 2**), and reconstitute with 100 μL of dichloromethane–methanol (60:40, v/v) containing 0.1% (v/v) formic acid and 10 mM ammonium formate (lipophilic extract) (*see* **Note 3**).

3.1.2 Extraction of Urine Samples

Urine samples must be diluted before MS analysis with the aim of reducing matrix effects [30]. To this end, urines are thawed on an ice bath and then centrifuged at 4000 × *g* for 10 min at 4 °C. Finally, the supernatant is tenfold diluted with methanol/water (1:1, v/v).

3.1.3 Extraction of Tissue Samples

Tissue samples, including hippocampus, brain cortex, cerebellum, striatum, olfactory bulbs, liver, kidney, spleen, and thymus, are extracted in accordance with the methodology previously optimized by González-Domínguez et al. [32, 33, 35–37].

1. Cryohomogenize tissues during 30 s at rate of 10 strokes per second using a Freezer/Mills 6770 homogenizer (*see* **Note 4**).
2. Weigh 30 mg of homogenized tissue in a 1.5 mL Eppendorf tube. Keep frozen until the addition of the extraction solvent.
3. Add 10 μL mg^{-1} of precooled 0.1% (v/v) formic acid in methanol (−20 °C).
4. Inside an ice bath, use a pellet mixer (VWR International) during 2 min for extracting tissues.
5. Centrifuge at 10,000 × *g* for 10 min at 4 °C.
6. Transfer the supernatant to a new tube for further analysis (polar extract).
7. Add 10 μL mg^{-1} of precooled chloroform–methanol (2:1, v/v), containing 0.1% (v/v) formic acid and 10 mM ammonium formate (−20 °C), to the pellet obtained in **step 5** (*see* **Note 5**).
8. Repeat **steps 4–6** to obtain lipophilic extracts.

3.1.4 Preparation of Quality Control Samples

Prepare quality control (QC) samples by pooling equal volumes of each individual sample for each biological matrix. These samples are analyzed at the start of the run and at intermittent points throughout the sequence in order to equilibrate the analytical system and to monitor system stability [39].

3.1.5 Derivatization of Sample Extracts

Polar extracts obtained from serum and tissue samples as previously described are derivatized following the two-step methodology developed by González-Domínguez et al. for subsequent GC-MS analysis [19, 32, 35–36] (*see* **Note 6**).

1. Take 50 μL of the polar extract into an Eppendorf tube and dry under nitrogen stream (*see* **Note 2**).
2. Add 50 μL of 20 mg mL^{-1} methoxyamine hydrochloride in pyridine to carry out the methoxymation (*see* **Note 7**).
3. Vortex the sample and incubate at 80 °C for 15 min using a water bath.

4. Add 50 μL of N-methyl-N-trimethylsilyl-trifluoroacetamide (MSTFA) to carry out the silylation (*see* **Note 7**).
5. Vortex and incubate at 80 °C for 15 min in a water bath.
6. Centrifuge at 4000 × *g* for 1 min and collect the supernatant for analysis.

3.2 Metabolomic Analysis

3.2.1 Direct Mass Spectrometry Fingerprinting

Extracts are directly infused into the QTOF-MS system by using an integrated syringe pump operating at 5 μL min^{-1} flow rate. Mass spectra are obtained by electrospray ionization (ESI) in positive and negative ionization modes in separate runs, as described elsewhere [17, 27, 33, 37]. Full scan spectra are acquired for 0.2 min in the *m/z* range 50–1100 Da, with 1.005 s scan time. High-purity nitrogen is used as curtain and nebulizer gas at flow rates of 1.13 L min^{-1} and 1.56 L min^{-1}, respectively. The source temperature is maintained at 60 °C, while the ion spray voltage (IS), declustering potential (DP) and focusing potential are fixed at 3300/−4000 V, 60/−100 V and 250/−250 V in positive and negative ion modes, respectively. To acquire MS/MS spectra, nitrogen is used as collision gas.

3.2.2 Ultrahigh-Performance Liquid Chromatography–Mass Spectrometry

A reversed-phase Hypersil Gold C18 column (2.1 × 50 mm, 1.9 μm) is used for chromatographic separation [28, 32, 35–36]. Methanol (solvent A) and water (solvent B), both containing 10 mM ammonium formate and 0.1% formic acid, are delivered at 0.5 mL min^{-1} flow rate following this elution gradient: 0–1 min, 95% B; 2.5 min, 25% B; 8.5–10 min, 0% B; 10.1–12 min, 95% B. The column is thermostated at 50 °C, using an injection volume of 5 μl. Full scan spectra are acquired in positive and negative polarities within the m/z range 50–1000 Da, with 1.005 s scan time. High-purity nitrogen is used as the curtain, nebulizer, and heater gas at flow rates of 1.48 L min^{-1}, 1.56 L min^{-1}, and 6.25 L min^{-1}, respectively. The source temperature is maintained at 400 °C, while the ion spray voltage, declustering potential (DP), and focusing potential are set at 5000/−2500 V, 100/−120 V, and 350/−350 V, respectively. To acquire MS/MS spectra, nitrogen is used as the collision gas.

3.2.3 Gas Chromatography–Mass Spectrometry

A Factor Four capillary column (VF-5MS 30 m × 0.25 mm, 0.25 μm film thickness) is used for GC-based separations [28, 32, 35–36]. The oven temperature gradient is as follows: 100 °C for 0.5 min, 15 °C per min until reaching 320 °C, and 320 °C for 2.8 min. The injector temperature is kept at 280 °C, working in the splitless mode (injection volume: 1 μL). Helium is used as carrier gas at a constant flow rate of 1 mL min^{-1}. Full scan MS spectra are acquired in the m/z range 35–650 Da by electron ionization at 70 eV. The ion source temperature is set at 200 °C.

3.3 Data Processing

3.3.1 DMS Data

Metabolomic raw data obtained by DMS fingerprinting are submitted to peak detection using the Markerview™ software (Applied Biosystems). For this, peaks above the noise level (10 counts, determined empirically from experimental spectra) are selected and binned in intervals of 0.1 Da.

3.3.2 UHPLC-MS and GC-MS Data

Raw data obtained by UHPLC/GC-MS analysis is preprocessed following the pipeline described by González-Domínguez et al., which proceeds through multiple stages including feature detection, alignment of peaks, and normalization [40]. UHPLC-MS files are converted into the mzXML format using the msConvert tool (ProteoWizard), while GC-MS files are converted into netCDF using the Thermo File Converter tool (Thermo Fisher Scientific). Then, the freely available software XCMS is used for data processing. First, data are extracted using the matchedFilter method by applying 0.1 Da as fixed step size, S/N threshold 2, and full width at half-maximum (fwhm) 10 or 3, for UHPLC and GC respectively. Subsequently, peak alignment is accomplished in three iterative cycles by descending bandwidth (bw) from 10 to 1 s for UHPLC-MS data, and descending bw from 5 to 1 s for GC-MS. Finally, imputation of missing values and data normalization are performed using the fillPeaks algorithm and the locally weighted scatter plot smoothing (LOESS) normalization method, respectively. These preprocessed data are exported as a .csv file for further data analysis.

3.4 Data Analysis

Statistical analysis of preprocessed data is carried out by using the MetaboAnalyst 3.0 web tool (http://www.metaboanalyst.ca/) and the SIMCA-P™ software (version 11.5, UMetrics AB, Umeå, Sweden). A conventional pipeline for metabolomic data analysis involves the following steps.

1. Data prefiltering: selection of metabolomic features present in at least 50% of samples.
2. Imputation: estimation of missing values by means of the k-nearest neighbor (KNN) algorithm (*see* **Note 8**).
3. Data filtering: application of the interquartile range (IQR) to remove variables with little variance.
4. Data normalization according to the total area sum (*see* **Note 8**).
5. Pareto scaling and logarithmic transformation.
6. Application of multivariate statistical techniques to visualize general trends and detect discriminant patterns: principal component analysis (PCA), partial least squares discriminant analysis (PLS-DA).

7. Application of univariate statistical techniques to identify discriminant metabolites: *t*-test or one-way analysis of variance (ANOVA) (*see* **Note 9**), with false discovery rate (FDR) correction for multiple testing.

4 Notes

1. To obtain plasma instead of serum, blood must be collected in anticoagulant-containing tubes.
2. Sample extracts can also be taken to dryness by using vacuum concentrators (e.g., SpeedVac), if available.
3. Sample extracts can be directly analyzed without performing a preconcentration step if MS sensitivity allows it.
4. Small organs (e.g., hippocampus, striatum, olfactory bulbs) are directly extracted without prior cryohomogenization.
5. This second extraction step can be omitted when brain tissue is analyzed, as previously reported [33, 40].
6. Derivatize only polar extracts containing low molecular weight metabolites, neutral lipids present in lipophilic extracts are not readily analyzable by GC-MS.
7. Derivatization reagents (i.e., methoxyamine and MSTFA) are very sensitive to moisture. Daily prepare the methoxyamine solution in pyridine, and use sealed glass ampoules of MSTFA that should be used within the day of opening.
8. These steps must only be performed on data from DMS fingerprinting. Imputation and normalization of UHPLC and GC data is already accomplished in XCMS (*see* Subheading 3.3.2).
9. Nonparametric methods must be used when variables do not show normal distribution (checked by normal probability plots) and variances are not homogeneous (checked by Levene's test).

References

1. González-Domínguez R, Sayago A, Fernández-Recamales Á (2017) Direct infusion mass spectrometry for metabolomic phenotyping of diseases. Bioanalysis 9:131–148
2. González-Domínguez R, Sayago A, Fernández-Recamales Á (2018) High-throughput direct mass spectrometry-based metabolomics to characterize metabolite fingerprints associated with Alzheimer's disease pathogenesis. Metabolites 8:52
3. Rainville PD, Theodoridis G, Plumb RS, Wilson ID (2014) Advances in liquid chromatography coupled to mass spectrometry for metabolic phenotyping. TrAC - Trends Anal Chem 61:181–191
4. Pasikanti KK, Ho PC, Chan ECY (2008) Gas chromatography/mass spectrometry in metabolic profiling of biological fluids. J Chromatogr B Anal Technol Biomed Life Sci 871:202–211
5. González-Domínguez Á, Durán-Guerrero E, Fernández-Recamales Á, Lechuga-Sancho AM, Sayago A, Schwarz M, Segundo C, González-Domínguez R (2017) An overview on

the importance of combining complementary analytical platforms in metabolomic research. Curr Top Med Chem 17:3289–3295

6. González-Domínguez R, Sayago A, Fernández-Recamales Á (2017) Metabolomics in Alzheimer's disease: the need of complementary analytical platforms for the identification of biomarkers to unravel the underlying pathology. J Chromatogr B Anal Technol Biomed Life Sci 1071:75–92
7. Blennow K, de Leon MJ, Zetterberg H (2006) Alzheimer's disease. Lancet 368:387–403
8. Maccioni RB, Munoz JP, Barbeito L (2001) The molecular bases of Alzheimer's disease and other neurodegenerative disorders. Arch Med Res 32:367–381
9. González-Domínguez R, García-Barrera T, Gómez-Ariza JL (2014) Characterization of metal profiles in serum during the progression of Alzheimer's disease. Metallomics 6:292–300
10. Xu J, Begley P, Church SJ et al (2016) Graded perturbations of metabolism in multiple regions of human brain in Alzheimer's disease: snapshot of a pervasive metabolic disorder. Biochim Biophys Acta Mol basis Dis 1862:1084–1092
11. Inoue K, Tsutsui H, Akatsu H, Hashizume Y, Matsukawa N, Yamamoto T, Toyo'oka T (2013) Metabolic profiling of Alzheimer's disease brains. Sci Rep 3:2364
12. Graham SF, Chevallier OP, Roberts D, Hölscher C, Elliott CT, Green BD (2012) Investigation of the human brain metabolome to identify potential markers for early diagnosis and therapeutic targets of Alzheimer's disease. Anal Chem 85:1803–1811
13. Kaddurah-Daouk R, Zhu H, Sharma S et al (2013) Alterations in metabolic pathways and networks in Alzheimer's disease. Transl Psychiatry 3:e244
14. Ibanez C, Simo C, Martin-Alvarez PJ, Kivipelto M, Winblad B, Cedazo-Minguez A, Cifuentes A (2012) Toward a predictive model of Alzheimer's disease progression using capillary electrophoresis-mass spectrometry metabolomics. Anal Chem 84:8532–8540
15. Czech C, Berndt P, Busch K, Schmitz O, Wiemer J, Most V, Hampel H, Kastler J, Senn H (2012) Metabolite profiling of Alzheimer's disease cerebrospinal fluid. PLoS One 7: e31501
16. Ibáñez C, Simó C, Barupal DK, Fiehn O, Kivipelto M, Cedazo-Mínguez A, Cifuentes A (2013) A new metabolomic workflow for early detection of Alzheimer's disease. J Chromatogr A 1302:65–71
17. González-Domínguez R, García-Barrera T, Gómez-Ariza JL (2014) Using direct infusion mass spectrometry for serum metabolomics in Alzheimer's disease. Anal Bioanal Chem 406:7137–7148
18. González-Domínguez R, García-Barrera T, Gómez-Ariza JL (2015) Application of a novel metabolomic approach based on atmospheric pressure photoionization mass spectrometry using flow injection analysis for the study of Alzheimer's disease. Talanta 131:480–489
19. González-Domínguez R, García-Barrera T, Gómez-Ariza JL (2015) Metabolite profiling for the identification of altered metabolic pathways in Alzheimer's disease. J Pharm Biomed Anal 107:75–81
20. González-Domínguez R, Rupérez FJ, García-Barrera T, Barbas C, Gómez-Ariza JL (2016) Metabolomic-driven elucidation of serum disturbances associated with Alzheimer's disease and mild cognitive impairment. Curr Alzheimer Res 13:641–653
21. Graham SF, Chevallier OP, Elliott CT, Hölscher C, Johnston J, McGuinness B, Kehoe PG, Passmore AP, Green BD (2015) Untargeted metabolomic analysis of human plasma indicates differentially affected polyamine and L-Arginine metabolism in mild cognitive impairment subjects converting to Alzheimer's disease. PLoS One 10:e0119452
22. Greenberg N, Grassano A, Thambisetty M, Lovestone S, Legido-Quigley C (2009) A proposed metabolic strategy for monitoring disease progression in Alzheimer's disease. Electrophoresis 30:1235–1239
23. Li N, Liu W, Li W, Li S, Chen X, Bi K, He P (2010) Plasma metabolic profiling of Alzheimer's disease by liquid chromatography/mass spectrometry. Clin Biochem 43:992–997
24. Cui Y, Liu X, Wang M et al (2014) Lysophosphatidylcholine and amide as metabolites for detecting Alzheimer disease using ultrahigh-performance liquid chromatography-quadrupole time-of-flight mass spectrometry-based metabonomics. J Neuropathol Exp Neurol 73:954–963
25. Liang Q, Liu H, Zhang T, Jiang Y, Xing H, Zhang A (2015) Metabolomics-based screening of salivary biomarkers for early diagnosis of Alzheimer's disease. RSC Adv 5:96074–96079
26. Liang Q, Liu H, Li X, Zhang A-H (2016) High-throughput metabolomics analysis discovers salivary biomarkers for predicting mild cognitive impairment and Alzheimer's disease. RSC Adv 6:75499–75504

27. González-Domínguez R, García-Barrera T, Vitorica J, Gómez-Ariza JL (2015) Application of metabolomics based on direct mass spectrometry analysis for the elucidation of altered metabolic pathways in serum from the APP/PS1 transgenic model of Alzheimer's disease. J Pharm Biomed Anal 107:378–385
28. González-Domínguez R, García-Barrera T, Vitorica J, Gómez-Ariza JL (2015) Deciphering metabolic abnormalities associated with Alzheimer's disease in the APP/PS1 mouse model using integrated metabolomic approaches. Biochimie 110:119–128
29. González-Domínguez R, García-Barrera T, Vitorica J, Gómez-Ariza JL (2015) Metabolomic research on the role of interleukin-4 in Alzheimer's disease. Metabolomics 11:1175–1183
30. González-Domínguez R, Castilla-Quintero R, García-Barrera T, Gómez-Ariza JL (2014) Development of a metabolomic approach based on urine samples and direct infusion mass spectrometry. Anal Biochem 465:20–27
31. Peng J, Guo K, Xia J, Zhou J, Yang J, Westaway D, Wishart DS, Li L (2014) Development of isotope labeling liquid chromatography mass spectrometry for mouse urine metabolomics: quantitative metabolomic study of transgenic mice related to Alzheimer's disease. J Proteome Res 13:4457–4469
32. González-Domínguez R, García-Barrera T, Vitorica J, Gómez-Ariza JL (2014) Region-specific metabolic alterations in the brain of the APP/PS1 transgenic mice of Alzheimer's disease. Biochim Biophys Acta Mol basis Dis 1842:2395–2402
33. González-Domínguez R, García-Barrera T, Vitorica J, Gómez-Ariza JL (2015) Metabolomic screening of regional brain alterations in the APP/PS1 transgenic model of Alzheimer's disease by direct infusion mass spectrometry. J Pharm Biomed Anal 102:425–435
34. Salek RM, Xia J, Innes A, Sweatman BC, Adalbert R, Randle S, McGowan E, Emson PC, Griffin JL (2010) A metabolomic study of the CRND8 transgenic mouse model of Alzheimer's disease. Neurochem Int 56:937–947
35. González-Domínguez R, García-Barrera T, Vitorica J, Gómez-Ariza JL (2015) Metabolomic investigation of systemic manifestations associated with Alzheimer's disease in the APP/PS1 transgenic mouse model. Mol BioSyst 11:2429–2440
36. González-Domínguez R, García-Barrera T, Vitorica J, Gómez-Ariza JL (2015) Metabolomics reveals significant impairments in the immune system of the APP/PS1 transgenic mice of Alzheimer's disease. Electrophoresis 36:577–587
37. González-Domínguez R, García-Barrera T, Vitorica J, Gómez-Ariza JL (2015) High throughput multiorgan metabolomics in the APP/PS1 mouse model of Alzheimer's disease. Electrophoresis 36:2237–2249
38. Wu J, Fu B, Lei H, Tang H, Wang Y (2016) Gender differences of peripheral plasma and liver metabolic profiling in APP/PS1 transgenic AD mice. Neuroscience 332:160–169
39. Sangster T, Major H, Plumb R, Wilson AJ, Wilson ID (2006) A pragmatic and readily implemented quality control strategy for HPLC-MS and GC-MS-based metabonomic analysis. Analyst 131:1075–1078
40. González-Domínguez R, González-Domínguez Á, Sayago A, Fernández-Recamales Á (2018) Mass spectrometry-based metabolomic multiplatform for Alzheimer's disease research. In: Perneczky R (ed) Biomarkers for Alzheimer's disease drug development, vol 1750. Humana Press (Springer), New York, pp 125–137

Chapter 16

GC-MS Nontargeted Metabolomics of Neural Tissue

Carolina Gonzalez-Riano, Mª. Fernanda Rey-Stolle, Coral Barbas, and Antonia García

Abstract

GC-MS is the most mature MS-based analytical technique. Besides, any GC-MS nontargeted metabolomics study of neural tissue will allow for obtaining information about the broad range of metabolites involved in the central metabolic pathways, including amino acids, some of them neurotransmitters, sugars, free fatty acids, polyalcohol, and cholesterol. The main disadvantage is the multiple steps required for sample treatment, but fortunately, the use of in-house and commercial spectrum libraries facilitates extraordinarily the identification of metabolites. Here, in this chapter detailed protocols describing sample and data treatment are described. Moreover, sample treatment for two different type of samples optic nerve and brain tissue is included.

Key words Hippocampus, Neural tissue, Nontargeted metabolomics, GC-MS, Spectrum libraries

1 Introduction

Metabolomics refers to the systematic analysis of the complete set of intra and extracellular metabolites (low-molecular-weight, typically <1500 Da, intermediates) in a given biological system, including microbial, plant, and mammalian systems. Depending on the objectives of metabolomics study, two analytical strategies can be applied: targeted and nontargeted approach. The classical approach, targeted metabolomics, focuses on the analysis of a small set of known compounds, while the nontargeted metabolomics is selected with the main objective of simultaneously measuring as many metabolites as possible in the biological sample without bias, within a single experimental run. These hypothesis-free studies allow for defining novel and previously unobserved changes in metabolome that can be linked to the biological function of an organism and mechanisms underlying altered metabolic changes related to the development of a condition or disease state. Finally, this process leads to the generation of a new hypothesis that should

Paul L. Wood (ed.), *Metabolomics*, Neuromethods, vol. 159, https://doi.org/10.1007/978-1-0716-0864-7_16,

be verified in the subsequent validation study, including targeted approach.

Gas chromatography-mass spectrometry (GC-MS) is the most mature MS analytical platform for metabolomics. It is especially suited for identifying and quantitating small molecular metabolites (<650 Da) with the highest chromatographic efficacy and resolution along with repeatability (in detector response and retention time).

Moreover, it offers the advantage of libraries for metabolites identification: commercially available spectrum libraries (NIST and Wiley libraries), and targeted libraries with dual information (spectra plus retention time under a selected GC-MS method) as Fiehn's library or any in-house library. These databases are especially useful in nontargeted metabolomics for identification of hundreds of metabolites found in the sample after deconvolution of the metabolite profile (*see* Subheading 3.4).

The main disadvantage of this analytical approach is the requirement of chemical derivatization of the biological sample before its analysis converting the metabolites into volatile derivatives (usually more than one per metabolite) lowering their boiling temperature, increasing their thermal stability, and improving chromatographic peak symmetry.

The neural o nervous tissue includes a great variety of cell types (among others: astrocytes, neurons, and oligodendrocytes), what makes one of the most complex tissues in the human body. It includes the brain, spinal cord, and other types of nerves. The optic nerve is the main myelinated neural tract of the central nervous system (CNS). It is formed by the projecting axons from retinal ganglion cells (RGCs) together with glia [1]. Regarding its composition, the optic nerve presents a high lipid content, comprised between 50% to 60% of its dry weight.

Despite the fact that the brain is the organ presenting the second highest lipid content behind the adipose tissue, a GC-MS–based metabolomics study is the most suitable approach to analyse the broad range of metabolites involved in the central metabolic pathways, including amino acids, some of them neurotransmitters such as Glu and Asp, along with sugars, free small, medium and long chain fatty acids, as well as polyalcohols and cholesterol that is the major architectural component of compact myelin.

2 Materials

2.1 Chemical Reagents

1. Internal Standards: Pentadecanoic acid and methyl stearate (C18:0 methyl ester) GC grade (Sigma Aldrich, Steinheim, Germany).

2. 25 ppm (μg/mL) Pentadecanoic acid (Internal Standard) solution in methanol. Prepare a 500 ppm stock solution in methanol and dilute it to 25 ppm with the same solvent.
3. 10 ppm C18:0 methyl ester (Internal standard) solution in heptane. Prepare a 1000 ppm stock solution in heptane and dilute it to 10 ppm with the same solvent.
4. *N,O*-Bis(trimethylsilyl)trifluoroacetamide (BSTFA) plus 1% trimethylchlorosilane (TMCS) (Supelco, Bellefonte, PA, USA) (*see* **Notes 1** and **2**).
5. *O*-methoxyamine hydrochloride GC grade (Sigma-Aldrich). Prepare a solution containing 15 mg/mL in pyridine. This solution is recommended to be freshly prepared for every experiment (*see* **Notes 1** and **2**).
6. Prepare a 100 fold dilution in dichloromethane from the grain fatty acid methyl esters mix (C8:0 – C22:1) (FAMEs) Ref. CRM47801 (Supelco, Sigma-Aldrich Chemie GmbH, Steinheim, Germany, 10 mg/mL in dichloromethane). The commercial solution contains a mixture of methyl esters of caprylic acid (C8:0) 1.9%, capric acid (C10:0) 3.2%, lauric acid (C12:0) 6.4%, tridecanoic acid (C13:0) 3.2%, myristic acid (C14:0) 3.2%, myristoleic acid (C14:1, cis-9) 1.9%, pentadecanoic acid (C15:0) 1.9%, palmitic acid (C16:0) 13.0%, palmitoleic acid (C16:1, cis-9) 6.4%, heptadecanoic acid (C17:0) 3.2%, stearic acid (C18:0) 6.5%, oleic acid (C18:1, cis-9) 19.6%, elaidic acid (C18:1, trans-9) 2.6%, linoleic acid (C18:2, cis-9,12) 13.0%, linolenic acid (C18:3, cis-9,12,15) 6.4%, arachidic acid (C20:0) 1.9%, cis-11-eicosenoic acid (C20:1) 1.9%, behenic acid (C22:0) 1.9%, and erucic acid (C22:1, cis-13) 1.9%. Keep the diluted and commercial mixture at (−20 °C).
7. The mixture of *n*-alkane calibration standards (C8 – C40) Ref. 40,147-U (Supelco, Sigma Aldrich, Steinheim, Germany) in dichloromethane contained 500 μg/mL of octane, nonane, decane, undecane, dodecane, tridecane, tetradecane, pentadecane, hexadecane, heptadecane, octadecane, nonandecane, prostane, eicosane, phytane, docosane, tricosane, tetracosane, pentacosane, heneicosane, hexacosane, heptacosane, octacosane, nonacosane, triacontane, hentriacontane, dotriacontane, tritriacontane, tetratriacontane, oentatriacontane, hexatriacontane, heptatriacontane, octatriacontane, nonatriacontane, tetracontane; and 5000 μg/mL of carbon disulfide. Keep at (−20 °C) and prepare a 50-fold dilution the day of the analysis with the same solvent.
8. Silylation-grade pyridine (VWR International BHD Prolabo, Madrid, Spain).
9. HPLC-grade 2-propanol (Sigma-Aldrich), HPLC-grade methanol (Sigma-Aldrich), HPLC-grade n-Heptane (VWR International BHD Prolabo, Madrid, Spain), HPLC-grade

dichloromethane (Carlo ERBA Reagents, Sabadell, Barcelona) and HPLC-grade methyl tert-butyl ether (MTBE) (Sigma-Aldrich).

10. All the solutions employed for the analyses have to be prepared using reverse-osmosed ultrapure water obtained from a Millipore Milli-Q Plus 185 water system (Millipore, Billerica, MA, USA).
11. The individual stock solutions are made and stored at −20 °C. The intermediate solutions are prepared from the stock solutions and stored at 4 °C during the working week. Finally, the intermediate standard solutions are diluted according to the concentration required on the day of the analysis.

2.2 Samples

1. Mouse optic nerve tissue.
2. Mouse left hippocampus tissue.
3. Hamster right brain hemisphere: the entire right hemisphere (300 mg approx.) is analyzed to reduce possible biological variability due to the brain region employed.

2.3 Apparatus

1. TissueLyser LT homogenizer (Qiagen, Germany).
2. Analytical balance Ohaus explorer.
3. Vortex mixer Fisherbrand (Fisher Scientific, Pittsburgh, USA).
4. Micropipettes HTL Labmate 2–20 μL, 20–200 μL, and 100–1000 μL.
5. Eppendorf tubes of 1.5 mL and 2 mL (Hamburg, Germany).
6. Speedvac concentrator (Thermo Fisher Scientific, Waltham, MA).
7. Centrifuge Heraeus instruments Megafuge 1.0R.
8. Eppendorf centrifuge 5415 R (Hamburg, Germany).
9. 2 mL GC-MS Chromacol glass vials with conical narrow opening inserts.
10. Ultrasonic bath (e.g., J.P. Selecta® Ultrasons-HD).

2.4 GC-MS Equipment

1. GC-Q-MS: auto sampler (Agilent Technologies 7693) and gas chromatograph (Agilent Technologies 7890A GC with split/splitless or multimode injector) coupled to a Mass Selective Detector (Agilent Technologies 5975 inert MSD with quadrupole mass analyzer and a triple-axis detector.
2. 10 m J&W precolumn (Agilent Technologies) integrated with a 122–5532G column: DB5-MS 30 m length, 0.25 mm i.d., and 0.25 μm film consisting of 95% dimethyl/5% diphenyl polysiloxane (Agilent Technologies).

3. Ultrainert low pressure drop liner with wool GC liner (Agilent Technologies cat.no. 5190-3165).

2.5 Software

1. Agilent MSD ChemStation software (Agilent Technologies).
2. Fiehn GC-MS Metabolomics RTL Library 2013.
3. National Institute of Standards and Technology (NIST) mass spectra library 2017.
4. Agilent MassHunter Qualitative Analysis version B.07.00 or subsequent (Agilent Technologies).
5. Agilent MassHunter Unknowns Analysis Tool version 7.0 or subsequent (Agilent Technologies).
6. MassProfiler Professional version B.12.1 or subsequent (Agilent Technologies).
7. Agilent MassHunter Quantitative Analysis version B.07.00 or subsequent (Agilent Technologies).
8. MATLAB version R2015a (MathWorks, Natick, MA).
9. SIMCA P+ version 14.0 (Umetrics, Umeå, Sweden).

3 Methods

After collection of the tissues, all the samples must be stored at −80 °C to avoid any postmortem ongoing metabolic processes. Defrost the samples the day of the analysis.

3.1 Metabolite Fingerprinting of Mouse Hippocampus and Hamster Brain Tissue Samples

This protocol was developed and validated for a metabolomics study based on lung tissue samples [2] and then was successfully applied for the analysis of different biological samples from brain tissue (*see* **Note 3**) [3, 4]. The major advantage of this methodology is that it can be used to carry out a multiplatform metabolomics analysis based on not only GC-MS but also on LC-MS and CE-MS. It consists of a single-phase extraction method that associates polar and aqueous solvents with a nonpolar organic solvent to extend the metabolite coverage of the analysis. At the end of the sample treatment, a monophasic extraction solvent made of MeOH–H_2O–MTBE will be obtained allowing for the performance of high reproducible analysis (*see* Fig. 1).

1. Add cold (−20 °C) MeOH–H_2O 50% (*v/v*) to each sample with a weight-to-volume ratio 1:10 for metabolites extraction. For example, for a 20 mg tissue add 200 μL of the mixture of solvents.
2. To achieve the tissue homogenization, perform 2 cycles of 3 min of TissueLyser LT homogenizer using 2.8 mm particle size steel bead. Alternative homogenizers, including the previously mentioned ones can be used for this process.

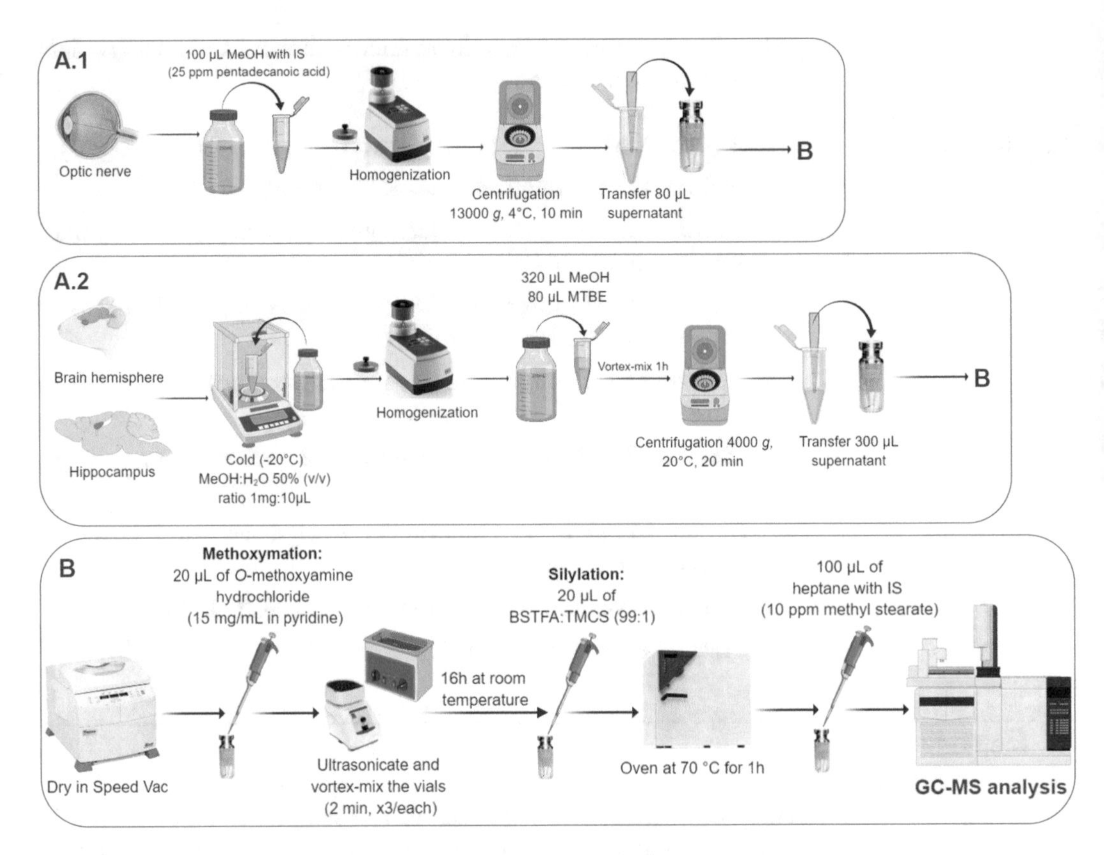

Fig. 1 General scheme of metabolite extraction from optic nerve (**A.1**) and brain tissue (**A.2**) samples, following the GC-MS derivatization process (**B**). Image created with BioRender

3. Vortex-mix 100 μL of tissue homogenate with 320 μL of cold (−20 °C) MeOH for 2 min.
4. Add 80 μL of methyl tert-butyl ether (MTBE) to each sample for the extraction of nonpolar metabolites and vortex the samples for 1 h at room temperature. The ratio 4:1 of MetOH: MTBE allows for the precipitation of proteins.
5. Centrifuge the extracted samples at 4000 × *g* at 20 °C for 20 min to separate the precipitated pellets.
6. Transfer 300 μL of the supernatant containing the extracted metabolites to a 0.3 mL-crimp-top-fixed insert Chromacol vial of borosilicate clear glass Ref. 10798345 Fisher Scientific (Loughborough, UK) and evaporate the samples to dryness with the SpeedVac Concentrator System at 35 °C (*see* **Note 4**).
7. Add 20 μL of *O*-methoxyamine hydrochloride (15 mg/mL in pyridine) to each sample to protect ketones and aldehydes functional groups. Close the vial and vortex-mix vigorously for 5 min (*see* **Note 2**).

8. Ultrasonicate the vials for 2 min and vortex-mix for 2 min. Then, repeat the process 3 times to guarantee the complete dissolution of compounds.
9. Incubate the vials for 16 h at room temperature in the dark covered with aluminium foil to avoid compounds hydrolysis due to humidity or light.
10. Add 20 μL of BSTFA with 1% TMCS (catalyst) to replace the active hydrogen atoms by an alkylsilyl group. Close the vial and vortex-mix the samples for 5 min (*see* **Note 2**).
11. For silylation, place the GC vials into an oven for 1 h at 70 °C.
12. Cool down the samples for 1 h at room temperature in the dark to avoid the loss of extremely volatile compounds.
13. Prior injection, add 100 μL of n-heptane containing 10 ppm C18:0 methyl ester (IS) to each vial as instrumental IS and then vortex-mix again the samples for 1 min to dissolve derivatives (*see* **Note 5**).
14. Centrifuge the vials at 2000 × *g* at 4 °C for 15 min before the GC-MS analysis (*see* **Note 6**).

3.2 Metabolite Fingerprinting in Optic Nerve Tissue Samples

The following analytical method has proven to be suitable for addressing two important challenges when treating mouse optic nerve tissue samples for metabolomics analysis. On the one hand, the limitation of matrix size that might hinder the performance of metabolomics study with such a small sample of tissue. On the other hand, the high lipid composition and the actin cytoskeleton of the optic nerve that might complicate the extraction and homogenization of the tissue [4].

The protocol consists on the addition of MeOH as the principal extracting solvent, which has proven to be a highly efficient and reproducible procedure for metabolomics-based approaches. The aim of the derivatization process is the same as the previously described one for the brain tissues but with slight modifications (*see* Fig. 1).

1. For metabolite extraction and protein precipitation, add 100 μL of cold methanol (−20 °C) containing 25 ppm pentadecanoic acid (IS) to the optic nerve samples (*see* **Note 5**). The pentadecanoic acid was previously used as an IS for both targeted and nontargeted GC-MS-based metabolomics studies [5].
2. To achieve the tissue disruption, use a 2 mm particle size glass beads for the TissueLyser LT homogenizer (Qiagen, Germany) and place the samples on ice for 30 s in between of TissueLyser cycles, 6 times each. Alternative homogenizers, such as liquid nitrogen-cooled mortar and pestle, electric tissue homogenizers, or Precellys 24-bead–based homogenizer can be used.

3. Centrifuge the homogenized samples at 13,000 × *g* at 4 °C for 10 min with the microcentrifuge to separate the precipitated pellets.
4. Transfer 80 μL of the supernatant to 0.3 mL crimp-top-fixed-insert Chromacol vial of borosilicate clear glass Ref. 10798345 Fisher Scientific (Loughborough, UK) and evaporate the samples to dryness with the SpeedVac Concentrator System at 35 °C (*see* **Note 4**).
5. Add 20 μL of *O*-methoxyamine hydrochloride (15 mg/mL in pyridine) to each sample to protect ketones and aldehydes functional groups and vortex-mix vigorously for 5 min (*see* **Note 2**).
6. Ultrasonicate the vials for 2 min and vortex-mix for 2 min. Then, repeat the process 3 times to guarantee the complete dissolution of compounds.
7. Incubate the vials for 16 h at room temperature in the dark covered with aluminium foil to avoid compounds hydrolysis due to humidity or light.
8. Add 20 μL of BSTFA with 1% TMCS (catalyst) to replace the active hydrogen atoms by an alkylsilyl group and vortex-mix the samples for 5 min (*see* **Note 2**).
9. For silylation, place the GC vials into an oven for 1 h at 70 °C.
10. Cool down the samples for 1 h at room temperature in the dark to avoid the loss of extremely volatile compounds.
11. Prior injection, add 100 μL of heptane containing 10 ppm C18:0 methyl ester (IS) to each vial as instrumental IS and then vortex-mix again the samples for 1 min to dissolve derivatives (*see* **Note 5**).
12. Centrifuge the vials at 2000 × *g* at 4 °C for 15 min before the GC-MS analysis (*see* **Note 6**).

3.3 GC-MS Analysis

1. Inject 2.0 μL of derivatized optic nerve, hippocampus, or brain samples in split mode (split ratio 1:10) into the GC-MS system (*see* **Note 7**).
2. Set a constant flow rate of the helium carrier gas at 1 mL/min through the column.
3. Perform the lock of the retention time (RTL) relative to the C18:0 methyl ester (IS) peak at 19.66 min, modifying consequently the flow rate.
4. Set the injector and the detector transfer line temperatures at 250 °C and 280 °C, respectively.
5. Programme the column oven temperature gradient as follows: set initial temperature at 60 °C, maintain it for 1 min, and then raise the temperature by 10 °C/min until reach 325 °C, and then hold at this temperature for 10 min before cooling down.

6. Operate the electron ionization (EI) source at −70 eV and set the filament source temperature at 230 °C (*see* **Note 8**).
7. Run the mass spectometer in scan mode only over the mass range of 50–600 *m/z* at a scan rate of 2 spectra/s.

3.4 Metabolite Identification

An important step in the analytical process is the characterization of the chromatographic profile obtained by GC-MS platform by using the standard "Fiehn metabolomics retention time lock (RTL)" method.

The identification is based on the comparison of both the RT and the mass spectrum of the compound already deconvoluted and the ones collected in an in-house, Fiehn's and NIST library (*see* Fig. 2). At the end of the process, 94 and 46 peaks were assigned to the optic nerve and hippocampus profiles, respectively. Carboxylic and dicarboxylic acids, organic nitrogen compounds, cyclic polyalcohols, cholesterol, amino acids, and fatty acids, among other metabolite classes, are included in the profiles (Figs. 3 and 4 and Tables 1 and 2). Cholesterol, myoinositol, and lactic acid can be observed as the highest peaks in both Total Ion chromatograms, TICs. Cholesterol is present at high levels in brain tissue, counting for 23% of total body cholesterol in humans. It is ubiquitously distributed within the brain and is an essential compound for a correct synaptic function and plasticity, allowing for synaptogenesis and the release of neurotransmitters [6]. Cholesterol has been also characterized as the main core of dendritic spines, since the alteration of its levels or inhibition of its synthesis led to the loss of these structures in cultured hippocampal neurons [7]. The dendritic spines are located in the pyramidal neurons of the hippocampus, as well as in neurons at different brain areas, and they provide biochemical compartments to control synaptic inputs within the CNS [8, 9]. Myoinositol plays multiple critical roles in brain, participating in major physiological functions. It can act either as an osmolyte or as neurotransmitter, and it is the precursor of phosphatidylinositols, an important lipid class especially abundant in brain tissue, corresponding up to a 10% of lipids content in this organ [10]. Regarding lactic acid, it has been proposed as a fuel metabolite since it can cross the blood–brain barrier (BBB) thanks to the presence of transporters that belong to the monocarboxylate carrier (MCT) family, which not only allows the passage of lactic acid but also of pyruvic acid and ketone bodies. MCT1-4 are the principal responsible transporters of lactic acid uptake and are located at different points in the brain. In fact, the transport of lactic acid between astrocytes and neurons has been connected with long-term memory formation since the astrocytic glycogenolysis is upregulated while the cognitive processes that take place in the hippocampus reach their highest activity point. This process leads

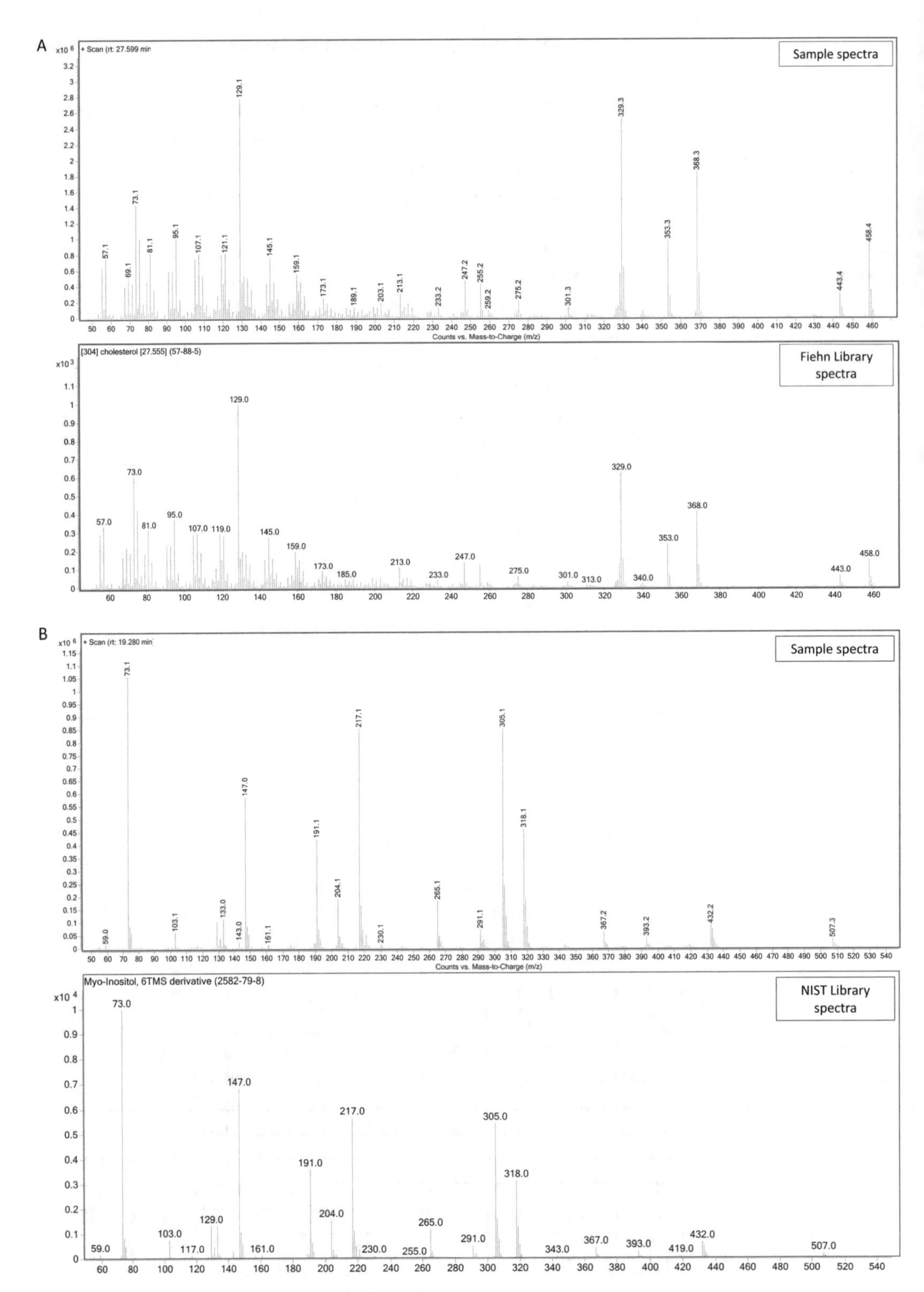

Fig. 2 Panel A. Identification of the pure spectrum, after deconvolution of the sample chromatogram, of the peak eluting at 27.599 min against the Fiehn library (cholesterol & Rt: 27.599 min). **Panel B.** Identification of the pure spectrum, after deconvolution of the sample chromatogram, of the peak eluting at 19.280 min against the NIST library (Myo-inositol 6TMS derivative)

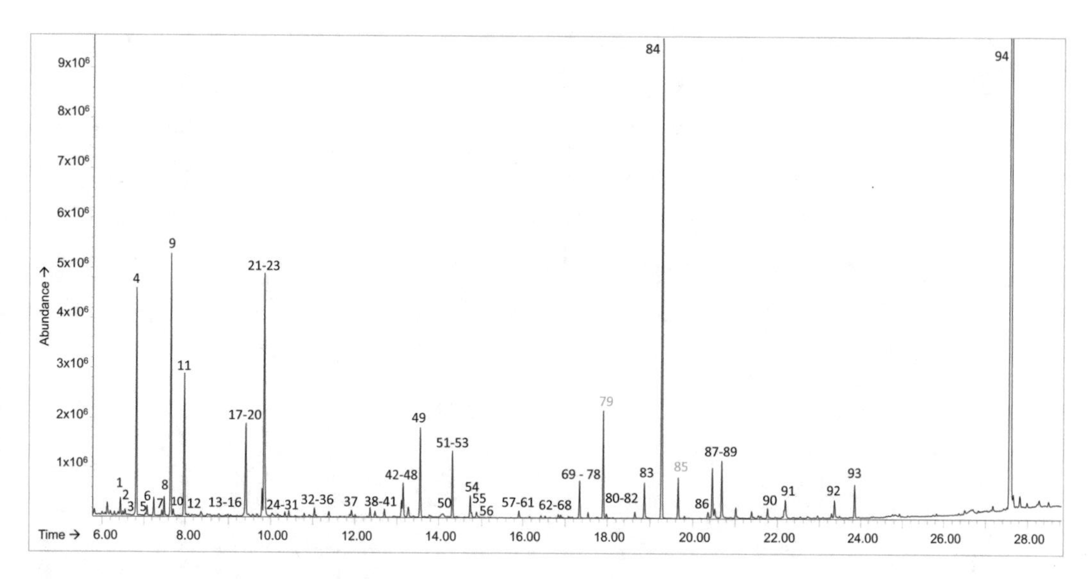

Fig. 3 Total Ion Chromatogram (TIC) of optic nerve profile obtained by GC-MS. The numbers assigned to each peak correspond to metabolites identified along the chromatographic profile and the numbers colored in light blue belong to the two IS spiked to the samples (Table 1)

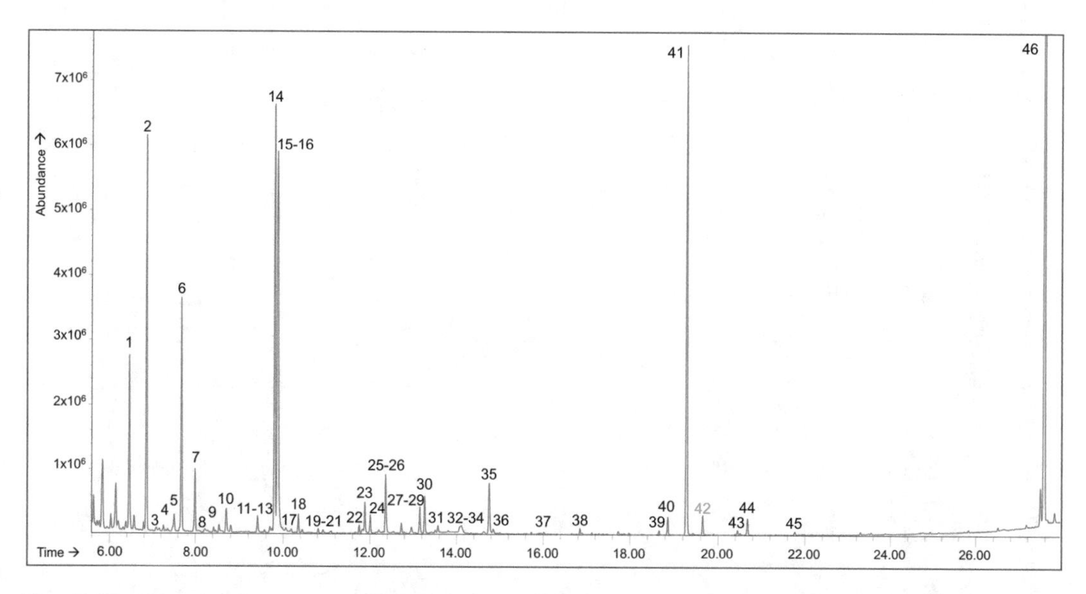

Fig. 4 TIC of hippocampus tissue profile obtained by GC-MS. The numbers assigned to each peak correspond to metabolites identified along the chromatographic profile and the number colored in light green belongs to the IS spiked to the samples prior to the analysis (Table 2)

Table 1
Compounds assigned across the TIC of the optic nerve tissue obtained by GC-MS analysis

Compound name	RT (min)	HMDB ID	Compound name	RT (min)	HMDB ID	Compound name	RT (min)	HMDB ID
1. Ethanolamine 2	6.450	HMDB00149	33. Fumaric acid	10.935	HMDB00134	65. Dehydroascorbic acid 1	16.846	HMDB01264
2. GC-MS reagent 1	6.554	–	34. Serine 2	11.060	HMDB00187	66. Tagatose 2	17.090	HMDB03418
3. Pyruvic acid	6.669	HMDB00243	35. Acetamide	11.060	HMDB31645	67. Fructose 1	17.090	HMDB00660
4. Lactic acid	6.832	HMDB00190	36. Threonine 2	11.398	HMDB00167	68. Sorbose 2	17.184	HMDB01266
5. Hexanoic acid	7.051	HMDB00535	37. Aspartic acid 1	11.923	HMDB00191	69. Tyrosine 1	17.297	HMDB00158
6. Glycolic acid	7.057	HMDB00115	38. 3-aminoisobutyric acid 2	12.380	HMDB03911	70. Galactose 1	17.322	HMDB00143
7. Alanine 1	7.451	HMDB00161	39. Aminomalonic acid	12.499	HMDB01147	71. Altrose 1	17.353	–
8. 3-methyl-2-oxobutanoic acid 1	7.504	HMDB0000019	40. Nicotinamide 2	12.712	HMDB01406	72. Talose 1	17.353	–
9. Acetoacetic acid 1	7.657	HMDB00060	41. Malic acid	12.718	HMDB00156	73. Glucose	17.415	HMDB00122
10. Glycine 2	7.726	HMDB00123	42. Aspartic acid 2	13.124	HMDB00191	74. Talose 2	17.559	–
11. Sarcosine	7.989	HMDB00271	43. Trans-4-hydroxy-L-proline 2	13.143	HMDB00725	75. Allose 2	17.559	HMDB01151
12. Leucine 1	8.252	HMDB00687	44. Glutamic acid 3 (dehydrated)	13.156	HMDB00148	76. Lysine 2	17.641	HMDB00182
13. Mimosine 1	8.790	HMDB15188	45. 4-Aminobutanoic acid	13.269	HMDB00112	77. Mannitol	17.778	HMDB00765
14. 3-aminoisobutyric acid 1	9.002	HMDB03911	46. 4-guanidinobutyric acid 2	13.275	HMDB03464	78. Tyrosine 2	17.810	HMDB00158

15. Valine 2	9.065	HMDB00883	47. Glutamic acid 1	13.287	HMDB00148	79. Pentadecanoic acid (IS)	17.900	HMDB00826
16. Beta-alanine 3	9.365	HMDB00056	48. Threonic acid	13.393	HMDB00943	80. Glucopyranose 5	18.154	HMDB0000516
17. Urea	9.422	HMDB00294	49. Creatinine	13.569	HMDB00562	81. Mannose 5	18.154	HMDB0000169
18. Benzoic acid	9.565	HMDB01870	50. Hypotaurine	14.119	HMDB00965	82. Scyllo-inositol 6	18.667	HMDB06088
19. Serine 1	9.672	HMDB00187	51. Glutamic acid 2	14.332	HMDB00148	83. Palmitic acid	18.879	HMDB00220
20. Ethanolamine 3	9.803	HMDB00149	52. Ornithine	14.332	HMDB00214	84. Myoinositol	19.286	HMDB00211
21. Phosphoric acid	9.853	HMDB02142	53. Phenylalanine 2	14.632	HMDB00159	85. Methyl stearate (IS)	19.661	–
22. Leucine 2	9.859	HMDB00687	54. N-acetyl-aspartic acid 1	14.751	HMDB00812	86. Tryptophan 2	20.349	HMDB00929
23. Glycerol	9.872	HMDB00131	55. Pyrophosphate	14.788	HMDB00250	87. Linoleic acid	20.412	HMDB00673
24. Isoleucine 2	10.153	HMDB00172	56. Lyxose 2	14.97	HMDB03402	88. Oleic acid	20.441	HMDB00207
25. Threonine 1	10.184	HMDB00167	57. Glycerol 1-phosphate	15.921	HMDB00126	89. Stearic acid	20.693	HMDB00827
26. Proline 2	10.234	HMDB00162	58. N-acetyl-L-phenylalanine 1	15.977	HMDB00512	90. Arachidonic acid	21.776	HMDB01043
27. 4-Aminobutanoic acid 2	10.281	HMDB00112	59. Glutamine 3	16.064	HMDB00641	91. Arachidic acid	22.370	HMDB02212
28. Beta-alanine 2	10.347	HMDB00056	60. Allothreonine 1	16.138	HMDB04041	92. Inosine	23.373	HMDB00195
29. Glycine	10.353	HMDB00123	61. O-phospho-colamine	16.177	HMDB00224	93. 1-monopalmitin	23.504	HMDB31074
30. Succinic acid	10.441	HMDB00254	62. Hypoxanthine	16.434	HMDB00157	94. Cholesterol	27.605	HMDB00067
31. GC-MS reagent 2	10.460	–	63. Citric acid	16.540	HMDB00094			
32. Uracil	10.760	HMDB00300	64. Arabitol 5	16.771	HMDB00568			

Table 2
Compounds assigned across the TIC of the hippocampus tissue obtained by GC-MS analysis

Compound name	RT (min)	HMDB ID	Compound name	RT (min)	HMDB ID	Compound name	RT (min)	HMDB ID
1. Ethanolamine 2	6.450	HMDB00149	17. Threonine 1	10.185	HMDB00167	33. Glutamic acid 2	14.338	HMDB00148
2. Lactic acid	6.857	HMDB00190	18. Glycine	10.353	HMDB00123	34. Phenylalanine 2	14.610	HMDB00159
3. Glycolic acid	7.076	HMDB00115	19. Succinic acid	10.441	HMDB00254	35. N-acetyl-L-aspartic acid 1	14.751	HMDB00812
4. Alanine 1	7.456	HMDB00161	20. Unknown 1	10.816	–	36. Ribose	15.001	HMDB0000283
5. 3-methyl-2-oxobutanoic acid 1	7.562	HMDB0000019	21. Fumaric acid	10.923	HMDB00134	37. Glutamine 3	16.021	HMDB00641
6. Acetoacetic acid 1	7.682	HMDB00060	22. Unknown 2	11.792	–	38. Myristic acid	16.896	HMDB0000806
7. Oxalic acid	7.970	HMDB0002329	23. Biogenic amine	11.880	–	39. Scyllo-inositol 6	18.667	HMDB06088
8. 3-Hydroxybutyric acid	8.214	HMDB0000442	24. Cadaverine 3	12.005	HMDB0002322	40. Palmitic acid	18.860	HMDB00220
9. Leucine 1	8.264	HMDB00687	25. Putrescine 4	12.361	HMDB0001414	41. Myoinositol	19.274	HMDB00211
10. Isoleucine 1	8.576	HMDB0000172	26. Nicotinamide 2	12.724	HMDB0001406	42. Methyl stearate (IS)	19.646	–
11. Urea	9.415	HMDB00294	27. Malic acid	12.731	HMDB00156	43. Oleic acid	20.449	HMDB00207
12. Benzoic acid	9.565	HMDB01870	28. Cadaverine 4	12.962	HMDB0002322	44. Stearic acid	20.681	HMDB00827
13. Serine 1	9.685	HMDB00187	29. Glutamic acid 3 (dehydrated)	13.156	HMDB00148	45. Arachidonic acid	21.760	HMDB01043
14. Ethanolamine 3	9.790	HMDB00149	30. 4-Aminobutanoic acid	13.271	HMDB00112	46. Cholesterol	27.580	HMDB00067
15. Phosphoric acid	9.859	HMDB02142	31. Creatinine	13.569	HMDB00562			
16. Glycerol	9.872	HMDB00131	32. Unknown 3	14.088	–			

to lactic acid formation and is transferred to neurons as fuel substrate when glucose levels are not enough [11].

3.5 Tips for Minimizing the Analytical Variability: Setting up and Controlling the Experiment

3.5.1 Equipment Maintenance

A regular maintenance of the equipment is essential to avoid bias, contamination, minimize fluctuations in equipment performance and, concurrently, increase its useful life time.

Given that samples are usually a complex matrix, subsequent interactions may result in a change in sensitivity and reproducibility. In addition, aggregates or precipitates can be formed in the GC injector and in the ionization source, respectively. Consequently, preliminary procedures in addition to specific maintenance instructions provided by the manufacturer must be executed before starting any analysis. Prior to any experiment or batch, in the case of large-scale studies, a thorough manual cleaning of the injection syringe with isopropanol and the liner and septum replacements should be performed.

Whether the chromatographic peaks show asymmetry, 10–20 cm of the guard column have to be cut to avoid contamination.

To facilitate subsequent identification of the metabolites, the analysis method must involve retention time locking (RTL). This approach allows for the employment of RTL Databases without distortion of peak elution time and preliminary analyte identification is possible without standards. Fiehn's method has been followed locking the RT of methyl stearate at 19.662 min, by setting the analytical flow in the chromatographic system. Furthermore, commercial mixtures of FAME (Fatty acid methyl esters) and n-alkanes with a wide coverage of elution times are injected prior to the analysis to build both fitting curves of retention index (RI) and retention time (RT) in order to apply Fiehn's and NIST libraries, respectively, for identification.

3.5.2 Analytical Variability

Nontargeted metabolomics entails minimizing analytical variability to attest the detection of the biological one. There are three different sources of variability: the *preanalytical* due to the collection, storage and preparation of samples, the *intra-analytical* as a result of the incorrect functioning of the equipment and the *post-analytical* that might be introduced by the researcher during data processing. A multifunctional autosampler and sample preparation robot is utilized to reduce the main source of preanalytical variability: the silylation and o-methoximation processes enclosed in the sample treatment. Another strategy to monitor the analytical variability is to use quality control (QC) samples, being the best approach, preparing the QCs as a pool of the same set of samples. QC samples are analyzed at the beginning and at the end of an analytical run as well as at regular intervals through the analysis to monitor the stability and reproducibility of the analytical process. Besides the correct maintenance of the equipment, the intra-analytical variability can be monitored by the visual inspection of the chromatograms and the use of blanks, internal standards and QC samples [12, 13] (*see* **Note 9**).

The blanks are included to determine the levels of contamination associated with the manufacture, extraction and analysis of the samples. In GC-MS based metabolomics, blanks will contain all volatile compounds coming from the laboratory air, containers, solvent, reagents, etc., therefore never should be skipped (*see* **Note 10**).

Methyl stearate has been used as the internal standard (IS) by reason of his stability and reproducible and repeatable RT. A solution of the IS was added to each sample at the end of the sample treatment to monitor sample injection and to correct the intrabatch fluctuation of the RT of the metabolites. Besides, the addition of the pentadecanoic acid at the beginning of the sample preparation allows for controlling both the whole sample treatment process and the system performance since it is also influenced by the equipment and analytical conditions.

The QCs are incorporated to investigate the individual metabolites variation within the analysis.

3.5.3 Setting Up the Analysis Sequence

Therefore, according to what have been previously exposed, a sequence of GC/MS analysis starts with several injections of the IS (10 ppm in n-heptane) to assure the equipment is under the optimal conditions to face the experiment. The abundance has to match the commonly obtained value for that concentration and the RT to remain unvarying. The commercial solutions of FAME and n-alkanes are the subsequence analyses in order to change the experimental RT to the RI of Fiehn's and NIST libraries, respectively. Blanks have to be analyzed successively for background subtraction. The equilibration of the equipment is achieved after several injections of one of the QC thus saturation the liner. At this point, the analysis of the samples, cases and controls (ideally in equal number), may start in a randomized order, running as well QCs at regular intervals, commonly, one per five injections. The sequence concludes with blanks, guaranteeing hence that additional contamination that may have emerged during the experiment is considered.

3.5.4 Monitoring the Sequence Run

While the acquisition of these results, the inspection of the chromatograms is essential. Chromatogram checking allows for the detection of discrepant profiles due to different factors such as acquisition problems, failed injection and incorrect derivatization, among others. As GC-MS data has a 3D nature including both *chromatogram* (analytes in the sample are physically separated due to their selective interaction in the analytical column and elute at different retention times) and a *spectrum* (each component separated by their different mass to charge ratios and abundances by the mass analyser and detector respectively).

3.6 Data Treatment

Once the sequence run has ended and the inspection of the chromatograms performed confirming the goodness of the response variability, raw data must be treated. Data treatment includes data preprocessing and data processing to improve signal quality and reduce possible biases present.

3.6.1 Data Preprocessing

Data preprocessing includes all the actions over raw data. This step includes (1) noise reduction, (2) peak detection and deconvolution, (3) identification, and (4) alignment.

Commonly used software integrates the filtering and peak detection algorithms into a single function. Deconvolution is the process of computationally separating coeluting components and creating a pure spectrum for each component by extracting ions from a complex total ion chromatogram (TIC), even with the target compound signal at trace levels. The identification is based on the comparison of both the RT and the mass spectra of the compound already deconvoluted and the ones collected in an in-house, Fiehn's and NIST library (*see* Subheading 3.4).

Spectral alignment (matching peak by *m/z* and RT among all samples) is required in MS-based studies because of the peak shifts observed across the RT axis. Data can be aligned before peak detection or across samples.

3.6.2 Data Processing

That includes all steps after data preprocessing and before the statistical analysis: (1) data integration, (2) normalization, and (3) data filtration.

The abundance of every compound is determined by integration of an abundant, discriminant, and characteristic ion (target ion).

After integration, the resulting abundances have to be normalized at least by the internal standard response.

Data filtering is necessary to discard nonrobust metabolites. It is usually performed by presence, based on the percentage: at least 80% in one of the sample group and by quality assurance criteria, according to the value of the QCs. CV for every metabolic feature detected in the QC samples acts as an indicator of the reproducibility and the repeatability of the analysis. Acceptance of CV limits, lower than 30% will allow for filtrating the complex data matrix and removing metabolites showing unacceptable variation. In addition, multivariate analysis methods such as principal components analysis, PCA can show how QCs cluster in the scores plot or how their spread is a signal of analytical variability.

Summing up the steps followed for data preprocessing and data processing with the corresponding software:

1. Unknowns Analysis Tools of Agilent is used for noise reduction, peak detection, deconvolution, and identification after library search.
2. Mass Profiler Professional (MPP, from Agilent Technologies) is the platform selected for the alignment process.
3. The quantification is performed using Agilent Quantitative Analysis, integrating an ion carefully selected for every potential metabolite to be discriminant enough.
4. The integration must be revise checking the retention time and qualifiers.
5. Normalize with the internal standard by MPP or by Microsoft Excel (Microsoft Office).
6. The filtration is performed by MPP or by Microsoft Excel (Microsoft Office).

This same workflow can be followed using open file platforms, for instance *Workflow4Metabolomics*, W4M. First of all MS files, preferably centroid format (XMCS centWave read only this format), must be converted into an open format, such as .mzML, . mzXML, .mzData, and .cdf. If the commercial software (usually supplied with the equipment) does not include an exporting tool, other open source software as *ProteoWizard* (http://proteowizard. sourceforge.net/) can be used for this raw data conversion. The visual inspection of the chromatograms can be performed by the software provided by the MS Instrument supplier but there are also open source alternatives, such as *InsilicosViewer*, a free viewer for displaying mzXML MS data with most of the functionalities offered by its commercial counterparts. This tool includes mzXML/ mzData reading software developed by the Institute of System Biology (http://www.systemsbiology.org).

Then for the preprocessing of the data, they have to be uploaded with GALAXY interface. There are two methods for data upload: with a set of single files (recommended as you will be able to launch several xcmsSet in parallel and use "xcms.xcmsSet Merger" before "xcms.group" or creating first a .zip file, containing all your conditions as sub-directories.

Deconvolution is performed running the R-package MetaMS, specifically MetaMS.RunGC, it includes XCMS peak detection, definition of pseudospectra, and compound identification by comparison to a database of standards. The annotation can be executed using in-house databases or by a managed service provider (MSP), PeakSpectra.MSP that exports the deconvoluted compound readable by NIST or Golm Metabolome.

4 Notes

1. Keep inside a desiccator at room temperature the reagents BSTFA + TMCS (99:1) and *O*-methoxyamine hydrochloride to prevent humidification, which impairs derivatization.
2. *O*-methoxyamine and BSTFA:TMCS are severe irritant agents; and pyridine is an extremely hazardous reagent. Take the appropriate precautions and work always under the chemical fume hood.
3. It is highly recommended whenever it is possible, to analyze always the same brain region to avoid region-specific metabolic differences. If not, analyze the whole brain hemisphere taking into account that maybe some metabolites only present in one region might be diluted.
4. It is crucial to assure that the vials are completely dry after using the speedvac to bypass the hydrolysis of the derivatization agents employed in the next sample treatment steps since their efficiency relies on the preservation of an anhydrous environment.
5. The C18:0 methyl stearate, which is added at the end of the sample treatment prior to the GC-MS analysis, ensures the system stability and the analytical process together with the QC samples. Therefore, it allows for adjusting systematic errors that may affect the analytical accuracy, besides variations in the exact volume of sample introduced in the chromatographic system. On the other hand, the addition of the pentadecanoic acid at the beginning of the sample preparation allows for controlling both the whole sample treatment process and the system performance since it is also influenced by the equipment and analytical conditions. Consequently, the addition of the IS at the beginning of sample preparation is recommended whenever is possible.
6. The final centrifugation step guarantees that any solid suspension material presented in the vial will be placed at the bottom of the insert and will not enter into the system when injecting the samples.
7. The amount of injected sample should be commensurate, because if it is overly elevated, the peaks will show a significant tailing, causing a poorer separation. This parameter can be easily controlled with the split mode. In addition, most detectors are relatively sensitive and do not need a lot of sample volume to produce a detectable signal.
8. The application of a high-energy electron impact (70 eV) will provide an extensive and reproducible fragmentation of the molecules. This will offer clues about the molecular structure,

which can be later identified by matching the spectrum of the unknown metabolite to the spectra of a commercial (Fiehn or NIST) or in-house spectral library.

9. The use of quality control samples (QCs) together with the internal standards is highly recommended to monitor sources of variation affecting the feasibility of the GC-MS analysis. Prepare the QCs by pooling equal volumes from each sample of the study. Then, treat them identically to the rest of the study samples in order to obtain representative QCs of the study. Analyze the QCs intermittently throughout the analytical run, between 5 to 10 sample injections, depending of the total number of samples in the study.
10. Prepare at least two blank solutions along with the rest of the samples. Analyze them at the beginning and at the end of the worklist to identify undesirable contaminants and artifact features.

Acknowledgement

Authors want to express their gratitude to the financial support received from the FEDER Program 2014-2020 of the Community of Madrid (Ref. S2017/BMD3684) and the Spanish Ministry of Science, Innovation and Universities (RTI2018-095166-B-I00).

References

1. Osborne N, Chidlow G, Layton C, Wood J, Casson R, Melena J (2004) Optic nerve and neuroprotection strategies. Eye 18:1075–1084
2. Naz S, Garcia A, Barbas C (2013) Multiplatform analytical methodology for metabolic fingerprinting of lung tissue. Anal Chem 85:10941–10948
3. Gonzalez-Riano C, Tapia-González S, García A, Muñoz A, DeFelipe J, Barbas C (2017) Metabolomics and neuroanatomical evaluation of post-mortem changes in the hippocampus. Brain Struct and Funct 222:2831–2853
4. Gonzalez-Riano C, León-Espinosa G, Regalado-Reyes M, García A, DeFelipe J, Barbas C (2019) Metabolomic study of hibernating Syrian hamster brains: in search of Neuroprotective agents. J Proteome Res 18:1175–1190
5. Naz S, Moreira dos Santos DC, Garcia A, Barbas C (2014) Analytical protocols based on LC-MS, GC-MS and CE-MS for nontargeted metabolomics of biological tissues. Bioanalysis 6:1657–1677
6. Smiljanic K, Vanmierlo T, Djordjevic AM, Perovic M, Ivkovic S, Lütjohann D, Kanazir S (2014) Cholesterol metabolism changes under long-term dietary restrictions while the cholesterol homeostasis remains unaffected in the cortex and hippocampus of aging rats. Age 36:9654
7. Hering H, Lin C-C, Sheng M (2003) Lipid rafts in the maintenance of synapses, dendritic spines, and surface AMPA receptor stability. J Neurosci 23:3262–3271
8. Sorra KE, Harris KM (2000) Overview on the structure, composition, function, development, and plasticity of hippocampal dendritic spines. Hippocampus 10:501–511
9. Bourne JN, Harris KM (2008) Balancing structure and function at hippocampal dendritic spines. Annu Rev Neurosci 31:47–67
10. Fisher SK, Novak JE, Agranoff BW (2002) Inositol and higher inositol phosphates in neural tissues: homeostasis, metabolism and functional significance. J Neurochem 82:736–754
11. Proia P, Di Liegro CM, Schiera G, Fricano A, Di Liegro I (2016) Lactate as a metabolite and

a regulator in the central nervous system. Int J Mol Sci 17:1450

12. Mastrangelo A, Ferrarini A, Rey-Stolle F, Garcia A, Barbas C (2015) From sample treatment to biomarker discovery: a tutorial for untargeted metabolomics based on GC-(EI)-Q-MS. Anal Chim Acta 900:21–35
13. Dudzik D, Barbas-Bernardos C, García A, Barbas C (2017) Quality assurance procedures for mass spectrometry untargeted metabolomics. A review. J Pharm Biomed Anal 147:149–173

Chapter 17

GC-MS of Pentafluourobenzyl Derivatives of Phenols, Amines, and Carboxylic Acids

Paul L. Wood

Abstract

The pentafluorobenzyl (PFB) derivatives of phenols, amines, and carboxylic acids demonstrate excellent electron capture properties for detection by negative chemical ion (NCI) mass spectrometric detection, combined with superior gas chromatographic (GC) separation due to their volatility. NCI-GC-MS quantitation of PFB derivatives is applicable to a variety of biological matrices.

Key words Negative ion mass spectrometry, Metabolomics, Pentafluorobenzyl, Amines, Carboxylic acids

1 Introduction

Analytical platforms for amines and carboxylic acids are valuable for both nontargeted metabolomics studies and targeted analyses [1]. In the case of targeted analyses the availability of analytical and stable isotope standards for these biomolecules allows for the development of absolute quantitation assays. In our experience the PFB derivatives of amines and carboxylic acids are valuable for the following reasons:

- The derivatization protocol is simple and robust.
- The derivatives are volatile and demonstrate excellent separation by gas chromatography.
- The pentafluro substitution is amenable to NCI-MS detection with high sensitivity.
- The derivatization procedure results in the addition of -CH_2-C_6F_5 (+180) for each -SH, Ar-OH, -COOH, -NH_2, and -NH functional group.

Paul L. Wood (ed.), *Metabolomics*, Neuromethods, vol. 159, https://doi.org/10.1007/978-1-0716-0864-7_17,

This derivatization is one of the most sensitive for free fatty acids [2], amino acids [3], and amines [4]. We have built an in-house database in Excel (Microsoft) of PFB derivatives. A number of key metabolites are presented in Table 1.

2 Materials

2.1 Buffers and Standards

- Extraction Buffer: acetonitrile–methanol–formic acid (800:200:2.5).
- Standards and Internal Standards are dissolved in ACN–MeOH–6 N-HCl (1:1:0.2).
- Pentafluorobenzyl bromide solution: (50 μL PFBB +950 μL dimethylformamide).
- Diisopropylethylamine.
- Stable isotope standards are purchased from CDN Isotopes and Cambridge Isotope Labs.

2.2 Analytical Equipment and Supplies

- GC-MS or GC-MS/MS.
- The GC column is a 30 m HP-5MS (0.25 mm ID; 0.25 μm film).
- Reagent gas: 10% ammonia in methane.
- Centrifugal dryer (Eppendorf Vacufuge Plus).
- Refrigerated microfuge capable of 30,000 xg (Eppendorf 5430R).
- Sonicator (Thermo Fisher FB50).
- 1.5 mL screw-top microfuge tubes (Thermo Fisher).

3 Methods

3.1 Sample Preparation

- Plasma or serum (100 μL), in 1.5 mL microfuge tubes, is vortexed with 1 mL of cold Extraction Buffer and 50 μL of the Internal Standard solution.
- Tissues (500–600 mg), in 1.5 mL microfuge tubes, are sonicated in 1 mL of cold Extraction Buffer and 50 μL of the Internal Standard solution.
- Next samples are centrifuged at 30,000 × *g* and 4 °C for 30 min.
- 750 μL of the supernatant is transferred to a clean 1.5 mL microfuge tube.
- The samples are next dried by vacuum centrifugation.

Table 1
PFB derivatives

Metabolite [Int. Std.]	MW	180	Deriv.	Calc. $[M-H]^-$	Calc. $[M-181]^-$	RT (min)	Obs. Anion (Int. Std.)
Geranoic acid	168	1	348	347	167	5	167
p-cresol	108	1	288	287	107	5.33	107
3-Methylthiopropionic acid	120	1	300	299	119	5.37	119
2-ethylphenol	122	1	302	301	121	5.5	121
Octanoic acid	144	1	324	323	143	5.52	143 (146)
Guaiacol [D4]	124	1	304	303	123	5.58	123 (127)
N-acetyl-alanine [D4]	131	1	311	301	130	5.6	130 (134)
4-ethylphenol	122	1	302	301	121	5.75	121
N-acetyl-glycine	117	1	297	296	116	5.78	116
Phenylacetic acid [D5]	136	1	316	315	135	5.81	135 (140)
Nicotinic acid [D4]	123	1	303	302	122	5.88	122 (126)
N-Ac-β-alanine	131	1	311	310	130	5.89	130
Thymol [D13]	150	1	330	329	149	5.9	149 (162)
Picolinic acid [D4]	123	1	303	302	122	5.97	122 (126)
Citronellic acid	170	1	350	349	169	6.0	169
Capric acid	172	1	352	351	171	6.27	171
Alanine [D4]	89	2	449	448	268	6.3	268 (272)
Dimethylglycine [D6]	103	1	283	282	102	6.38	282 (288)
Sarcosine [D3]	89	1	449	448	268	6.47	268 (271)
4-acetamidobutyric acid	145	1	325	324	144	6.6	144
Valine	117	2	477	476	296	6.67	296
β-Alanine [D4]	89	2	449	448	268	6.8	268 (272)
Norvaline	117	2	477	476	296	6.82	296
Cinnamic acid	148	1	328	327	147	6.65	147
Leucine [D3]	131	2	491	490	310	6.9	310 (313)
Lauric acid [D5]	200	1	380	379	199	6.94	199 (204)
Ibuprofen [D3]	206	1	386	385	205	6.95	205 (208)
Isoleucine	131	2	491	490	310	6.95	310
Norleucine	131	2	491	490	310	7.09	310
GABA [D6]	103	2	463	462	282	7.1	282

(continued)

Table 1
(continued)

Metabolite [Int. Std.]	MW	180	Deriv.	Calc. $[M-H]^-$	Calc. $[M-181]^-$	RT (min)	Obs. Anion (Int. Std.)
Succinic acid [D4]	118	2	478	477	297	7.1	297 (301)
Proline [D7]	115	2	295	294	114	7.12	294 (301)
N-acetyl-methionine [D6]	191	1	371	370	190	7.16	190 (196)
Glycinamide	74	2	434	433	253	7.17	394 [M-40]
N-acetyl-alanine	147	2	507	506	326	7.2	326
N-acetyl-aspartic acid [D3]	175	2	535	534	354	7.27	354 (357)
Pipecolic acid [D9]	129	1	309	308	128	7.3	308 (317)
O-acetyl-serine	147	2	507	506	326	7.31	326
4-thiaisoleucine	149	2	509	508	328	7.5	328
Resorcinol [D6]	110	2	470	469	289	7.59	289 (293)
Hippuric acid	179	1	359	358	178	7.74	178
Methionine [D4]	149	2	509	508	328	7.79	328 (332)
Phenyl-acetyl-glycine	193	1	373	372	192	7.79	192
Pyroglutamate [D5]	129	2	489	488	308	7.79	308
Farnesoic acid	236	1	416	415	235	7.8	235
3-methyl-adipic acid	160	2	520	519	339	7.8	339
Hypotaurine [D4]	109	2	469	468	288	7.86	288 (292)
Adipic acid	146	2	506	505	325	7.87	325
Salicylic acid [D4]	138	2	498	497	317	7.88	317 (321)
Hippuric acid	179	1	359	358	178	7.94	178
Uracil [N2]	112	1	472	471	291	8.0	291 (293)
n-acetyl-putrescine	130	2	490	489	309	8.02	430 [M-60]
Thymine [D4]	136	2	486	485	305	8.04	305 (309)
Lipoic acid	206	1	386	385	205	8.1	205
Phenylalanine [D5]	165	2	525	524	344	8.1	344 (349)
Palmitic acid [C16]	256	1	436	435	255	8.12	255 (271)
Cysteamine [D4]	77	3	617	616	436	8.18	577 (581) [M-40]
Prenylcysteine [D4]	189	2	549	548	368	8.36	368 (372)
3-methoxytyramine [D4]	167	2	527	526	346	8.4	507 (511) [M-20]

(continued)

Table 1 (continued)

Metabolite [Int. Std.]	MW	180	Deriv.	Calc. [M-H]⁻	Calc. [M-181]⁻	RT (min)	Obs. Anion (Int. Std.)
Putrescine [D4]	88	3	628	627	447	8.43	568 (572) [M-60]
10-methylhexadecanoic acid	270	1	450	449	269	8.48	269
Quinolinic acid	167	2	527	526	346	8.48	346
Phytanic acid	312	1	492	491	311	8.63	311
Vanillic acid	168	2	528	347	346	8.66	347
Adenine	135	2	495	494	314	8.7	455 [M-40]
Aspartic acid [D3]	133	3	673	672	492	8.7	492 (495)
Stearic acid [C18]	284	1	464	463	283	8.7	283 (301)
Cysteine [D4]	121	3	661	660	480	8.8	480 (484)
Glutamic acid [D5]	147	3	687	686	505	9.0	506 (511)
p-Aminobenzoic acid	137	2	497	496	316	9.01	316
Arachidonic acid	304	1	484	483	303	9.14	303
Aminolevulinic acid [D4]	131	3	671	670	490	9.14	490 (494)
Eicosapentanoic acid	302	1	482	481	301	9.17	301
Gabapentin [D4]	171	3	711	710	530	9.35	530 (534)
N-methyl-lysine	160	2	520	519	339	9.39	519
Aminoadipic acid [D4]	161	3	701	700	520	9.4	520 (523)
p-coumaric acid	164	2	524	523	343	9.47	343
Citric acid [D4]	192	3	732	731	551	9.52	551 (555)
2-methylcitric acid [D3]	206	3	746	745	565	9.5	565 (568)
Cytosine	111	2	471	470	290	9.65	470
Cycothionine	205	3	745	744	564	9.69	564
Docosahexaenoic acid	328	1	508	507	327	9.87	327
Kyneurenic acid [D5]	189	2	549	548	368	9.86	368 (373)
Ferulic acid [D3]	194	2	554	553	373	10.0	373 (376)
Behenic acid [D4]	340	1	520	519	339	10.06	339 (343)
Dopamine [D4]	153	3	693	692	512	10.1	667 (671)
Tryptophan [D5]	204	2	564	563	383	10.12	383 (387)
Histidine [D3]	155	2	515	514	334	10.2	514 (517)

(continued)

Table 1
(continued)

Metabolite [Int. Std.]	MW	180	Deriv.	Calc. [M-H]$^-$	Calc. [M-181]$^-$	RT (min)	Obs. Anion (Int. Std.)
Phenyl-acetyl-glutamine	264	2	624	623	443	10.46	443
Ornithine [D4]	132	3	672	671	491	10.5	671 (677)
Tricosanoic acid	354	1	534	533	353	10.5	353
Tyrosine [D7]	181	3	721	720	540	10.6	540 (547)
Cysteinyl-glycine	178	3	718	717	537	10.89	537 and 497 [−40]
Nervonic acid	366	1	546	545	365	10.95	365
Lignoceric acid [D4]	368	1	548	547	367	11.0	367 (371)
N-acetyl-spermidine	187	3	727	726	546	11.0	546 and 506 [−40]
Lysine [D4]	146	3	686	685	505	11.1	685 (689)
Hexadecanediooic acid [D28]	286	2	646	645	465	11.2	465 (493)
Dihydrolipoic acid	208	3	748	747	567	11.77	567 and 547 [−40]
Farnesylcysteine	325	2	685	684	504	11.77	504
Cerebronic acid	384	1	564	563	383	11.9	383
Thialysine	164	3	704	703	523	11.99	703
Cystathionine [D4]	222	4	942	941	761	14.6	761 (765)
Cystine [D4]	240	4	960	959	779	15.7	779 (783)

MW molecular weight, *Deriv.* PFB derivative, *RT* retention time, *Int. Std.* internal standard

3.2 Derivatization

- To the dried samples are added 50 μL of pentafluourobenzyl bromide solution and 10 μL of diisopropylethylamine as catalyst.
- The reaction is conducted with shaking at 80 °C for 1 h and then vortexed with 200 μL of hexane–ethyl acetate (3:2).
- The tubes are next centrifuged at 30,000 × g for 10 min to precipitate salts.
- The supernatants are transferred to autosampler vials for GC-MS analyses.

3.3 GC-NCI-MS

- The PFB derivatives are analyzed using 10% ammonia in methane as the reagent gas for NCI-GC-MS.
- The GC column is held at 100 °C for 1 min. Followed by a gradient of 30 °C/min, with a helium flow of 1.2 mL/min.

- The $[M-181]^{-}$ anions predominate for most biomolecules. However the molecular anion predominates in some cases, while some metabolites are characterized by the loss of 1 to 3 HF (−20) groups.
- The syringe is flushed between samples with acetonitrile–methanol (1:1).

3.4 Data Reduction

- There currently is no database for PFB derivatives such that in-house construction is required (Table 1).
- For absolute quantitation, the ratio of the peak area for the biological is divided by the peak area for its stable isotope internal standard and this ratio applied to a standard curve of at least 7 points.

4 Notes

1. All organic solvents should be of LC-MS grade.
2. The extraction buffer must be cold to stabilize metabolites from further metabolism.
3. Deuterated internal standards run slightly faster on the GC column than the endogenous compounds due to decreased hydrogen bonding. With more deuterium substitutions the greater this effect. ^{13}C and ^{15}N analogs are superior since they have identical retention times to the endogenous metabolites.
4. Very-long-chain fatty acids (≥24 carbons) demonstrate a large GC column ghost effect.

References

1. Wood PL (2014) Mass spectrometry strategies for targeted clinical metabolomics and lipidomics in psychiatry, neurology, and neuro-oncology. Neuropsychopharmacology 39:24–33
2. Quehenberger O, Armando AM, Dennis EA (2011) High sensitivity quantitative lipidomics analysis of fatty acids in biological samples by gas chromatography-mass spectrometry. Biochim Biophys Acta 1811:648–656
3. Smith T, Ghandour MS, Wood PL (2011) Detection of N-acetyl methionine in human and murine brain and neuronal and glial derived cell lines. J Neurochem 118:187–194
4. Wood PL, Khan MA, Smith T, Goodenowe DB (2011) Cellular diamine levels in cancer chemoprevention: modulation by ibuprofen and membrane plasmalogens. Lipids Health Dis 10:214

Chapter 18

GC-MS of *tert*-Butyldimethylsilyl (tBDMS) Derivatives of Neurochemicals

Paul L. Wood

Abstract

tBDMS derivatization, prior to GC-MS, is extremely versatile in that a wide variety of functional groups are derivatized. These include –COOH, $–NH_2$, –SH, $–SO_2H$, –OH, –C=O, and –NH–OH groups. The advantages of this GC-MS methodology include: (a) simple sample preparation; (b) a single derivatization step; (c) direct GC-MS analysis of the reaction mix; (d) high precision as a result of isotopic dilution analyses; (e) high sensitivity and specificity as a result of strong $[MH]^+$ ions with ammonia reagent gas; (f) no hydrolysis of glutamine to glutamate or asparagine to aspartate; (g) stability of the derivatives to aqueous contamination; and (h) applicability to a wide range of neurochemicals.

With 10% ammonia in methane as the reagent gas, all amino acids, polyamines, and urea yield strong $[MH]^+$ ions with little or no fragmentation. In the case of carboxylic acids, $[M + 18]^+$ ions predominate.

Key words Positive ion mass spectrometry, Metabolomics, tBDMS, Amino acids

1 Introduction

Analytical platforms for amino acids are valuable for both nontargeted metabolomics studies and targeted analyses [1]. In the case of targeted analyses the availability of analytical and stable isotope standards for these biomolecules allows for the development of absolute quantitation assays. We have utilized the tBDMS derivatization procedure to monitor amino acid release in brain slices and tissue culture polyamine levels [2]. In our experience the tBDMS derivatives of neurochemicals are valuable for the following reasons:

- The derivatization protocol is simple and robust.
- The derivatives are volatile and demonstrate excellent separation by gas chromatography and optimal detection with ammonia positive chemical ionization (PCI) GC-MS.
- The derivatization procedure results in the addition of a tBDMS group (+114) to each –COOH, $–NH_2$, –SH, $–SO_2H$, and –OH

Paul L. Wood (ed.), *Metabolomics*, Neuromethods, vol. 159, https://doi.org/10.1007/978-1-0716-0864-7_18,

functional group. In addition, –C=O (keto) and –NH–OH (hydroxylamine) functional groups are derivatized.

- The tBDMS derivatives, in contrast to TMS derivatives, are stable in aqueous media, thereby limiting their degradation during refrigerated storage.

This derivatization provides a sensitive PCI assay for amino acids, carboxylic acids, and keto compounds [1]. We have built an in-house database in Excel (Microsoft) of tBDMS derivatives. A number of key metabolites are presented in Table 1.

2 Materials

2.1 Buffers and Standards

- Extraction Buffer: acetonitrile–methanol–formic acid (800:200:2.5) (*see* **Notes 1** and **2**).
- Standards and Internal Standards are dissolved in ACN–MeOH–6 N HCl (1:1:0.2).
- *N*-(*tert*-butyldimethylsilyl)-*N*-methyltrifluroacetamide with 1% *tert*-butyldimethylchlorosilane.
- Acetonitrile.
- Stable isotope standards are purchased from CDN Isotopes and Cambridge Isotope Labs (*see* **Note 3**).
- Custom-synthesized stable isotope variants of dipeptides are purchased from Creative Peptides.

2.2 Analytical Equipment and Supplies

- GC-MS or GC-MS/MS.
- The GC column is a 30 m HP-5MS (0.25 mm ID; 0.25 μm film).
- Reagent gas: 10% ammonia in methane.
- Dry block heater with shaking.
- Centrifugal dryer (Eppendorf Vacufuge Plus).
- Refrigerated microfuge capable of 30,000 × *g* (Eppendorf 5430R).
- Sonicator (Thermo Fisher FB50).
- 1.5 mL screw-top microfuge tubes (Thermo Fisher).

3 Methods

3.1 Sample Extraction

- Plasma or serum (100 μL), in 1.5 mL microfuge tubes, is vortexed with 1 mL of cold Extraction Buffer and 50 μL of the Stable Isotope Internal Standard solution.

Table 1
Cations for tBDMS derivatized neurochemicals

Compound	MW	# of 114	Deriv.	RT	SIM $[MH]^+$ (Int.Std.)
Sarcosine	89	1	203	3.32	204
N-acetyl-β-alanine	131	1	245	3.39	246
Dimethylglycine	103	1	217	3.54	218
N-acetyl-putrescine	130	1	244	3.58	245
3-acetyl-1-propanol	102	1	216	3.8	217
2-pyrrolidone	85	1	199	4.2	200
Pipecolic acid	129	1	243	4.6	244
Picolinic acid	123	1	237	4.85	238
N-benzyl-hydroxylamine	123	1	237	4.76	238
Alanine [D4	89	2	317	4.8	318 (322)
Ethanolamine [D4]	61	2	289	4.87	290 (294)
Lactate [D3]	90	2	318	4.9	319 (322)
2-amino-isobutyric acid	103	2	331	5	332
3-amino-1-propanol	75	2	303	5.1	304
Norleucine	130	2	358	5.16	359
Ketoisovaleric acid	116	2	344	5.2	345
β-alanine [D4]	89	2	317	5.26	318 (322)
Leucine [D3]	131	2	359	5.3	360 (363)
Cysteamine	77	2	305	5.37	306
N-acetyl-methionine [D6]	191	1	305	5.39	306 (312)
Norleucine	131	2	359	5.42	360
4-amino-1-butanol	89	2	317	5.45	318
Isoleucine	131	2	359	5.47	360
Urea [C2N]	60	2	288	5.5	289 (292)
Uracil [D2]	112	2	340	5.69	341 (343)
Succinic acid	118	2	346	5.7	347
Isoputreanine	160	2	388	5.73	389
Oxoisoleucine	130	2	358	5.8	359
GABA [D6]	103	2	331	5.8	332 (338)
Putrescine [D4]	88	2	316	5.85	317 (321)
Aspartic acid [D3]	133	3	475	5.9	476 (479)

(continued)

Table 1 (continued)

Compound	MW	# of 114	Deriv.	RT	SIM $[MH]^+$ (Int.Std.)
Proline [D7]	115	2	343	5.9	344 (361)
Leucinamide	130	2	358	5.89	359
Thymine	126	2	354	5.98	355
Glutaric acid [D4]	132	2	360	6	361(365
Isoputreanine lactam	142	1	256	6.1	257
Cytosine	111	2	339	6.38	340
Adipic acid	146	2	374	6.4	375/392
Pyroglutamate	129	2	357	6.4	358/375
Adenine	135	1	249	6.5	250
N-methyl-lysine	160	2	388	6.59	389
Threonine	119	3	461	6.6	462
Serine [C2N]	105	3	447	6.7	448 (351)
Methionine [D4]	149	2	377	6.8	378 (382)
N-acetyl-cysteine	163	2	391	6.98	392
Ketoadipate (Trp, Lys)	160	3	502	6.9	502
α-ketoglutarate	146	3	488	7.2	489/506
Spermidine	145	2	373	7.2	374
Hydroxyproline	131	3	463		474
Cysteine	121	3	463	7.18	464 (468)
Cystamine	152	2	380	7.2	381
3-methoxytyramine [D4]		2	395	7.45	396 (400)
Glutamic acid [D5]	147	3	489	7.5	490 (495)
Ornithine [D4]	132	3	474	7.37	475 (481)
Homocysteine	135	3	477	7.46	478
Lysine	146	3	488	7.66	489

- Tissues (500–600 mg), in 1.5 mL microfuge tubes, are sonicated in 1 mL of cold Extraction Buffer and 50 μL of the Stable Isotope Internal Standard solution.
- Next samples are centrifuged at 30,000 × g and 4 °C for 30 min.
- 750 μL of the supernatant is transferred to a clean 1.5 mL microfuge tube.
- The samples are next dried by vacuum centrifugation.

3.2 Ion Exchange Purification

- Most extracts can be derivatized and analyzed directly. However, in cases where amino acids require some level of purification we utilize cation exchange chromatography.
- Tissue culture media, tissue superfusates, CSF, or plasma samples are transferred to 18 mm × 100 mm glass tubes containing appropriate stable isotope internal standards and mixed with 0.5 mL of cation-exchange resin, Dowex AG 50 W-X8 (200–400 mesh, hydrogen form, Bio-Rad, Hercules, CA).
- The tubes are shaken for 5 min, the resin allowed to settle, and the supernatant aspirated.
- The resin was next washed three times via brief vortexing with 4 mL of water. The washes are aspirated each time after the resin settles.
- The acids and polyamines are eluted with 1 mL of 8 N NH_4OH.
- 700 μL is transferred to 1.5 mL screw-top microfuge tubes and dried overnight in a Savant concentrator.

3.3 Derivatization

- To the Savant-dried samples are added 75 μL of acetonitrile and 75 μL of MTBSTFA containing 1% *t*-butyldimethylchlorosilane (TBDMS-Cl) as a catalyst.
- The tubes are capped and heated at 80 °C for 1 h in a dry block with shaking.
- The samples are centrifuged at 30,000 × *g* for 10 min in a microfuge to precipitate any salts.
- The supernatants are transferred to autosampler vials for GC–MS analyses.

3.4 GC-NCI-MS

- The GC column is held at 100 °C for 1 min and then placed on a gradient of 30 °C per min to 300 °C with carrier gas (He) flow of 1.2 mL/min.
- The injector, interface, and source temperatures are 250 °C, 320 °C, and 150 °C, respectfully.
- The tBDMS derivatives are analyzed using 10% ammonia in methane as the reagent gas for PCI-GC-MS.
- Amino acids, polyamines, and urea yield strong $[MH]^+$ ions with little or no fragmentation. In the case of carboxylic acids, $[M + 18]^+$ ions predominate.
- The syringe is flushed between samples with acetonitrile–methanol (1:1).

3.5 Data Reduction

- There currently is no database for t-BDMS derivatives such in-house construction is required (Table 1).
- For absolute quantitation, the ratio of the peak area for the biological is divided by the peak area for its stable isotope internal standard and this ratio applied to a standard curve of at least 7 points.

4 Notes

1. All organic solvents should be of LC-MS grade.
2. The extraction buffer must be cold and kept cold to stabilize metabolites from further metabolism.
3. Deuterated internal standards run slightly faster on the GC column than the endogenous compounds due to decreased hydrogen bonding. The more deuterium substitutions the greater this effect. ^{13}C and ^{15}N analogs have identical retention times to the endogenous metabolites.

References

1. Wood PL, Khan AK, Moskal JK (2006) Neurochemical analysis of amino acids, polyamines and carboxylic acids: GC-MS quantitation of tBDMS derivatives using ammonia positive chemical ionization. J Chromatogr B 831:313–319
2. Wood PL, Khan MA, Kulow SR, Mahmood SA, Moskal JR (2006) Neurotoxicity of reactive aldehydes: the concept of "aldehyde load" as demonstrated by neuroprotection with hydroxylamines. Brain Res 1095:190–199

Chapter 19

UHPLC-MS/MS Method for Determination of Biologically Important Thiols in Plasma Using New Derivatizing Maleimide Reagent

Dominika Olesova, Andrej Kovac, and Jaroslav Galba

Abstract

Although direct quantification of reactive oxygen species (ROS) is challenging, their interacting partners can readily be used for assessment of redox status in living organisms. Biothiols are molecules directly involved in scavenging of ROS, and their levels reflect the amount of oxidative burden. However, even analysis of these molecules has its obstacles. Due to their high reactivity, proper derivatization protocol is needed. Here, we present a protocol for LC-MS/MS analysis of the four most significant thiols in plasma using new N-phenylmaleimide derivatization reagent. Both analytical method and sample preparation were thoroughly optimized to provide a method that is fast and straightforward and easily applicable in biomedical research.

Key words Thiols, Liquid chromatography, Mass spectrometry, Derivatization, N-phenylmaleimide

1 Introduction

Biologically active thiols are sulfur-containing molecules important in maintaining redox homeostasis and detoxication of dangerous side products and xenobiotics. They also play a crucial regulatory role in nitric oxide and prostaglandin metabolism [1]. Redox homeostasis is carefully regulated, but under pathological conditions it is usually compromised. Metabolites indicating these alterations can be readily detected in biofluids and quantification of oxidative stress markers could be useful for tracking the progression of diseases [2]. Many diseases are connected with some level of oxidative stress such as cardiovascular diseases [3–5], diabetes mellitus [6], chronic kidney disease [7], neurodegenerative diseases [8–10], and cancer in general [11]. Due to their high reactivity and short half-life, direct quantification of reactive oxygen species (ROS) is challenging. Thus, the focus is rather directed to stable molecules that are generated by interaction with ROS

Paul L. Wood (ed.), *Metabolomics*, Neuromethods, vol. 159, https://doi.org/10.1007/978-1-0716-0864-7_19,

[12]. Although ROS reacts with all molecules, their major targets are unsaturated lipids and thiols. Moreover, thiols serve as cofactors for enzymes metabolizing the oxidized lipids. This makes thiols both targets and the defense system against oxidative stress [13]. The most abundant thiol in cells is tripeptide glutathione (GSH). Besides its role as ROS scavenger and catabolism of toxic compound, it is also involved in protein and DNA synthesis. Cysteinyl-glycine is a product of glutathione catabolism by enzyme γ-glutamyltransferase [14, 15]. Thiol group–containing amino acids cysteine and homocysteine are essential components of metabolism on many levels. Homocysteine is produced from methionine by demethylation. The connection between homocysteine and endothelial dysfunction has been found. Also, it is recognized as a risk factor for cardiovascular diseases and correlates with cognitive decline in dementia patients [16, 17].

The quantification of biothiols can be helpful in the determination of the relationship between redox status and ongoing pathology. Over the past 20 years, several methods have been developed using different separation and detection techniques. Most used separation methods are liquid chromatography and capillary electrophoresis [18–21]. For detection of thiols mostly fluorescence detector (FLD) [22, 23] or mass spectrometer (MS) is used [18, 24, 25]. Although florescence detection provides an easy and sensitive approach, low selectivity, long derivatization procedures, and instability of fluorescent reagents favor the use of mass spectrometry. Due to their high reactivity, appropriate derivatization is needed to stabilize thiol groups [10]. Many of the reagents used in FLD can also be used for MS detection (i.e., maleimides, SBD-F, ABD-F, monobromobimane) [2, 18, 22, 26, 27]. Here, we present a protocol for selective and sensitive determination of four aminothiols, namely, cysteine (Cys), homocysteine (Hcy), cysteinylglycine (CG), and glutathione (GSH) in plasma. The proposed method can also be adapted for other compounds containing thiol group which widens its application range in biomedical analysis (Fig. 1).

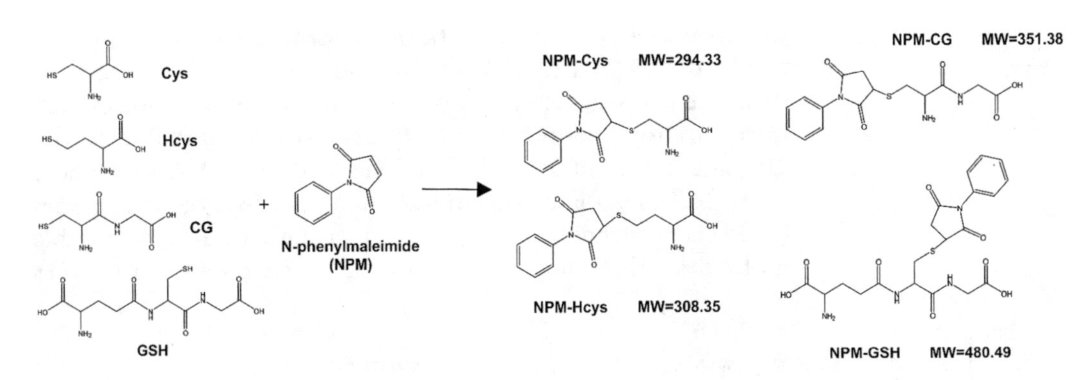

Fig. 1 Derivatization scheme of thiols with N-phenylmaleimide derivatization reagent

2 Materials

All solutions for sample preparation are prepared using LC-MS grade reagents and deionized water.

Analytical standards L-cysteine, homocysteine, cysteinyl-glycine are purchased from Sigma-Aldrich, glutathione from FLUKA. Isotopically labeled internal standards D3-cysteine, D4-homocysteine and 2-^{13}C, ^{15}N-glutathione are purchased from Cambridge Isotope Laboratories and D2-cysteinyl-glycine from EZBiolab.

2.1 Derivatization Procedure Reagents

1. *Reduction reagent: tris(2-carboxyethyl)phosphine hydrochloride (TCEP) 10 g/L:* Dissolve 0.1 g of TCEP in 10 mL of deionized water. Prepare 200 μL aliquots. Store at −20 °C (*see* **Note 1**).
2. *Trichloroacetic acid (TCA) 2 mol/L:* Dissolve 32.6 g of TCA in 100 mL of deionized water. Mix well. Store at room temperature (RT).
3. *Derivatization reagent—N-phenyl maleimide (NPM) 20 mmol/L:* Dissolve 3.46 mg NPM in 1 mL of acetonitrile (*see* **Note 2**).
4. *Potassium phosphate dibasic (K_2HPO_4) 0.5 mol/L:* Dissolve 8.7 g K_2HPO_4 in 100 mL of deionized water. Mix well. Store at RT (*see* **Note 3**).

2.2 Solid Phase Extraction Materials

1. Oasis PRiME HLB 96-well Plate, 30 mg sorbent per well (*see* **Note 4**) 2 mL 96-well collection plate and waste plate.
2. Positive Pressure-96 Processor (Waters).
3. *Solvents*: deionized water, 5% methanol, 100% methanol, 80% acetonitrile.

2.3 LC Solvents

1. *A stock solution of ammonium formate (AF) 1 mol/L:* Dissolve 6.305 g of ammonium formate in 100 mL of deionized water. Mix well by stirring with a magnetic bar. Filter through 0.2 μm filter and store at 4–8 °C (*see* **Note 5**).
2. *Mobile phase A—50 mmol/L ammonium formate + 0.25% formic acid:* Prepare mobile phase A by diluting 20 volumes of the AF (1 mol/L) with 380 volumes of deionized water and add 1 volume of formic acid. Mix well. Filter through 0.2 μm filter and store at 4 °C (*see* **Note 6**).
3. *Mobile phase B*—100% acetonitrile.
4. *Wash solvent:* 50% acetonitrile (*see* **Note 7**).

2.4 LC-MS Instrumentation

1. *Chromatographic apparatus:* ACQUITY UPLC I-class system (Waters) with a binary gradient pump, autosampler, and column thermostat.
2. *Column*: UPLC BEH Amide 2.1 × 100 mm, 1.7 μm.
3. *Mass spectrometer:* XEVO TQD triple-quadrupole mass detector (Waters) with Electrospray ion source (ESI).

2.5 Data Acquisition and Processing

1. Data are acquired under the control of MassLynx v4.1 software.
2. Quantification processing is performed using TargetLynx SE software.

3 Methods

3.1 Preparation of Standard Solution

1. *The stock solutions of standards and internal standards* are prepared by dissolving 1.0 mg of the reference standard in 1 mL of 0.1% hydrochloric acid, aliquoted by 50 μL and stored at −20 °C.
2. The *mixed working solution of standards* is prepared by adding 48 μL of Cys (400 nmol/mL), 3.56 μL of Cys-Gly (20 nmol/mL), 12.4 μL of glutathione (40 nmol/mL), and 2.7 μL of HCys (20 nmol/mL) into 930 μL of deionized water.
3. The *mixed working solution of internal standards* is prepared by adding 24 μL of D3-Cys, 3.56 μL of D2-Cys-Gly, 12.4 μL of 2-13C, 15 N-GSH, and 2.7 μL of D4-HCys in 260 μL deionized water.

3.2 Preparation of Calibration Standards

The calibration standards ranging in the intervals of 5–400 nmol/mL for Cys, 0.25–25 nmol/mL for CG and Hcy, and 0.5–40 nmol/mL for GSH are prepared from the working solution by dilution with deionized water (*see* Table 1).

3.3 Reduction and Protein Precipitation

1. Add 100 μL standard solution or plasma, 10 μL solution of internal standards, and 10 μL of TCEP (10 g/L) into 100 μL of deionized water.
2. Incubate for 30 min at RT.
3. Add 25 μL of TCA (2 mol/L), immediately vortex and centrifuge for 10 min at 30,000 × *g*.
4. Transfer the supernatants into separate 1.5 mL tubes.

3.4 Derivatization

1. Add 60 μL of K_2HPO_4 (0.5 mol/L) into 100 μL of the supernatants.
2. Add 60 μL of N-phenylmaleimide (20 mmol/L).

Table 1
Calibration standard preparation guide

Level	The volume of working solution (μL)	The volume of deionized water (μL)	Conc. Cys (nmol/mL)	Conc. CysGly (nmol/mL)	Conc. GSH (nmol/mL)	Conc. HCys (nmol/mL)
1	0	400	0	0	0	0
2	5	395	5	0.25	0.5	0.25
3	10	390	10	0.5	1	0.5
4	25	375	25	1.25	2.5	1.25
5	50	350	50	2.5	5	2.5
6	100	300	100	5	10	5
7	200	200	200	10	20	10
8	400	0	400	20	40	20

3. Incubate for 10 min at RT, then dilute with 800 μL of deionized water.

3.5 Solid-Phase Extraction Clean-Up

1. Set up an OASIS HLB PRIME well plate in Positive pressure 96 SPE instrument (Waters).
2. Condition wells by adding 1 mL of deionized water.
3. Load 1 mL of the diluted sample.
4. Wash with 1 mL of 5% methanol in water (v/v).
5. Insert collection plate and elute with 1 mL of methanol.
6. Transfer 900 μL of eluate into 1.5 mL tube and evaporate the solvent in vacuum concentrator Savant SpeedVac or Eppendorf™ Concentrator.
7. Redissolve samples in 60 μL of 80% acetonitrile, vortex immediately (20 s), and centrifuge for 10 min at 30,000 × *g*.
8. Transfer 50 μL of the supernatants into analytical vials (*see* **Note 8**) and store in autosampler at 10 °C until LC-MS analysis (*see* **Note 9**).

3.6 Liquid Chromatography

1. Set up LC acquisition method according to chromatographic parameters summarized in Table 2 (*see* **Note 10**).

3.7 Mass Spectrometry

1. XEVO TQD triple-quadrupole mass detector is operated in positive electrospray ion mode (ESI+). The capillary voltage is set at 2.0 kV; source temperature is 150 °C and nitrogen is used as desolvation gas with flow rate 600 L/h and temperature 350 °C.

Table 2
Chromatographic conditions

Mobile phase A	50 mmol/L ammonium formate +0.25% formic acid in water
Mobile phase B	100% acetonitrile
Column temperature	40 °C
Sample temperature	6 °C
Flow rate	0.5 mL/min
Injection volume	10 μL

Gradient table		
Time (min)	**%A**	**%B**
0.00	10	90
0.60	10	90
Feb-40	60	40
Mar-50	60	40
Mar-55	10	90
5.00	10	90

2. Analytes are detected in selected reaction monitoring mode (SRM). Cone voltage and collision energy are optimized to achieve the best intensity of precursors and fragment ions. Transitions with the highest intensity are used as quantifiers, whereas less intensive transitions are used as qualifiers (Fig. 2). Optimized parameters used for SRM analysis are summarized in Table 3 (*see* **Note 11**).

3.8 Data Processing

1. Exact quantification is performed by calculating the peak area ratios of samples and internal standards and comparing those with the calibration curve in TargetLynx SE (*see* **Note 13**).

4 Notes

1. The solution is stable for 6 months at −20 °C.
2. Solution should be prepared freshly before the procedure.
3. It is recommended to check the pH of K_2HPO_4 (0.5 mol/L) solution before further use. If the pH does not equal 9.5, adjustment with potassium hydroxide is necessary. This solution is stable at RT for 1 week.

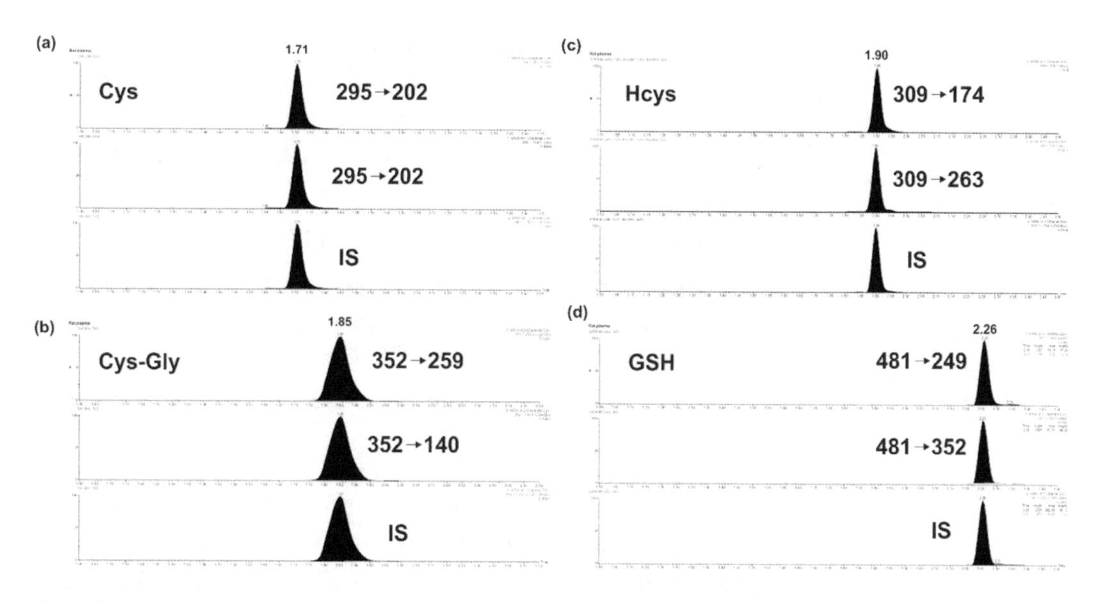

Fig. 2 Representative MRM ion chromatograms of rat plasma sample (**a**) Cys, (**b**) CG, (**c**) Hcys, and (**d**) GSH, and respective internal standards (IS)

Table 3
MS/MS transitions and corresponding optimized SRM conditions

Compound name	Parent ion (m/z)	Daughter ion (m/z)	Dwell (s)	Cone voltage (V)	Collision energy (eV)	
Cysteine	295.0 295.0	184.1 202.2	0.034 0.034	20 20	20 15	 Quantifier
D3-cysteine	298.2	205.1	0.034	20	15	
Cysteinyl-glycine	352.0 352.0	140.1 259.2	0.029 0.029	20 20	25 20	 Quantifier
D2-Cysteinyl-glycine[a]	354.2	261.2	0.029	20	20	
Glutathione	481.0 481.0	249.2 352.3	0.050 0.050	30 30	20 20	 Quantifier
2-^{13}C, ^{15}N-glutathione	484.0	355.2	0.029	30	20	
Homocysteine	309.0 309.0	174.1 263.2	0.029 0.029	30 30	20 10	Quantifier
D4-Homocysteine	313.2	174.1	0.029	30	20	

[a]*See* **Note 12**

4. 96-well plates in combination with Positive pressure 96 system is ideal for high-throughput analysis. Although more labor demanding, SPE cartridges can also be used.
5. Solution is stable at refrigerator temperature for 6 months.
6. Solution is stable for 2 weeks at RT.
7. 50% acetonitrile can be replaced by 50% methanol.
8. 2-mL glass vials with 250 μL plastic inserts are recommended.
9. Storage longer than 12 h is not recommended due to acetonitrile evaporation and limited GSH stability.
10. Longer equilibration time is needed in HILIC separation modes.
11. These parameters will vary between different instruments to some extent. It is necessary to check all these parameters and find a setting that would be optimal for your instrument.
12. D2-Cysteinyl-glycine can be replaced by D4-Homocysteine.
13. After method establishment, short validation procedure should be performed in terms of precision and accuracy.

References

1. Wu G, Fang Y-Z, Yang S, Lupton JR, Turner ND (2004) Glutathione metabolism and its implications for health. J Nutr 134:489–492
2. Isokawa M, Shimosawa T, Funatsu T, Tsunoda M (2016) Determination and characterization of total thiols in mouse serum samples using hydrophilic interaction liquid chromatography with fluorescence detection and mass spectrometry. J Chromatogr B Anal Technol Biomed Life Sci 1019:59–65
3. Banos-Gonzalez MA et al (2012) Lipoprotein (a) and Homocysteine potentiate the risk of coronary artery disease in male subjects. Circ J 76:1953–1957
4. Humphrey LL, Fu R, Rogers K, Freeman M, Helfand M (2008) Homocysteine level and coronary heart disease incidence: a systematic review and meta-analysis. Mayo Clin Proc 83:1203–1212
5. Hobbs CA, Cleves MA, Zhao W, Melnyk S, James SJ (2005) Congenital heart defects and maternal biomarkers of oxidative stress. Am J Clin Nutr 82:598–604
6. Al-Maskari MY, Waly MI, Ali A, Al-Shuaibi YS, Ouhtit A (2012) Folate and vitamin B12 deficiency and hyperhomocysteinemia promote oxidative stress in adult type 2 diabetes. Nutrition 28:e23–e26
7. Noce A et al (2014) Erythrocyte glutathione transferase activity: a possible early biomarker for blood toxicity in uremic diabetic patients. Acta Diabetol 51:219–224
8. Zoccolella S et al (2009) Hyperhomocysteinemia in levodopa-treated patients with Parkinson's disease dementia. Mov Disord 24:1028–1033
9. Puertas MC et al (2012) Plasma oxidative stress parameters in men and women with early stage Alzheimer type dementia. Exp Gerontol 47:625–630
10. Forgacsova A et al (2019) Ultra-high performance hydrophilic interaction liquid chromatography – triple quadrupole tandem mass spectrometry method for determination of cysteine, homocysteine, cysteinyl-glycine and glutathione in rat plasma. J Pharm Biomed Anal 164:442–451
11. Stabler S et al (2011) Serum methionine metabolites are risk factors for metastatic prostate cancer progression. PLoS One 6:1–9
12. Ho E, Karimi Galougahi K, Liu C-C, Bhindi R, Figtree GA (2013) Biological markers of oxidative stress: applications to cardiovascular research and practice. Redox Biol 1:483–491
13. Baba SP, Bhatnagar A (2018) Role of thiols in oxidative stress. Curr Opin Toxicol 7:133–139
14. Toyo'oka T (2009) Recent advances in separation and detection methods for thiol compounds in biological samples. J Chromatogr B Anal Technol Biomed Life Sci 877:3318–3330

15. Paolicchi A, Dominici S, Pieri L, Maellaro E, Pompella A (2002) Glutathione catabolism as a signaling mechanism. Biochem Pharmacol 64:1027–1035
16. Tyagi N et al (2005) Mechanisms of homocysteine-induced oxidative stress. Am J Physiol Circ Physiol 289:H2649–H2656
17. Herrmann W, Obeid R (2011) Homocysteine: a biomarker in neurodegenerative diseases. Clin Chem Lab Med 49:435–441
18. Moore T et al (2013) A new LC-MS/MS method for the clinical determination of reduced and oxidized glutathione from whole blood. J Chromatogr B Anal Technol Biomed Life Sci 929:51–55
19. Nolin TD, McMenamin ME, Himmelfarb J (2007) Simultaneous determination of total homocysteine, cysteine, cysteinylglycine, and glutathione in human plasma by high-performance liquid chromatography: application to studies of oxidative stress. J Chromatogr B Anal Technol Biomed Life Sci 852:554–561
20. Bayle C, Issac C, Salvayre R, Couderc F, Caussé E (2002) Assay of total homocysteine and other thiols by capillary electrophoresis and laser-induced fluorescence detection: II. Pre-analytical and analytical conditions. J Chromatogr A 979:255–260
21. Guo XF, Arceo J, Huge BJ, Ludwig KR, Dovichi NJ (2016) Chemical cytometry of thiols using capillary zone electrophoresis-laser induced fluorescence and TMPAB-o-M, an improved fluorogenic reagent. Analyst 141:1325–1330
22. Sawuła W et al (2008) Improved HPLC method for total plasma homocysteine detection and quantification. Acta Biochim Pol 55:119–125
23. McDermott GP, Terry JM, Conlan XA, Barnett NW, Francis PS (2011) Direct detection of biologically significant thiols and disulfides with manganese(IV) chemiluminescence. Anal Chem 83:6034–6039
24. Liem-Nguyen V, Bouchet S, Björn E (2015) Determination of sub-nanomolar levels of low molecular mass thiols in natural waters by liquid chromatography tandem mass spectrometry after derivatization with p-(hydroxymercuri) benzoate and online preconcentration. Anal Chem 87:1089–1096
25. Gori SS et al (2014) Profiling thiol metabolites and quantification of cellular glutathione using FT-ICR-MS spectrometry. Anal Bioanal Chem 406:4371–4379
26. Sun Y, Yao T, Guo X, Peng Y, Zheng J (2016) Simultaneous assessment of endogenous thiol compounds by LC–MS/MS. J Chromatogr B 1029–1030:213–221
27. McMenamin ME, Himmelfarb J, Nolin TD (2009) Simultaneous analysis of multiple aminothiols in human plasma by high performance liquid chromatography with fluorescence detection. J Chromatogr B Anal Technol Biomed Life Sci 877:3274–3281

Chapter 20

Untargeted Metabolomics Determination of Postmortem Changes in Brain Tissue Samples by UHPLC-ESI-QTOF-MS and GC-EI-Q-MS

Carolina Gonzalez-Riano, Antonia García, and Coral Barbas

Abstract

Metabolomics is a well-established method that allows for the screening of a broad range of metabolic shifts, capturing the global state of a complex system. Postmortem biochemical processes induce significant metabolic changes within the brain, hindering later the proper interpretation of the results. Consequently, one of the main challenges when facing a metabolomics study based on brain tissue samples is dealing with such alterations induced by tissue degradation and the hypoxic/ischemic state generated in this organ after death. Generally speaking, metabolomics experiments can be addressed following a discovery-orientated untargeted approach or an aim-dependent targeted analysis. Here, we describe a protocol to carry out untargeted metabolomics studies based on brain tissue samples by liquid chromatography (LC) and gas chromatography (GC) coupled to mass spectrometry (MS) aiming to gain a deeper knowledge of the biochemical changes that occur in the brain tissue following death. We also provide some recommendations to avoid postmortem-induced changes in brain samples.

Key words Postmortem interval, Multiplatform metabolomics, Untargeted lipidomics, LC-MS analysis, GC-MS analysis, MS^1 annotation

1 Introduction

One of the most challenging and critical tasks to deal with not only in research studies but also in forensic sciences is the postmortem time interval (PMI) [1]. This period elapsed between the death of an individual and the collection of the brain represents a problem of stability since multiple and complex variations at morphological and biochemical levels take place in this organ after death [2]. Postmortem human brain samples are critical for neuroscience research. The future of this science depends on the availability of high-quality tissue samples in the biobanks. In consequence, several guidelines and standardized protocols have been proposed for the correct acquisition and storage of samples [3, 4]. However, postmortem time (PT) delay is absolutely unavoidable. Rodent models are also

Paul L. Wood (ed.), *Metabolomics*, Neuromethods, vol. 159, https://doi.org/10.1007/978-1-0716-0864-7_20,

invaluable for the research landscape in neurodegenerative disorders and neurobiology. It must be kept in mind that once the body circulation stops, the sample collection step plays a decisive role since autolysis (caused by enzymes) or putrefaction (induced by the action of microorganisms) can produce structural and composition modifications [5]. The identification of the metabolic alterations generated by the postmortem time delay within the brain is a critical point to ensure the quality of a research study based on brain tissue samples.

In this regard, metabolomics aims the analysis of the entire set of small-molecule, metabolism products or metabolites that are present in a biological sample, providing mechanistic insights into the physiological or pathological status of the system. The measurement of abnormal levels of metabolites in response to internal or external stimuli is achieved through the combination of different analytical approaches. Nuclear magnetic resonance (NMR) spectroscopy and mass spectrometry (MS) detection are the two methods that have formed the central axis for metabolomics. Especially, chromatography-MS coupled techniques, including liquid (LC), and gas (GC) chromatography are the most popular combinations [6]. Although NMR has some unique advantages, MS is arguably the most widely used analytical tool for projects in metabolomics due to its selectivity and sensitivity [7, 8]. Recent improvements in MS have broadened the range of metabolites that can be analyzed in different types of biological samples, and have stimulated great interest in the application of metabolomics to explore the central nervous system (CNS) to detect and quantify neuroactive metabolites. GC-MS analysis is intended to measuring low-molecular-weight metabolites (50–650 Da), including amino acids, short and long-chain free fatty acids, neurotransmitters, amines, and sugars, among others. The main features that render GC-MS advantageous for metabolomics studies rely on its high separation capability, robustness, sensitivity, sensitivity, and excellent reproducibility. Reversed-phase (RP) LC-MS is a well-suited technique that allows the determination of non-volatile and low to nonpolar compounds (mainly lipids) across a broad molecular weight range, from 50 to more than 1500 Da with high sensitivity [9]. Hence, these high-throughput analytical techniques can shed light on the biochemical modifications occurring after death.

The extraction method proposed in this chapter was developed to unveil the significant metabolic changes that take place in the hippocampus tissue at three PMI: t_0, t_{2h}, and t_{5h} [10]. The protocol covers the extraction of the main lipids classes presented in brain tissue for their subsequent analysis by UHPLC-ESI-QTOF-MS, including phosphatidylcholines (PC), phosphatidylethanolamines (PE), lysophosphatidylcholines (LPC), lysophosphatidylethanolamines (LPE), phosphatidylserines (PS), phospatydylglycerides (PG), phosphatydic acids (PA), carnitines, diacylglycerides

(DAG), monoacylglycerides (MAG), ceramides (Cer), and sphingomyelins (SM). Regarding the polar compounds, several aminoacids, short chain organic acids, biogenic amines, sugars, free fatty acids, and cholesterol can be extracted from the hippocampus tissue samples for their analysis by GC-EI-Q-MS. In this chapter, we propose several recommendations to obtain animal brain tissue samples, since a rapid dissection of the preselected brain region/s and effective methods to halt the tissue metabolism to prevent the alteration of the levels of important metabolites are imperative. Finally, we describe the general data treatment workflow for multiplatform untargeted metabolomics studies, including metabolites and lipids annotation based on MS^1 data.

2 Materials

2.1 Equipment and Supplies

1. Analytical balance.
2. Homogenizer, such as TissueLyser LT homogenizer (Qiagen, Germany).
3. Multitube vortex mixer.
4. Ultrasonic bath.
5. Calibrated micropipettes of 2–20 μL, 20–200 μL, and 100–1000 μL.
6. Eppendorf tubes of 1.5 mL and 2 mL (Hamburg, Germany).
7. SpeedVac concentrator system.
8. Centrifuge.
9. Microcentrifuge.
10. 2 mL GC-HPLC glass vials with 300 μL conical narrow opening inserts.
11. Laboratory oven.
12. Water purification system, such as Milli-Qplus185 system (Millipore).

2.2 Solvents and Chemical Reagents

1. Ultrapure water.
2. HPLC-grade methanol (Sigma-Aldrich).
3. HPLC-grade 2-propanol (Sigma-Aldrich).
4. HPLC-grade methyl tert-butyl ether (MTBE) (Sigma-Aldrich).
5. HPLC-grade n-Heptane (VWR International BHD Prolabo, Madrid, Spain).
6. Analytical grade formic acid (Sigma-Aldrich).
7. Analytical grade ammonia hydroxide (30% ammonium in high purity water) (Panreac Quimica SA).

8. Silylation-grade pyridine (VWR International BHD Prolabo, Madrid, Spain).
9. *N,O*-Bis(trimethylsilyl)trifluoroacetamide (BSTFA) plus 1% trimethylchlorosilane (TMCS) (Supelco, Bellefonte, PA, USA).
10. *O*-Methoxyamine hydrochloride GC grade (Sigma-Aldrich). Prepare a solution containing 15 mg/mL in pyridine the day of the analysis.
11. Internal Standard: methyl stearate (C18:0 methyl ester) GC grade (Sigma Aldrich, Steinheim, Germany).
12. Standard mix containing fatty acid methyl esters (C8:0-C22:1, n9) (FAMEs) (Ref. CRM47801—Supelco, Sigma-Aldrich Chemie GmbH).

2.3 LC-MS Equipment

1. UHPLC system, such as Agilent 1290 Infinity II (Agilent Technologies).
2. Quadrupole Time of Flight mass spectrometer, such as 6545 Q-TOF MS (Agilent Technologies).
3. UHPLC guard column, such as InfinityLab Poroshell 120 EC-C8 guard column (2.1 × 5 mm, 2.7 μm) (Agilent Technologies).
4. UHPLC sub-2-micron column, such as InfinityLab Poroshell 120 EC-C8 column (2.1 × 150 mm, 2.7 μm) (Agilent Technologies).

2.4 GC-MS Equipment

1. GC system, such as the Agilent Technologies 7890A GC with split/splitless or multimode injector, coupled to an autosampler such as the Agilent Technologies 7693.
2. Mass selective detector such as the Agilent Technologies 5975 inert MSD with a Quadrupole mass analyzer and a triple-axis detector.
3. 10 m J&W precolumn (Agilent Technologies) integrated with a 122–5332G column: DB5-MS 30 m length, 0.25 mm i.d., and 0.25 μm film consisting of 95% dimethyl and 5% diphenyl polysiloxane (Agilent Technologies).
4. Ultrainert low pressure drop liner with wool GC liner (Agilent Technologies cat.no. 5190-3165).

3 Methods

3.1 Important Preanalytical Aspects

Any metabolomics study based on brain tissue samples requires taking into account several considerations in the preanalytical step, since the shutting down of a complex system, like the brain, does not implicate the immediate cessation of its metabolism at the time of death. Foremost, the anatomical region that will be employed for the analysis and the postmortem time interval (PMI) should be well established before starting the experiment [11]. The brain possesses a complex metabolome and, therefore, exhibits a broad spectrum of metabolite patterns, that are in charge of specific functions within a particular brain area [12]. Additionally, there are multiple neurodegenerative disorders which pathophysiology is initially associated with a located dysfunction of cells within the brain, including Parkinson's disease (PD) (basal ganglia and substantia nigra), Huntington's disease (HD) (basal ganglia), or Alzheimer's disease (AD) (entorhinal cortex, cerebral cortex, and hippocampus), among others [13]. Although a brain section is more suitable than others depending on the scope of the study, it also should be contemplated the integration of multiple regions in the analysis because it might provide a broader view of the pathophysiological processes that take place during the development and progression of a disease [14] (*see* **Note 1**).

In general, lipids seem to be resistant to PMI effects, being stable up to 3 days postmortem [15]. It is, however, worth noting that lipids comprise a heterogeneous group in terms of physicochemical properties. Therefore, assuming that all the lipid classes will behave equally within a PT interval seems to be quite risky. That is the case of the oxidized lipids. This group of lipids is highly metabolically dynamic with short half-life rates and, consequently, postmortem delay can induce a metabolic instability problem.

The brain is highly sensitive to oxidative injury because of its elevated oxygen uptake, high polyunsaturated fatty acids (PUFA) content, and the weak presence of antioxidant defenses, which indeed dramatically decrease postmortem resulting in lipid oxidation and degradation [16]. Oxidized lipids are recognized as significant signaling mediators that provide quantitative readout relating to neuroinflammation and oxidative stress, which are involved in the development of many neurological disorders and neurodegenerative diseases [17]. Oxylipins can be defined as PUFA oxidation products that regulate the synaptic transmission, vasodilation, angiogenesis, neuronal morphology, and inflammatory signaling, among other brain activities [18]. It has been described that oxylipins can be generated artifactually ex vivo since it exists a time-dependent increase of their levels induced by the postmortem ischemic state in the brain tissue that activates the continuous release of PUFAs [19]. Moreover, the removal of the brain from the skull and the subsequent dissection of its areas can also modify the levels of many oxylipins [18]. Nevertheless, not only some lipid classes are vulnerable to the PMI effects in the brain tissue. The levels of several metabolites have also been reported as affected by PT delay in brain tissue, including cadaverine, pyroglutamic acid, succinic acid, glycine, ethanolamine, putrescine, GABA, *N*-acetyl-aspartic acid (NAA), creatinine, lactic acid, and nucleosides, among many others [10, 20, 21].

Therefore, in order to minimize as much as possible the postmortem modifications in the brain metabolome, carry out the following:

1. Select the most appropriate freezing method; add enough volume of isopentane or liquid nitrogen in a Dewar at least 10 min before starting. There are multiple methods for brain tissue freezing. However, if samples collected are intended for metabolomics, the preferred method for sample conservation is snap freezing the brain area previously dissected with isopentane or liquid nitrogen since it is a rapid and easy way to preserve the tissue stable.
2. Label all the material that will be employed for the sample storage; including tubes or sample bags.
3. Perform the brain dissection over ice. The recommended temperature to section the brain should be ranged between −9 and −19 °C to avoid tissue hardness.
4. Put dry ice into a foam to store temporarily the already frozen samples while continuing to freeze the rest of the samples.
5. Snap freeze the dissected brain region samples within 2 h postmortem.
6. Store the samples at −80 °C immediately after collection all the samples and keep them in these conditions until the day of the analysis.
7. Once the samples are stored at −80 °C, do not allow the temperature to exceed −60 °C and try to avoid the thaw of the samples while transporting them.
8. Minimize the sample's exposure to oxygen while in storage to halter the lipid oxidation processes (e.g., substitute air with a dry nitrogen gas atmosphere).

3.2 Metabolite Fingerprinting of Mouse Hippocampus

The extraction protocol described below has proven to be suitable to perform untargeted metabolomics analyses based on brain tissue samples by LC-MS and GC-MS techniques [10, 22]. The solvent mixture used in this method (H_2O–MeOH–MTBE5:37:8, v/v/v) allows the performance of a one-single phase extraction of both polar and nonpolar metabolites, maximizing the chemical range of metabolites that will be later detected. Furthermore, this method is simple to execute, efficient, reproducible, and safe since chloroform has been replaced by MTBE [23]. Moreover, only 30 mg of brain tissue is required for the total experiment. To correct the weight differences between the samples, the first solvent mixture consisted on methanol–water (1:1, v/v) will be added in a specific ratio (1 mg,10 μL) in order to ensure the metabolites extraction

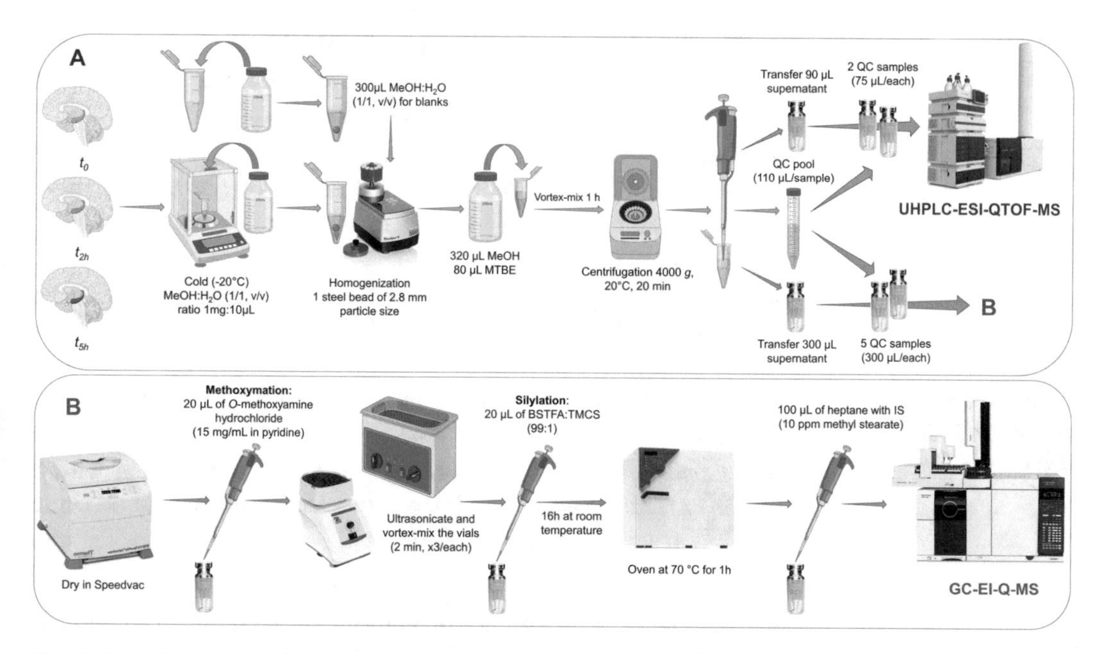

Fig. 1 General scheme of the sample treatment workflow for a multiplatform (LC-MS and GC-MS) untargeted metabolomics study based on hippocampus tissue collected at three post-mortem intervals (t_0, t_{2h}, t_{5h}). The **A** section corresponds to the metabolite extraction and the LC-MS sample treatment, while the **B** section describes the derivatization process required for the analysis by GC-MS.

normalization. A summary of the main steps in the sample treatment workflow is presented in Fig. 1.

3.2.1 Metabolite Extraction

1. On ice, add cold (−20 °C) methanol–water (1:1, *v*/*v*) to each sample with a weight-to-volume ratio 1:10 for metabolites extraction. For example, for a 30 mg tissue add 300 µL of the solvent mixture.
2. Pipette 300 µL of the cold mixture of methanol–water (1:1, *v*/*v*) in two empty Eppendorf tubes for blank samples. Treat blank solutions in parallel with the rest of the samples of the study (*see* **Note 2**).
3. For metabolite extraction of the tissue, add a 2.8 mm particle size steel bead to each sample and place them in the TissueLyser LT homogenizer. Perform two cycles of 3 min/each. Alternative homogenizers, such as liquid nitrogen-cooled mortar and pestle, electric tissue homogenizers, or Precellys 24 bead-based homogenizer can be used.
4. Pipet 100 µL of homogenate to an Eppendorf tube.
5. Add 320 µL of cold (−20 °C) methanol and vortex-mix samples 2 min.

6. Then, add 80 μL of methyl tert-butyl ether to each sample for the extraction of nonpolar metabolites and vortex-mix samples for 1 h at room temperature (*see* **Note 3**).
7. Centrifuge for 20 min at 4000 × *g* at 20 °C to concentrate the protein and tissue debris to the bottom of the sample tube. This supernatant will be used for the further steps of sample treatment.
8. Collect 110 μL of supernatant from each sample in order to prepare a QC pool sample (*see* **Note 4**).

3.2.2 Sample Preparation for LC-MS Analysis

1. Collect 90 μL of supernatant and transfer it to an LC vial equipped with a 300 μL insert.
2. Transfer 75 μL of QC pool to two LC vials equipped with a 300 μL insert (one QC for positive ionization mode and one QC for negative ionization mode).
3. Inject the samples directly into the LC-MS system.
4. Analyze the QC sample throughout the analytical run to provide a measurement of the stability and performance of the system.

3.2.3 Sample Preparation for GC-MS Analysis

1. Collect 300 μL of the supernatant to a GC vial equipped with a 300 μL insert.
2. Transfer 300 μL of the QC pool sample to five GC vials equipped with a 300 μL insert.
3. Evaporate the samples and the QCs to dryness by SpeedVac Concentrator System at 30 °C. Check to assure the vial is completely dry.
4. For methoximation, add 20 μL of *O*-methoxyamine hydrochloride (15 mg/mL) in pyridine to each GC vial to protect ketones and aldehydes functional groups.
5. Vortex-mix vigorously for 5 min.
6. Ultrasonicate the vials for 2 min and vortex for 2 min—repeat the process 3 times.
7. Cover the GC vials with aluminum foil and incubate at room temperature for 16 h under darkness to avoid compounds hydrolysis due to humidity or light.
8. For silylation, add 20 μL of BSTFA:TMCS (99:1) to each vial in order to replace the active hydrogen atoms by an alkylsilyl group.
9. Vortex-mix for 5 min.
10. Place the GC vials into an oven for 1 h at 70 °C.
11. Cool down the samples for at least 30 min at room temperature in the dark to minimize the loss of highly volatile compounds.

12. Add 100 μL of heptane containing 10 ppm C18:0 methyl ester (IS) to each GC vial.
13. Vortex-mix for 2 min to dissolve derivatives.
14. Centrifuge the vials at 2500 × *g* at 20 °C for 15 min before GC-MS analysis.
15. Analyze the QC sample throughout the analytical run to provide a measurement of the stability and performance of the system.

3.3 LC-MS Analytical Settings

1. Calculate the mobile phases volume needed for the entire run of each ion-mode, taking into account the run time, the mobile phases flow, and the total number of injections included in the analytical run (blanks, QCs, and samples) (*see* **Note 5**).
2. For positive ionization mode, prepare the previously calculated volume of 5 mM ammonium formate in Milli-Q water for mobile phase A; and 5 mM ammonium formate in methanol–isopropanol (85:15, v/v) for mobile phase B.
3. For negative ionization mode, prepare the volume calculated in advance of 0.1% formic acid in Milli-Q water for mobile phase A; and 0.1% formic acid in methanol/isopropanol (85:15, v/v) for mobile phase B.
4. Set the following elution gradient pattern for both ion modes: 0 min, 75% of B; 23.00 min - 31.00, 96% of B; 31.50–32.50 min, 100% of B; 33.00–40.00 min, 75% of B, at a flow rate of 0.5 mL/min.
5. Set the column temperature at 60 °C, the rack temperature at 15 °C, and the injection sample volume at 1.5 μL (*see* **Notes 6** and **7**). Set the mass range window at 50–1500 *m/z* at scan rate of 1.00 spectra/s.
6. If your equipment includes the multiwash function, prepare a multiwash solution consisted of the initial conditions of your analytical gradient. For this method, mix 25% of mobile phase A with 75% of mobile phase B.
7. If your system pump presents the seal wash function, prepare a 10% isopropanol in Milli-Q water solution.
8. For the purpose of mass accuracy performance verification, prepare fresh reference mass solution containing the reference masses required for both ion-modes. Check the intensity and the nebulization of the reference masses by extracting their corresponding *m/z* values.

3.4 GC-MS Analytical Settings

1. Prepare a 10 ppm C18:0 methyl ester (Internal standard) solution in heptane. First, prepare a 1000 ppm stock solution in heptane and dilute it to 10 ppm with the same solvent. Inject the IS at least 5 times prior to the analysis in order to observe the signal intensity ($\geq 5 \times 10^5$) and its RT.

2. *O*-Methoxyamine hydrochloride GC grade (Sigma-Aldrich). Prepare a solution containing 15 mg/mL in pyridine.
3. Prepare a 100 fold dilution in dichloromethane from the grain fatty acid methyl esters mix (C8:0–C22:1) (FAMEs) Ref. CRM47801 (Supelco, Sigma-Aldrich Chemie GmbH, Steinheim, Germany, 10 mg/mL in dichloromethane). Inject it at the beginning of the analytical run.
4. Prepate a 50-fold dilution of the mixture of *n*-alkanes calibration standards (C8–C40) Ref. 40,147-U (Supelco, Sigma Aldrich, Steinheim, Germany) in dichloromethane. Inject it at the beginning of the analytical run, after the grain fatty acid methyl esters mix.
5. All the solutions are recommended to be freshly prepared for every experiment the day of the analysis.
6. Set the injection volume at 2 μL, with a split ratio 1:10.
7. Set the next temeperatre gradient for the column oven: 0 min, 60; 1 min, raise the temperature 10 °C/min until 325 °C; held at this temperature for 10 min before cooling down.
8. Set the injector and the transfer line temperatures at 250 °C and 280 °C, respectively.
9. Configure the electron impact ionization operating parameters as follows: 230 °C for the filament source, and 70 eV of electron ionization energy.
10. Set the mass range window at 50–600 *m/z* at scan rate of 2.00 spectra/s.

4 Data Processing Pipeline

Data processing can be defined as the computational method to transform raw MS data into biological knowledge [24]. This is a multistep process within the metabolomics workflow that will provide reliable information leading to accurate identification of the significant metabolites. The main steps in the data analysis workflow are (a) data preprocessing, (b) data pretreatment, and (c) data treatment.

4.1 Data Preprocessing

The raw data obtained after an MS-based metabolomics analysis consists of complex data with a 3D structure composed by the mass-to-charge ratio (*m/z*), retention time (RT), and chromatographic peak area. However, working directly with 3D data is not trivial; therefore, the raw data should be converted into a two-dimensional form to proceed with the next data processing steps [25, 26]. In order to reduce the complexity of raw data obtained and extracting the relevant features of the analysis, the principal steps in the data processing pipeline are as follows:

1. Opening the raw data: each vendor utilizes its own data format that can be utilized only with proprietary software; however, there are open data formats such as the mzXML, mzML, mzData, or NetCDF that are suitable for their reprocessing with open source tools (e.g., XCMS, MZmine 2, Metaboanalyst, Workflow4Metabolomics, MetAlign). One of the most commonly used software in metabolomics to convert MS data is ProteoWizard's MSConvert.
2. Raw data filtering, including reducing the noise or applying a baseline correction to improve the peak detection and avoid the introduction of unwanted features.
3. Peak detection: aiming to identify all the signals generated by true ions and avoid the undesirable false ions (*see* **Note 8**).
4. Deconvolution is an essential process that consists on grouping the *m/z* ions of a single component since the nontargeted metabolomics chromatograms are characterized by the presence of coeluting compounds that constitutes overlapped signals and multiple fragments originated from the same metabolite (*see* **Note 9**). The outcome of this step is a peak list of *m/z*@time entries for each sample analyzed.
5. Alignment: the retention time of each feature is corrected to integrate the data acquired from each separate sample in to a single data matrix. A peak list table containing aligned *m/z*@time variables is then obtained, allowing for the comparison between samples (*see* **Note 10**).
6. Missing values replacement; to fill the gaps occasioned by technical factors (e.g., instrument sensitivity threshold, matrix effect, or computational processing limitations), or biological factors (e.g., drug treatment) (*see* **Note 11**).

4.2 Data Pretreatment

1. Data filtration. The scope of this first step is to remove those signals that are not stable and present a coefficient variation higher than the predefined threshold (e.g., CV > 30% in the QC samples). Then, exclude signals presenting a high percentage of missing values within the sample groups (e.g., signal must be present in at least 50% of samples in at least one group).
2. Data normalization. This process aims to correct the technical variations introduced in the analysis that lead to peak intensities out of the trend between measurements while retaining the biological variation. Depending on the source of variation that wants to be corrected, different normalization strategies can be applied, such as the median fold change, MS "total useful

signal" (MSTUS), or Intra-batch effect correction using QC samples and support vector regression (QC-SVRC).

3. Data scaling. This approach divides each variable by a scaling factor, which is different for each variable, aiming to adjust for differences in fold change between metabolites. The most extended methods for scaling metabolomics data are the Auto-scaling, Pareto scaling, and Vast scaling.

4.3 Data Treatment

1. Perform a data quality check by performing a principal component analysis (PCA) on the data set previously obtained to observe if the QC samples are tightly clustered revealing the robustness of the analytical performance and the system stability.
2. Investigate the biological differences by carrying out a statistical analysis.
   (a) The univariate data analysis points out whether an alteration of a specific variable (peak or signal) found after the analytical stage differs significantly or not between groups. Best-suited statistical tests for datasets following normal distribution are unpaired *t*-test (compare two unpaired groups), paired t-test (compare two paired groups), and One-way ANOVA (compare more than two unmatched groups). On the other hand, the most appropriated test for data far from the normal curve are Mann–Whitney *U*-test (compare two unpaired groups), Wilcoxon signed-rank (compare two paired groups), and Kruskal-Wallis (compare more than two unmatched groups). Select those metabolites presenting a p-value <0.05.

   (b) The multivariate data analysis takes into consideration more than one variable at the same time. PCA scatter plot, which is an unsupervised method, is considered as a key tool to inspect the overall structure of the data and obtain a preliminary evaluation of natural clustering of the samples. Then, build the partial least squares regression (PLS-DA) and orthogonal projections to latent structures (OPLS-DA) supervised models to discriminate which variables are responsible of variation, set parameters to predict new data, and select potential biomarkers. Select the important variable based on their contribution to the model based of metabolites presenting a VIP score > 1.

   (c) Calculate the percentage of change of each metabolite between the study groups. Check that the biological variation is higher than the analytical variation expressed by the coefficient of variation. Exclude those metabolites that do not satisfy this criterion.

4.4 Metabolite Identification

4.4.1 GC-EI-Q-MS Metabolite Annotation

The annotation of metabolites obtained after LC/GC-MS analysis is a challenging task. However, the metabolite identification process for GC-EI-Q-MS experiments presents a major advantage, which is the widespread standardization of setting the electron energy applied to the system at 70 eV. This facilitates the comparison of fragmentation spectra with online or commercial libraries available for GC-MS metabolite ID. To achieve this goal, the analysis method must be performed using the standard "Fiehn metabolomics retention time lock (RTL)" method. This methodology consists on locking the RT of methyl stearate at 19.662 min, by setting the analytical flow in the chromatographic system. Furthermore, commercial mixtures of FAME (Fatty acid methyl esters) and *n*-alkanes with a wide coverage of elution times are injected prior to the analysis to build both fitting curves of retention index (RI) and retention time (RT) in order to apply Fiehn's and NIST libraries, respectively, for identification (Figs. 2 and 3) [27, 28].

4.4.2 LC-MS Metabolite Annotation

The correct annotation of the metabolites detected by LC-MS analysis is an essential step in order to give a biological explanation to the results obtained. Frequently, the assignation of a metabolite ID is selected among several other possible identities. Therefore, the metabolomics community has established some guidelines to

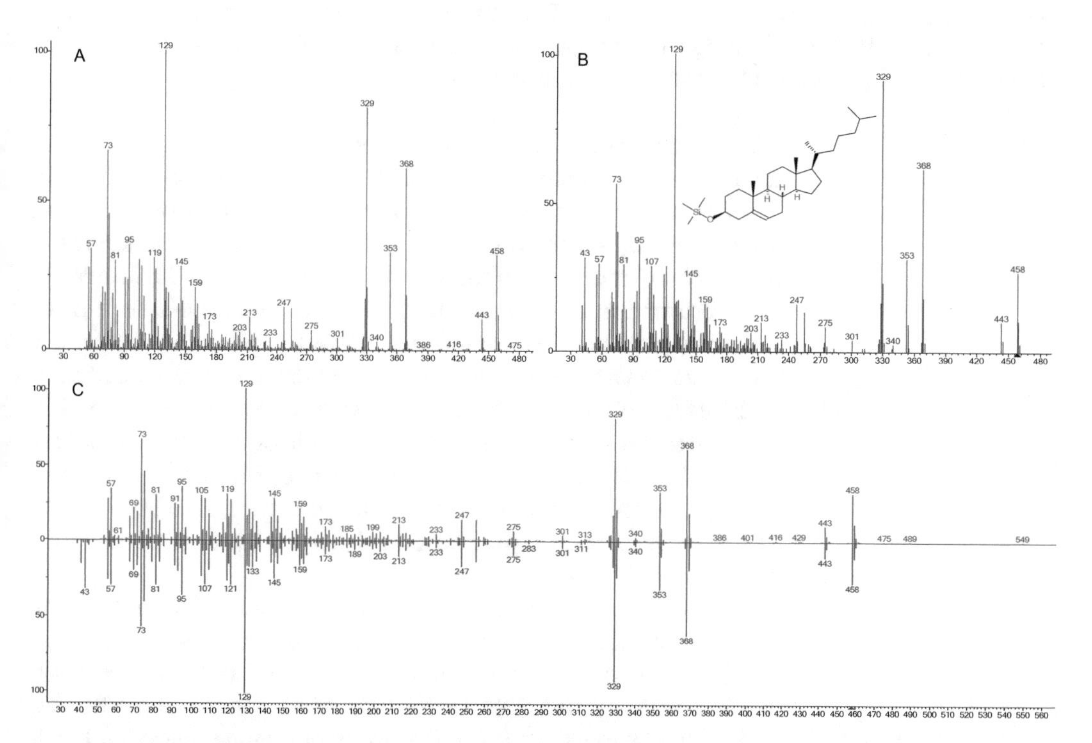

Fig. 2 Metabolite identification after GC-EI-Q-MS analysis based on RT and spectrum matching with a targeted library. Figure **A** displays the obtained spectrum, figure **B** shows the library spectrum of the cholesterol at 27.555 min, and figure **C** corresponds to the overloading spectra plot

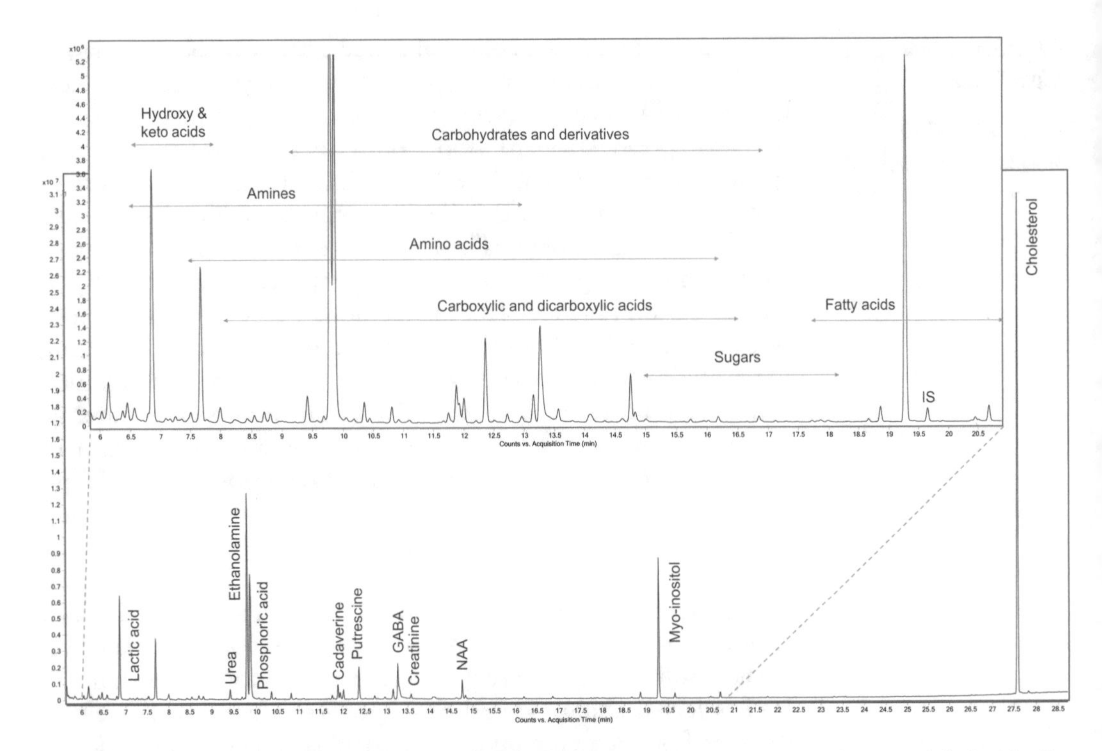

Fig. 3 TIC of hippocampus tissue profile obtained by GC-MS. The chromatogram regions are delimited by the metabolite class that eluted within that time window. The metabolites pointed out in the chromatogram are the ones that presented the highest changes in the hippocampus tissue due to the PMI. The peak marked as the internal standard (IS) belongs to the methyl stearate at 19.662 min

assess confidence when annotating metabolites. Recently, a confidence scale for the metabolite identification using high-resolution MS has been proposed [29]. The tentative annotation based on accurate mass, retention time, change, and adduct formation is the lowest confidence level in the identity confirmation scale. However, a first estimation of the possible metabolite ID is necessary prior to performing an MS^n experiment since it will help later to the proper interpretation of the spectra obtained. Additionally, the amount of available sample sometimes is not enough to carry out the MS/MS fragmentation or the commercial standards of certain metabolites are not in the market yet. In consequence, the MS^1 annotation might be considered as the stepping stone to the final identification of the metabolites, playing a vital role in any untargeted metabolomics analysis [30, 31].

Search the *m/z* of the statistically significant features in the online tool CEU Mass Mediator (http://ceumass.eps.uspceu.es/mediator/) [30, 31]. This search engine comprises the information available in Kegg, HMDB, LipidMaps, Metlin, an in-house library, and MINE. You can perform the query of a single mass or a batch of the significant experimental masses.

1. Set at 20 ppm the mass error tolerance allowed for the tentative annotation for the experimental masses introduced.
2. Select the databases to perform the search and then, select the type of metabolites to search. For metabolomics studies, select all metabolites except peptides.
3. Since the values introduced in the first step corresponded to the *m/z* values, select the *m/z* Masses option as the mass mode, and the ionization mode of the analysis through which those features were obtained.
4. Depending on the ion mode selected, choose the possible adducts formed during the ionization process, taking into account the mobile phases and the modifiers employed for the analysis. For the analytical conditions and the sample type used for the protocol described in this chapter select:
   (a) Positive ion mode: $[M + H]^+$, $[M + Na]^+$, $[M + K]^+$, $[M + NH_4]^+$, $[M + H\text{-}H_2O]^+$, $[M + 2H]^{2+}$, $[2\ M + H]^+$.
   (b) Negative ion mode: $[M\text{-}H]^-$, $[M + Cl]^-$, $[M + FA\text{-}H]^-$, $[M\text{-}H\text{-}H_2O]^-$, $[M\text{-}2H]^{2-}$, $[2\ M\text{-}H]^-$.
5. Select the best-suited candidate for the annotation considering the following tips:
   (a) Elution order of the compounds based on the chromatographic conditions: since the LC analysis has been performed in a reversed-phase (RP) mode, the polar metabolites will elute in the first minutes of the chromatogram, while the nonpolar (lipids) will be retained longer in the RP column and, consequently, will elute later. The RT in lipid classes depends on the length of the carbon chains of the fatty acids and the degree of unsaturation. The higher the number of carbon atoms, the greater the retention time; and the number of double bonds decreases the retention of the metabolites. Therefore, a LysoPC(16:0) will elute earlier than a LysoPC(18:0), and the RT of LysoPC(18:2) must be lower than the RT of LysoPC (18:0). Take also into account the LogP value of the possible candidates since in RP-LC, the higher the value of the LogP, the more hydrophobic will be the molecule, resulting in greater RT.

      Figure 4 displays the Total Compound Chromatograms (TCC) acquired following the previously described reversed-phase LC-MS-based strategy, pointing out the different areas throughout the chromatogram depending on the lipid class that elutes within that time window.
   (b) Ionization and adduct formation; considering the ionization mode selected to carry out the analysis, mobiles phases, modifiers, and the lipid class, the possibilities of adduct formation are described in Table 1. The order of

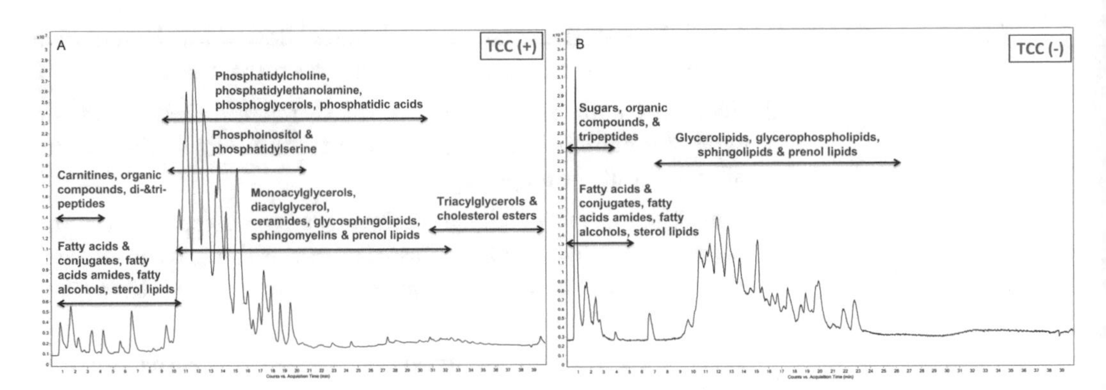

Fig. 4 Extracted compound chromatogram acquired by LC-MS in positive (**A**) and negative (**B**) ionization modes from the hippocampus tissue. The time regions of the chromatogram are delimited by the lipid class that eluted within that window of time

Table 1
Possible adduct formation and its intensity pattern for the main lipid classes that can be detected by LC-MS in samples collected from the CNS

Lipid class	Positive ion mode	Negative ion mode
Fatty acids	$[M+H]^+$, $[M+H\text{-}H_2O]^+$, $[M+Na]^+$	$[M\text{-}H]^-$, $[M\text{-}H\text{-}H_2O]^-$
PC	$[M+H]^+$, $[M+Na]^+$, $[M+K]^+$	$[M+FA\text{-}H]^-$, $[M+Cl]^-$
PE	$[M+H]^+$, $[M+Na]^+$, $[M+K]^+$	$[M\text{-}H]^-$, $[M+FA\text{-}H]^-$
PI	$[M+Na]^+$	$[M\text{-}H]^-$, $[M+FA\text{-}H]^-$, $[M+Cl]^-$
PG	$[M+H]^+$, $[M+NH_4]^+$	$[M\text{-}H]^-$, $[M+FA\text{-}H]^-$, $[M+Cl]^-$
PS	$[M+H]^+$	$[M\text{-}H]^-$, $[M+FA\text{-}H]^-$, $[M+Cl]^-$
PA	–	$[M\text{-}H]^-$, $[M+FA\text{-}H]^-$, $[M+Cl]^-$
MG	$[M+Na]^+$, $[M+K]^+$, $[M+NH_4]^+$, $[M+H]^+$	–
MGDG	$[M+NH_4]^+$, $[M+Na]^+$, $[M+K]^+$	$[M\text{-}H]^-$
DGDG	$[M+NH_4]^+$, $[M+Na]^+$, $[M+K]^+$	$[M\text{-}H]^-$
DG	$[M+Na]^+$, $[M+K]^+$, $[M+NH_4]^+$, $[M+H]^+$	–
CER	$[M+H]^+$, $[M+Na]^+$	$[M+FA\text{-}H]^-$, $[M+Cl]^-$, $[M\text{-}H]^-$
GlcCer/GalCer	$[M+H]^+$, $[M+Na]^+$	$[M\text{-}H]^-$, $[M+FA\text{-}H]^-$
SM	$[M+NH_4]^+$, $[M+H]^+$, $[M+Na]^+$, $[M+K]^+$	$[M+FA\text{-}H]^-$, $[M+Cl]^-$, $[M\text{-}H]^-$

PC glycerophosphatidylcholine, *PE* glycerophosphatidylethanolamine, *PI* glycerophosphatidylinositol, *PG* glycerophosphoglycerol, *PS* glycerophosphoserine, *PA* glycerophosphate, *MG* monoradylglycerols, *MGDG* monogalactosyldiacylglycerol, *DGDG* digalactosyldiacylglycerol, *DG* diradylglycerols, *CER* ceramide, *GlcCer/GalCer* glucosyl–/galactosylceramide, *SM* sphingomyelin

the possible adducts displayed in the table for each lipid class is based on the intensity pattern of adduct formation. For example, the intensity pattern for PCs is $[M + H]^+ > [M + Na]^+ > [M + K]^+$ (*see* **Notes 12** and **13**).

(c) Some lipid classes can suffer a neutral loss of water caused by in-source fragmentation, yielding to the adduct $[M + H\text{-}H_2O]^+$, including fatty acids, ceramides, monoacylglycerols, diacylglycerols, GlcCer, LysoPC, and LysoPE. On the other hand, PC and PE will never show neutral loss of water.

(d) Since it is not possible to distinguish the structure of the chains presented in a lipid at MS^1 level, an example of the standardized annotation format for lipids will be PX(A:a), being *A* the sum of the total number of carbons and *a* the total number of double bonds. (e.g., PC(18:0/16:2) = PC(34:2)). An instructive example that illustrates the tentative annotation process with CMM is described in Fig. 5.

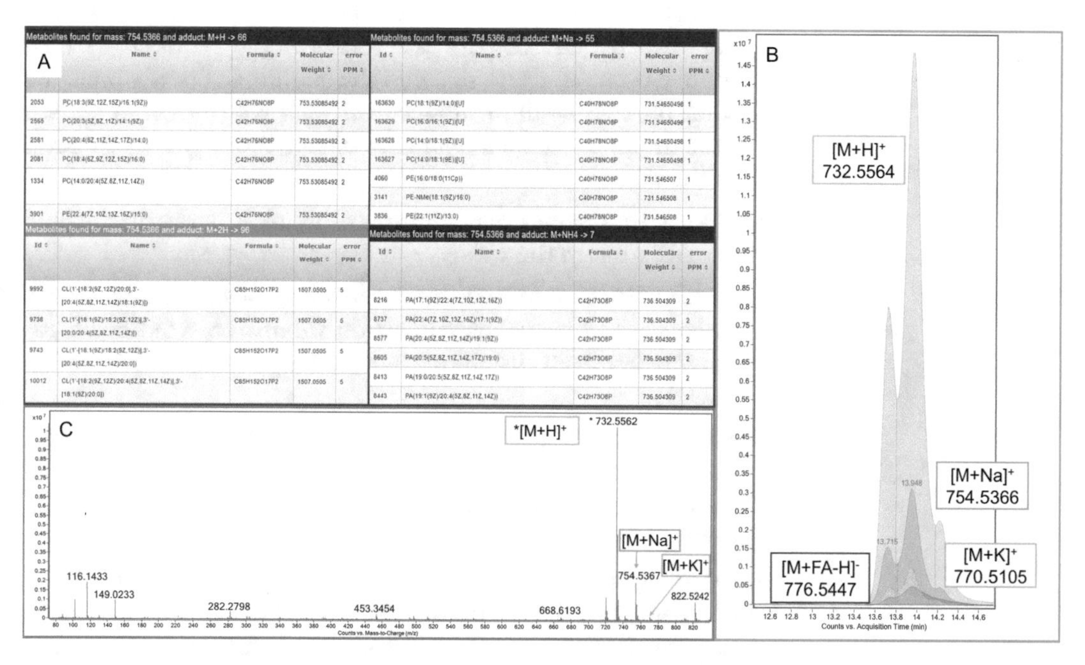

Fig. 5 Tentative annotation of a significant experimental *m/z* (754.5366) obtained by LC-ESI (+) MS analysis. (**A**) Possible candidates displayed by CMM after searching the experimental *m/z*. (**B**) EIC of the experimental *m/z* (blue), and the EICs of the *m/z* corresponding to the $[M + H]^+$ (green) and $[M + K]^+$ (orange) adducts, together with the presence of the $[M + FA\text{-}H]^-$ adduct in the same sample analyzed in negative ion mode. It can be observe the intensity pattern of the adducts is $[M + H]^+ > [M + Na]^+ > [M + K]^+$. (**C**) Saturation of the $[M + H]^+$ adduct signal, resulting in the $[M + Na]^+$ adduct as the statistically significant *m/z*. Since the $[M\text{-}H]^-$ adduct is not present and the RT and peak shape of the $[M + FA\text{-}H]^-$ adduct (**B**) match to the ones in positive ion mode, we can assume that the tentative annotation for this lipid should be PC(32:1)

5 Notes

1. If the brain samples available for your study cannot be dissected, keep in mind that the analysis of the whole brain might cause the loss of region-specific information.
2. It is highly recommended to prepare at least two blank samples for each analytical platform and treat them along the rest of the study samples in order to detect possible contaminations or undesirable artifacts. Analyze the blank samples at the beginning and at the end of the analytical run, but do not inject them in the middle of the sequence to avoid the breakdown of the system conditioning.
3. The ratio 4:1 of MetOH: MTBE allows for the precipitation of proteins and provides a monophasic solvent mixture.
4. The number of QCs required for the multiplatform analysis depends of the total number of samples included in the study. Therefore, the aliquot volume needed to prepare the QC pool must be calculated in advanced. It is recommended to analyze them intermittently throughout the run, between 5 and 10 injections.
5. Calculate carefully the volumes of mobile phases prior to starting the analysis. Prepare the total volume required of each mobile phase in order to avoid the addition of newly prepared solution during the run since this fact will induce retention time modifications and will later hinder the alignment process of the data.
6. Do not set the rack temperature below 15 °C to avoid the lipid precipitation in the vials.
7. Take into account the detector sensitivity of your equipment in order to set the adequate injection volume to avoid overloading the detector (saturation) but reaching the limit of detection (LOD).
8. Set a minimum intensity threshold for peak detection to select only the MS signals and discard the noise. This can be achieved selecting the regional peak maxima from small spectral windows [32].
9. Deconvolution process is especially important for GC-EI-Q-MS data due to the extensive fragmentation of molecular ions produced by the electron impact ionization (EI), which is a hard ionization technique [33].
10. The alignment process is more relevant for LC-MS analysis since GC-MS offers a higher reproducibility in terms of retention time [34].

11. The most popular imputation strategies for missing values (MV) replacement are zero, mean, median, k-nearest neighbors (kNN), and random forest imputation. The selection method chosen to replace MV is a crucial step in the preprocessing of data since it may have critical repercussion in the study outcome [35].
12. If the statistical differences obtained resulted from the $[M + Na]^+$ adduct might be due to the saturation of the most abundant signal ($[M + H]^+$). Then, the equipment will show the same abundance for that metabolite in both control and case groups although the real scenario is the opposite, and therefore, the significant differences will be detected for the adduct $[M + Na]^+$.
13. If CMM retrieves a PC or PE as possible candidates, they can be distinguished if negative ion mode data is available since PCs in negative mode will never form $[M\text{-}H]^-$ adduct.

Acknowledgments

Authors want to express their gratitude to the financial support received from the Spanish Ministry of Science, Innovation and Universities RTI2018-095166-B-I00, and the FEDER Program 2014–2020 of the Community of Madrid (Ref. S2017/BMD3684).

References

1. Mora-Ortiz M, Trichard M, Oregioni A, Claus SP (2019) Thanatometabolomics: introducing NMR-based metabolomics to identify metabolic biomarkers of the time of death. Metabolomics 15(3):37
2. Donaldson AE, Lamont IL (2013) Biochemistry changes that occur after death: potential markers for determining post-mortem interval. PloS One 8(11):e82011
3. Mighdoll MI, Hyde TM (2018) Brain donation at autopsy: clinical characterization and toxicologic analyses. Handb Clin Neurol 150:143–154
4. Lee K-H, Seo SW, Lim TS, Kim E-J, Kim B-C, Kim Y, Lee H-W, Jeon JP, Shim S-M, Na DL (2017) Proposal guidelines for standardized operating procedures of brain autopsy: brain bank in South Korea. Yonsei Med J 58 (5):1055–1060
5. Christensen AM, Passalacqua NV (2018) A laboratory manual for forensic anthropology. Academic Press, San Diego
6. Naz S, Moreira dos Santos DC, Garcia A, Barbas C (2014) Analytical protocols based on LC-MS, GC-MS and CE-MS for nontargeted metabolomics of biological tissues. Bioanalysis 6(12):1657–1677
7. Khamis MM, Adamko DJ, El-Aneed A (2017) Mass spectrometric based approaches in urine metabolomics and biomarker discovery. Mass Spectrom Rev 36(2):115–134
8. Haggarty J, Burgess KEV (2017) Recent advances in liquid and gas chromatography methodology for extending coverage of the metabolome. Curr Opin Biotechnol 43:77–85
9. Fujimori T, Sasaki K (2013) LC-MS-based metabolomics. Encyclopedia Syst Biol 2013:1109–1111
10. Gonzalez-Riano C, Tapia-González S, García A, Muñoz A, DeFelipe J, Barbas C (2017) Metabolomics and neuroanatomical evaluation of post-mortem changes in the hippocampus. Brain Struct Funct 222:2831–2853

11. Astarita G, Stocchero M, Paglia G (2018) Unbiased lipidomics and metabolomics of human brain samples. In: Biomarkers for Alzheimer's disease drug development. Springer, pp 255–269
12. Ivanisevic J, Epstein AA, Kurczy ME, Benton PH, Uritboonthai W, Fox HS, Boska MD, Gendelman HE, Siuzdak G (2014) Brain region mapping using global metabolomics. Chem Biol 21(11):1575–1584
13. Bayati A, Berman T (2017) Localized vs. systematic neurodegeneration: a paradigm shift in understanding neurodegenerative diseases. Front Syst Neurosci 11:62
14. Vasilopoulou CG, Margarity M, Klapa MI (2016) Metabolomic analysis in brain research: opportunities and challenges. Front Physiol 7:183
15. Samarasekera N, Salman RA-S, Huitinga I, Klioueva N, McLean CA, Kretzschmar H, Smith C, Ironside JW (2013) Brain banking for neurological disorders. Lancet Neurol 12 (11):1096–1105
16. Shichiri M (2014) The role of lipid peroxidation in neurological disorders. J Clin Biochem Nutr 54(3):151–160
17. de la Fuente AG, Traldi F, Siroka J, Kretowski A, Ciborowski M, Otero A, Barbas C, Godzien J (2018) Characterization and annotation of oxidized glycerophosphocholines for non-targeted metabolomics with LC-QTOF-MS data. Anal Chim Acta 1037:358–368
18. Hennebelle M, Metherel AH, Kitson AP, Otoki Y, Yang J, Lee KSS, Hammock BD, Bazinet RP, Taha AY (2019) Brain oxylipin concentrations following hypercapnia/ischemia: effects of brain dissection and dissection time. J Lipid Res 60(3):671–682
19. Bazán NG Jr (1970) Effects of ischemia and electroconvulsive shock on free fatty acid pool in the brain. Biochim Biophys Acta 218 (1):1–10
20. Kovács Z, Kékesi KA, Bobest M, Török T, Szilágyi N, Szikra T, Szepesi Z, Nyilas R, Dobolyi A, Palkovits M (2005) Post mortem degradation of nucleosides in the brain: comparison of human and rat brains for estimation of in vivo concentration of nucleosides. J Neurosci Methods 148(1):88–93
21. Fujii T, Hattori K, Miyakawa T, Ohashi Y, Sato H, Kunugi H (2017) Metabolic profile alterations in the postmortem brains of patients with schizophrenia using capillary electrophoresis-mass spectrometry. J Schizophr Res 183:70–74
22. Gonzalez-Riano C, León-Espinosa G, Regalado-Reyes M, García A, DeFelipe J, Barbas C (2019) Metabolomic study of hibernating Syrian hamster brains: in search of Neuroprotective agents. J Proteome Res 18:1175–1190
23. Sostare J, Di Guida R, Kirwan J, Chalal K, Palmer E, Dunn WB, Viant MR (2018) Comparison of modified Matyash method to conventional solvent systems for polar metabolite and lipid extractions. Anal Chim Acta 1037:301–315
24. Lämmerhofer M, Weckwerth W (2013) Metabolomics in practice: successful strategies to generate and analyze metabolic data. Wiley
25. Lamichhane S, Sen P, Dickens AM, Hyötyläinen T, Orešič M (2018) An overview of metabolomics data analysis: current tools and future perspectives. Comp Anal Chem 82:387–413
26. Li Z, Lu Y, Guo Y, Cao H, Wang Q, Shui W (2018) Comprehensive evaluation of untargeted metabolomics data processing software in feature detection, quantification and discriminating marker selection. Anal Chim Acta 1029:50–57
27. Mastrangelo A, Ferrarini A, Rey-Stolle F, Garcia A, Barbas C (2015) From sample treatment to biomarker discovery: a tutorial for untargeted metabolomics based on GC-(EI)-Q-MS. Anal Chim Acta 900:21–35
28. Garcia A, Barbas C (2011) Gas chromatography-mass spectrometry (GC-MS)-based metabolomics. In: Metabolic profiling. Springer, pp 191–204
29. Rochat B (2017) Proposed confidence scale and ID score in the identification of known-unknown compounds using high resolution MS data. J Am Soc Mass Spectrom 28 (4):709–723
30. de la Fuente AG, Godzien J, López MF, Rupérez FJ, Barbas C, Otero A (2018) Knowledge-based metabolite annotation tool: CEU mass mediator. J Pharm Biomed Anal 154:138–149
31. Gil-de-la-Fuente A, Godzien J, Saugar S, Garcia-Carmona R, Badran H, Wishart DS, Barbas C, Otero A (2018) CEU mass mediator 3.0: a metabolite annotation tool. J Proteome Res 18(2):797–802
32. Karaman I, Pinto RC, Graça G (2018) Metabolomics data preprocessing: from raw data to features for statistical analysis. Data analysis for omic sciences: methods and applications vol 82, pp 197–225
33. Domingo-Almenara X, Brezmes J, Vinaixa M, Samino S, Ramirez N, Ramon-Krauel M,

Lerin C, Díaz M, Ibáñez L, Correig X (2016) eRah: a computational tool integrating spectral deconvolution and alignment with quantification and identification of metabolites in GC/ MS-based metabolomics. Anal Chem 88 (19):9821–9829

34. Castillo S, Gopalacharyulu P, Yetukuri L, Orešič M (2011) Algorithms and tools for the preprocessing of LC–MS metabolomics data. Chemometr Intell Lab Syst 108(1):23–32
35. Gromski P, Xu Y, Kotze H, Correa E, Ellis D, Armitage E, Turner M, Goodacre R (2014) Influence of missing values substitutes on multivariate analysis of metabolomics data. Meta 4 (2):433–452

Chapter 21

Metallomics Imaging

Valderi Luiz Dressler, Graciela Marini Hiedrich, Vinicius Machado Neves, Eson Irineu Müller, and Dirce Pozebon

Abstract

This chapter focuses on bioimaging in metallomics where spectroscopic and nonspectroscopic techniques commonly applied in bioimaging are discussed as well as methods of sample preparation for bioimaging. Applications for metals and other important elements in animal tissues, cells, and subcellular units are presented. In general, bioimaging metallomics requires techniques that enable high resolution (nm to μm) to obtain qualitative and/or quantitative information about elements distribution in biological tissues and cells. The tendency is in vivo analysis to get more realistic information about the role of metals and key elements in the living beings. Bioimaging metallomics is a complex theme and conclusive results are in general obtained in a multidisciplinary way.

Key words Metallomics, Bioimaging, Techniques for bioimaging, Biological tissues, Cells

Abbreviations

2D	Two dimensional
3D	Three dimensional
AD	Alzheimer disease
AFM	Atomic force microscopy
AmD	Amyloid deposits
Ar	Argon
CCD	Charge-coupled device
CLMS	Confocal laser scanning microscopy
CRM	Certified reference material
CT	Computed tomography
CyTof-ICP-MS	Cytometry-ToF-ICP-MS
Da	Dalton
DNA	Deoxyribonucleic acid
DOTA	Tetraazacyclododecane-1,4,7,10-tetraacetic acid
DTPABMA	2-[bis[2-[carboxymethyl-[2-(methylamino)-2-oxoethyl]amino]ethyl] amino]acetic acid
EDFM	Enhanced dark-field microscopy
EDS	Energy dispersive X-ray spectrometry

Paul L. Wood (ed.), *Metabolomics*, Neuromethods, vol. 159, https://doi.org/10.1007/978-1-0716-0864-7_21,

EDX	Energy-dispersive X-ray spectroscopy
EDXS	Energy-dispersive X-ray spectroscopy
EELS	Electron energy loss spectroscopy
EFTEM	Energy-filtered transmission electron microscopy
EM	Electron microscopy
EPRI	Electron paramagnetic resonance imaging
FEG	Field-emission gun
fs	Femtosecond
FTIR	Fourier transform infrared
Gd-DTPABMA	2-[bis[2-[carboxylatomethyl-[2-(methylamino)-2-oxoethyl]amino]ethyl] amino]acetate;gadolinium(3+)
GE	Gel electrophoresis
HIS	Hyperspectral imaging
HPF	High-pressure freezing
ICP	Inductively coupled plasma
ICP-MS	Inductively coupled plasma mass spectrometry
ICP-QMS	Inductively coupled plasma quadrupole mass spectrometry
ID	Isotope dilution
IMCy	Imaging mass cytometry
IP	Ionization potential
IS	Internal standard
LA	Laser ablation
LA-ICP-MS	Laser ablation-inductively coupled plasma mass spectrometry
LA-MSI	Laser ablation mass spectrometry imaging
LMIG	Liquid-metal ion gun
LOD	Limit of detection
m/z	Mass-to-charge ratio
MeCAT	Metal-coded affinity tag
MRI	Magnetic resonance image
NanoSIMS	Nano-secondary ionization mass spectrometry
Nd:YAG	Neodymium–yttrium–aluminum garnet
nm	Nanometer
NPs	Nanoparticles
PAGE	Polyacrylamide gel electrophoresis
PET	Positron emission tomography
QDs	Quantum dots
ROI	Region of interest
ROS	Reactive oxygen species
RS	Raman spectroscopy
RSFs	Relative sensitivity factor
SEM	Scanning electron microscopy
SF-ICP-MS	Sector field-inductively coupled plasma mass spectrometry
SIMS	Secondary ionization mass spectrometry
SRM	Super resolution microscopy
STEM	Scanning transmission electron microscopy
STEM-EELS	Scanning transmission electron microscopy- electron energy loss spectroscopy
SXRF	Synchrotron-based X-ray fluorescence
SXRFM	Scanning X-ray fluorescence microscopy

TEM	Transmission electron microscopy
Tm(DOTA)	Thulium(tetraazacyclododecane-1,4,7,10-tetraacetic acid)
ToF	time-of-flight
ToF-ICP-MS	time-of-flight inductively coupled plasma mass spectrometry
ToF-SIMS	Time-of-flight secondary ionization mass spectrometry
VSOP	Very small iron oxide particles
WD	Wilson disease
WDS	Wavelength dispersive spectrometry
XAFS	X-ray absorption fine structure
XANES	X-ray absorption near edge structure spectroscopy
XAS	X-ray absorption spectroscopy
XEDS	Energy-dispersive X-ray spectroscopy
XRF	X-ray fluorescence
μFTIR	Micro-Fourier transformed infrared microscopy
μPIXE	Micro-proton-induced X-ray emission
μXANES	Micro-X-ray absorption near edge structure spectroscopy
μXRF	Micro-X-ray fluorescence

1 Introduction

Several metals and other relevant elements are present in a wide range of concentrations in living organisms and are crucial for life and involved in for many biological processes, including cellular functions, cell signaling, (macro)molecular structure formation, and (co)enzyme activation. About one-third of the proteins require metals to fulfil biological processes and half of the enzymes are associated to a particular metal. Twenty-eight chemical elements are essential for biological systems function, whereas abnormal concentrations of them are associated with physiological disturbances and diseases. These elements are heterogeneously distributed in organs, tissues and cells, bound to (macro)molecules or as free ions. Ions such as Mg^{2+}, Ca^{2+}, Zn^{2+}, and Mn^{2+} are involved in activation and stabilization of macromolecules while redox-active ions such as Fe^{2+}/Fe^{3+} and Cu^{+}/Cu^{2+} are essential for catalytic reactions and/or electron transfer. For example, higher amounts of Zn^{2+}, Cu^{+}/Cu^{2+}, and Ca^{2+} are present in more active brain regions and are associated with neuronal cell signaling through the action of electric current within and between cells. Other free metal ions are actively involved in immune responses, through catalysis of reactive oxygen species (ROS), activation of inflammasomes, and T cell activation or suppression. Thus, it is important to know the concentration of involved metals and metalloids, their chemical form (species) and distribution within the organism.

Many discoveries of function, speciation, location, and interactions of chemical elements in biological processes are due to the development of sensitive and selective instrumental techniques. Depending on the exact location of a given element, its function could be elucidated. Therefore, in recent years, efforts have been made to develop and launch techniques that allow obtaining images of spatial distribution (bioimaging) of chemical elements in biological tissues and cells (even at subcellular level). The spatial resolution achieved ranges from nm to μm, two- and three-dimensionally (2D and 3D, respectively). These techniques are based on absorption, emission, transmittance and reflectance of the electromagnetic radiation, mass spectrometry and particle beam microscopy, which are nowadays employed to investigate and determine local distribution and interaction of metals and metalloids in the human body. The following techniques have played an important role in bioimaging metallomic in cells and biological tissues: magnetic resonance (MR), Fourier transform infrared (FTIR), Raman spectroscopy (RS), confocal laser scanning microscopy (CLSM), super resolution microscopy (SRM), energy-dispersive X-ray spectroscopy (EDX, EDXS, XED), synchrotron-based X-ray fluorescence (SXRF), X-ray absorption spectroscopy (XAS), positron emission tomography (PET), laser ablation-inductively coupled plasma mass spectrometry (LA-ICP-MS), secondary ion mass spectrometry (SIMS), scanning electron microscopy (SEM), transmission electron microscopy (TEM) and scanning transmission electron microscopy (STEM).

Most of these techniques are based on the use of a high energy bean, which scans the solid sample when bioimaging is intended. The beam is composed of photons, ions, or electrons and the signal intensity of each pixel of the specific element is recorded and converted into an image representing the element in the specimen. Essential information can be obtained from biological tissue analysis using the cited techniques whose results are often complementary to each other and make possible a better understanding of metal and metalloids functions and behavior in biological processes. In general, the specimen cannot be analyzed directly, requiring some modification or an appropriate treatment in order to produce high-quality images. More information of the techniques above cited and their applications in metallomic bioimaging will be described in the following sections.

2 Mass Spectrometry–Based Techniques for Bioimaging

2.1 Laser Ablation Inductively Coupled Plasma Mass Spectrometry

Inductively coupled plasma mass spectrometry (ICP-MS) is a multielemental and quantitative analysis technique that offers large linear response, isotopic analysis capability, and high sensitivity. The limit of detection (LOD) for most elements is typically at the ng L^{-1} level when the ICP-MS instrument is operated in standard mode and samples in the form of solutions are introduced in the

ICP by pneumatic nebulization. The ICP-ion source is versatile to be associated with different sample introduction systems, including laser ablation (LA). Mass spectrometers are combined with the ICP ion source, and they differ mainly according to the mass filter; quadrupole (ICP-QMS), time of flight (ToF-ICP-MS), and sector field (SF-ICP-MS) are the most common.

Since the early applications of LA-ICP-MS for bioimaging [1, 2], the technique has grown substantially and remarkably studied [3, 4], including calibration strategies for quantitative bioimages, despite that qualitative bioimages can also provide useful information about elements distribution in the sample under analysis. The technique is suitable for determination of most elements in solid samples, with reasonable to good spatial resolution.

Three distinct processes are sequentially involved in LA-ICP-MS: (1) ablation of the sample material (sample located inside an ablation chamber); (2) ionization of the elements in the ICP source; and (3) separation/measurement of one or more isotopes of investigated elements by the mass spectrometer. The ions are separated according their mass-to-charge ratio (m/z) by the mass spectrometer and once reaching the detector an electric signal is generated and measured. The ablated material in the form of vapor or fine particles is carried to the ICP by an inert gas stream, usually Ar or Ar mixed with He or N_2. The use of Ar or Ar mixed gases depends mainly on the wavelength of the laser employed. For example, better results are obtained for lower wavelength laser (213 nm) when Ar plus He are used, whereas the carrier gas type does not have any influence on the results obtained when a 266 nm laser is employed. The gas stream flows through the LA chamber that is connected by a transfer line to the injector tube fitted to the ICP torch.

The laser pulses diameter ranges from one to several hundred μm, being the spatial resolution dependent of the spot size and number of spots in the scanned sample area. The sample ablation is software-controlled, allowing for accurate and precise laser shots at specific points on the sample. For bioimaging, all surface of the biological sample or a region of interest (ROI) is scanned line-by-line by laser pulses. The resulting time resolved data is then reconstructed into an image using a software such as MATLAB, ImageJ, Origin, R, or others.

In LA-ICP-MS the ICP operates on dry conditions and interference by hydrogen and oxygen polyatomic ions derived from water in the ICP are strongly reduced. This constitutes an advantage in comparison to nebulization systems, typically used in the analysis of solid samples that are decomposed and brought into aqueous solutions. However, polyatomic ions are still formed and they interfere on biologically relevant elements. These polyatomic ions are from major elements constituting the sample, plasma gas (Ar), and gas from the atmosphere, namely, CO_2, N_2, and O_2

[5]. That interference can be circumvented or eliminated by employing ICP-QMS instruments equipped with dynamic reaction cells or collision cells [5–7], or instruments with superior resolution as SF-ICP-MS and multicollector ICP-MS [8, 9].

A laser ablation system consists of a high-power laser source, an ablation chamber, and a CCD camera. The most common laser sources combined with ICP-MS instruments are the solid state neodymium–yttrium–aluminum garnet (Nd:YAG) and Ti:sapphire which generates laser pulses lasting (ns) and femtosecond (fs), respectively. Lasers with different wavelengths (from ultraviolet to infrared) have been hyphenated with ICP-MS; 1064 nm, 694 nm, 532 nm, 266 nm, 213 nm, and 193 nm. Shorter wavelength lasers as fs lasers have demonstrated better ablation performance and led to less matrix interference. The ablation performance is also dependent on the sample matrix; for soft biological tissues, ns laser (266 nm or 213 nm) can yield images with good spatial resolution. Even better spatial resolution (in nm) is achieved using fs laser [10], with vaporization of fg to pg of sample per laser pulse, making LA-ICP-MS a quasi-nondestructive technique.

Although the spatial resolution provided by LA-ICP-MS is lower than that of electron microscopy (EM), SIMS, and SXRF, the former is more sensitive and capable of isotopic analysis. In addition, LA-ICP-MS enables multielement images and can be combined with immunohistochemistry, isoelectric focusing electrophoresis, and polyacrylamide gel electrophoresis (PAGE) for proteins mapping in biological tissues [4].

The volume of the LA chamber where the sample is accommodated is about 3–100 cm^3. The volume and geometry of the LA chamber and the transfer line size (its length and internal diameter) influence the aerosol transport efficiency and the analyte signal profile because they affect the dispersion of the ablated material vapor and the image resolution as a consequence. High dispersion of the sample vapor due to long washout time of the ablation chamber worsen the bioimage resolution. Ablation chambers with different geometry for bioimaging were proposed for bioimaging by Malderen et al. [11]. When high resolution is required, as for element imaging in cells, LA chambers designed for rapid pulse response (<10 ms) yield better images. To detect such short signal time, ToF-ICP-MS instruments are more feasible since the ToF mass analyzer performs very fast mass scanning and quasi-simultaneous real-time full-elemental-mass range detection. However, the sensitivity of ToF-ICP-MS is about one order of magnitude lower than that of ICP-QMS. Even so, both instruments are feasible for multiprotein imaging at subcellular level, similar to mass cytometry [12]. If the LA chamber geometry allows a constant laminar flow and well-controlled delivery of the aerosol into the transfer line, the aerosol dispersion is reduced and images resolution improved due to better analyte signals. By employing this type of LA chambers and SF-ICP-MS instrument, a high-resolution image (1 μm) of Ho added to a cell biomarker in breast cancer tissue was obtained [13].

Laser ablation at temperature lower than 0 °C is also used for biological tissue analysis [14–18]; frozen sections of the sample can be analyzed directly, preserving the sample integrity, improving precision, and enabling the generation of bioimages with less blurring. The better resolution has been attributed to low water load and/or less energy dissipation/heating effects around the spot crater in the sample under ablation, with fewer effects on the ablation process/crater formation.

Laser ablation is also used for sampling in imaging mass cytometry (IMCy). In this case, the LA system is associated with a cytometry-ToF-ICP-MS instrument (CyToF) [19] that enables simultaneous detection of isotopes, at a high acquisition speed and resolution comparable to light microscopy. Such high-resolution was achieved using a low-dispersion LA chamber and 213 nm laser operated at 100–200 Hz, focused to produce 1 μm spot size [12]. Mass cytometry is based on the use of antibodies conjugated to polymers incorporating a specific chelating agent. Antibodies are labeled with stable isotopes primarily from the lanthanide elements, with additional nonlanthanide metal tags (Bi, Au, Pt), which sets an upper limit of 40 antibodies that can be combined in staining protocol. In addition to the metal-tagged antibody approach, there is potential to develop oligonucleotide probes for gene transcription analysis using IMCy [20]. By means of IMCy, analysis of human formalin-fixed, paraffin-embedded breast cancer and human mammary epithelial samples revealed substantial tumor microenvironment heterogeneity. Cells or tissue sections were labeled with antibodies that were selected to target proteins and protein modifications relevant to breast cancer. Before staining, antibodies were tagged with a lanthanide element. Then, the samples were submitted to LA and the ablated material was then transported to a CyToF instrument where the tagged element was ionized and detected. The signals of individual laser shots at 20 Hz were fully separated, and the LOD was approximately six ion counts that corresponded to ~500 molecules. IMCy was also used to investigate cisplatin (an anticancer drug) distribution in patient-derived xenografts, where it was observed extensive binding of Pt to collagen fibbers in cancer tumor, as well as in normal tissues. These results provided new information with respect to the cisplatin pharmacology [21].

Calibration has been one of the main topics studied in the development of LA-ICP-MS. Despite that and the large amount of work already conducted, calibration in quantitative analysis using LA-ICP-MS is still challenging [4]. Aiming a quantitative analysis, many strategies and approaches of calibration have been evaluated, including internal standardization, external calibration with matrix-matched standards, standard addition, isotope dilution (ID), film coating and printing. Elemental fractionation (sum of nonstoichiometric effects occurring in sample ablation, aerosol transport, and analyte ionization in the ICP) is the main responsible for impairing quantitative analysis using LA-ICP-MS. Laser wavelength, laser pulse duration (fs or ns), laser fluence, and spot size influence on fractionation. This phenomenon can be reduced by employing lower laser wavelength and shorter laser pulses because smaller aerosol particles are produced and thermal effects reduced when the laser interact with the sample. In this way, not only accuracy but also sensitivity and precision are improved [22]. In a study conducted by Claverie et al. [23] using fs laser at high repetition frequency, the ablation of polyacrylamide gels for Se determination in selenoproteins showed that up to 77% of the aerosol particles were below 1 μm, whereas transport of large particles coming from the deepest part of the sample was low. Thus, fractionation was greatly reduced since the aerosol particles were very small. Calibration with internal standardization can compensate fluctuations in laser output power, differences in sample thickness, sample mass ablated and transported, instabilities and drifts in the ICP source. Homogeneous distribution of the internal standard (IS) in the sample, similarity of the m/z of the IS and that of the analyte, as well as the respective ionization potentials (IP), and absence of spectral interference are the main recognized prerequisites for selecting an IS in the analysis of a solid sample by LA-ICP-MS. In the case of biological tissues, ^{13}C, ^{33}S and ^{34}S naturally present in the sample are commonly used as IS, although they do not meet the mentioned prerequisites; the ionization potentials of C and S are relatively high, S isotopes are subject to spectral interferences by polyatomic ions generated from the atmospheric ICP, and both elements are not always uniformly distributed in the sample material. Even so, accurate results can still be obtained, as was observed by Sela et al. [24], Pozebon et al. [25], and Luo et al. [26] in hair analysis using ^{34}S and ^{32}S as IS for analytes with m/z quite different from those of the S isotopes. In the case of hair, internal standardization with S isotopes is advantageous because that element is homogeneously distributed in hair. In addition, ^{13}C has been used as IS to cancel out variations in ablation, transport, and ionization efficiency in element quantification in animal tissues [14, 27–29]. Internal standardization is also recommended for qualitative bioimaging being images generated from the ratio analyte signal/IS signals. Thereby, qualitative bioimages of Mg, P, Ca, Mn, Fe, Cu, Zn, and Pb in kidney and brain were obtained using ^{13}C as IS [30]. The different water content among the sample and standards used for calibration can also be compensated by dividing the analyte signal to that of ^{13}C or S isotopes [31].

However, several authors mention that ^{13}C might not be an efficient IS in quantitative analysis by LA-ICP-MS [32–36] by considering that C is released from the sample as gaseous species (CO_2 and CO) and solid particles. Consequently, the transport of these species and the analyte to the ICP and their ionization would differ remarkably, which would affect accuracy and precision of the results. It should also be considered that the C and S IP are higher than those of most elements, and thus the degree of ionization of them in the ICP will be different. Calcium isotopes (^{43}Ca or ^{44}Ca), naturally present in tooth and bone, were used as IS in element quantification in tooth layers [37–40] and bone [41–43]. However, similarly to C and S, Ca is also not homogeneously distributed in tooth [39]. Therefore, there is not any element naturally present in biological tissues that fulfils the requirement of an ideal IS.

The IS element can also be added to the analyzed sample, as investigated by Austin et al. [44]. In a detailed study about calibration in LA-ICP-MS, ^{13}C, ^{52}Cr, ^{53}Cr, ^{89}Y, and ^{101}Ru were evaluated as IS and the authors concluded that any of these isotopes was effective to compensate for changes in mass ablated, but ^{13}C was in general less effective than the others. They observed the accuracy and precision were better when the *m/z* and IP of IS were close to those of the analyte and the IS signal at least 6% of that of the analyte. The IS addition can be carried out by spiking and homogenizing the sample with a defined amount of the IS element, printing the sample with an ink enriched with the IS or sputtering the IS over the sample. Following such approaches, in Pt bioimaging in kidney of rat treated with cytostatic Pt drugs (cisplatin, carboplatin, and oxaliplatin) Ir was used as IS by printing it on 4 μm sagittal sections of kidney by means of a conventional inkjet printer [45]. The bioimages obtained allowed a reliable comparison of the kidneys from rats treated with different Pt-based drugs. However, it was a pseudo-internal standardization because the IS remains located only on the sample surface and could not overcome differences in sample thickness. Instead of printing, Konz et al. [46] sputtered the IS (Au) on the sample surface (sliced eye tissue) for bioimaging of Cu, Fe, and Mg. The authors affirm that Au was a more reliable IS than ^{13}C because better precision was achieved using Au. An extension of this approach was proposed by O'Reilly et al. [47] for Fe bioimaging in sheep brain; slices of sheep brain fixed on glass slides were immersed in solutions containing Rh, that was used as IS. Through this internal standardization procedure, precision was improved twofold and accuracy similar to that obtained using μXRF, for specific brain regions. In summary, internal standardization usually improves precision and accuracy in LA-ICP-MS, but it is challenging to choose the most appropriate IS.

Matrix matching calibration is usually necessary for quantitative bioimaging using LA-ICP-MS, even with internal standardization. Certified reference materials (CRM) or homemade standards are used for calibration, but the CRMs and sample matrices should be as similar as possible to circumvent matrix effects. However, the number of CRMs available with matrix corresponding to that of the analyzed biological sample is limited. Therefore, homemade standards are quite often used for matrix matching calibration. In preparing the standards, the analyte is added to the sample followed by their mixing and homogenization. If the standard is from animal tissue, the homogenate is sliced and mounted on glass support for ablation. Pioneering works about element imaging in brain tissue conducted by Becker et al. [48, 49] were based on this calibration procedure.

Aliquots of homogenized brain tissue spiked with elements of interest and then encapsulated in a sol–gel matrix produced by tetraethyl orthosilicate [24] has been used as a homemade standard; chicken breast homogenate [5, 27] and egg yolk [50], both spiked with the analytes, are also examples of homemade standards. The egg yolk was spiked with the analyte (Tm) and heated up to 90 °C for 10 min to form a solid, which was subsequently frozen and cryocut, fixed on a glass slide, and then the sample was ablated. Quantitative images were obtained for tumor cells and macrophages labeled with Tm(DOTA), complementing information obtained by magnetic resonance imaging (MRI). Other procedures involving homemade standards for quantitative elemental imaging in animal tissue are as follows.

Whole blood or blood serum spiked with known amounts of Sr, Gd, and Pt were deposited on holes drilled in a carboxymethyl cellulose block that was used as support. Before LA-ICP-MS analysis, the block filled with the standards was cryocut into 20–60 μm slices thickness [51] and the slices deposited on glass slides. Cold-polymerizing resin poly-2-hydroxyethyl methacrylate (Technovit) was used to support Pt (as platinum acetylacetonate) and samples of cochlea, testis, and kidney of mice. The analyzed samples were washed with water/methanol/ethanol, incubated in isobutanol, infiltrated with the resin, and then polymerized and cured. The polymer block with the tissue or standards was sectioned and the slices placed on object slides for LA-ICP-MS analysis. Platinum distribution at different time intervals after cisplatin treatment of mice was determined quantitatively in the different tissues [52]. The same method was applied for Pt bioimaging in rat bone submitted to treatment with Pt drugs [53]. For quantitative bioimaging of Cu, Zn, Cd, Hg, and Pb in ancient Chilean mummies and contemporary Egyptian teeth, samples were mounted in epoxy resin and sliced longitudinally using a steel-bladed saw. Reference material of caprine bone [40] or certified bone meal [54] pressed into pellets were used to prepare standards for quantitative elements image in teeth, whereas Ca was used as IS. Hair strands enriched with the analyte [25, 55] or powdered hair standards placed on carbon tabs were used as standards [55] for quantitative determination of elements along the hair. Following this procedure, elements could be mapped in single hair strands, making possible to monitor absorption/elimination of nutrient or toxic elements in human body.

Continuing calibration studies in LA-ICP-MS, standards supported on appropriate material such as filter paper or gelatin were proposed. It is assumed that this strategy of calibration combines simplicity and easy handling, and standards can be prepared to encompass a wide analyte concentration range. When filter paper is used, the analyte solution is simply deposited onto the paper, which is then let stand for drying. In addition to the contribution by the paper constituents for matrix matching, ^{13}C naturally present in the paper can be used as IS, despite that other IS element in solution can be easily added to the filter paper. However, care must be taken when the solution is added to the paper in order to avoid chromatographic effect, which results in nonhomogeneous distribution of analytes and IS in the paper. Calibration with supported standards was proposed for Mn, Ni, Cu, Zn, and K imaging in human malignant mesothelioma biopsy [56] and Pt in human malignant pleural mesothelioma [57]. In both cases, Au that was used as pseudo-IS was sputtered onto the surface of the filter paper and samples. Calibration with dried-droplet standards deposited on filter paper was carried out by Shariatgorji et al. [58] for quantitative imaging of Mg, Ca, Mn, Cu, and Zn in mouse kidney and traumatic brain injury model tissue sections. The authors also extended this calibration strategy to laser ablation mass spectrometry imaging (LA-MSI) for Na and K imaging in such samples.

Gelatin has been used to prepare matrix-matched standards in single-cell analysis [59]. In this case, cells and standards in gelatin can be placed in microarray plates with small holes. For holes with 110 μm diameter, spaced 185 μm apart, it was possible to determine Cu in the cells. The images of Cu obtained by LA-ICP-MS were similar those obtained using SXRF.

Solution based calibration has been investigated for quantitative imaging of elements in rat brain; the dry aerosol from the LA was mixed with the desolvated aerosol produced by a nebulizer coupled to a desolvation system. Both aerosols were online mixed through a Y piece placed before the injector tube of the ICP torch. Matrix-matching calibration was based on nebulization of aqueous standards prepared in 2% v/v HNO_3 with simultaneous laser ablation of mouse brain homogenate. The ratio of the slope of calibration curve obtained with homemade mouse brain homogenate standard and that with aqueous standards was used to correct the difference of sensitivity among solution nebulization and LA [60]. For Th imaging in human brain, Becker et al. [48] adapted a total consumption micronebulizer directly to a cooled LA chamber. During ablation of human brain slices with thickness of 20 μm the standard solutions were nebulized inside the LA chamber. A correction factor was applied to compensate differences of sensitivity in nebulization-ICP-MS and LA-ICP-MS.

Isotope dilution has also been employed for quantitative bioimaging. ID is an absolute method of quantification, which consists of adding an enriched isotope to the sample and comparing the ratio of two isotopes of the analyte, being one of them enriched. In LA-ICP-MS the enriched isotope solution can be added directly to the solid sample and homogenized or online mixed with the dry aerosol that constitutes the ablated material. For online mixing, a total consumption nebulizer or a nebulizer with desolvation system can be adapted to the LA chamber or the aerosol from the solution nebulization can be added to the carrier gas after the LA chamber and before the ICP source. The almost dry aerosol generated by solution nebulization is comparable to that produced by LA, reducing water interferences, mainly by oxides and hydroxides. For quantitative Fe bioimaging in sheep brain tissues a solution of enriched ^{57}Fe was added by a total consumption nebulizer (the solution flow rate was 8 μL min^{-1}) to the aerosol from LA before entering the ICP. An SF-ICP-MS instrument operated at medium mass resolution ($R = 4000$) was employed for Fe detection and quantification whereas the ratio ^{57}Fe/^{56}Fe was measured [61].

Quantitative and qualitative bioimaging of nanoparticles (NPs) and quantum dots (QDs) by LA-ICP-MS are still not well established due to difficulties associated with the signal output, which appears in the form of pulses or spikes. The signal is dependent of particles size, particles aggregation, gravitational settling in the transport line, incomplete ionization in the ICP and number of them in the sample. Therefore, matrix-matched standards to obtain calibration curves for each specific type of NPs or QDs are necessary [62]. The incorporation of Au, Ag, and Al_2O_3 NPs in Danio rerio (D. rerio) embryos and Daphnia magna (D. magna) were evaluated by Böhme et al. [63]. Matrix-matched standards consisting of NPs of similar size embedded in agarose gel were used for calibration. Despite calibration with matrix-matched standards, size-dependent sensitivity was observed and it was related to particles segregation in the ablation process, changes in NPs size distribution, NPs agglomeration, and gravitational settling in the transport line or incomplete ionization in the ICP. Very small Fe oxide particles (VSOP) doped with Eu were injected in mice and further mapped in their liver and atherosclerotic plaques using LA-ICP-MS [64]. For matrix-matched calibration, a EuVSOP suspension was dropped onto deparaffinized mouse liver tissue sections and dried. The LA-ICP-MS analysis resulted in sensitive and specific detection of Eu in EuVSOP in liver and atherosclerotic plaques, allowing calculating Fe and EuVSOP concentration in tissue sections and correlating endogenous Fe in aortic roots sections.

Cadmium was mapped in zebrafish larvae exposed to SeCd/ZnS QDs, which is a semiconductor in the form of NPs. For LA-

ICP-MS analysis, frozen zebrafish larvae were placed over a polycarbonate plate previously covered with a thin Au layer to serve as IS. Ionic Cd was detected mainly in eye areas while QDs appeared to be more homogeneously distributed along the larvae surface. These results suggested that QDs were not taken by the zebrafish larvae and the accumulated Cd could correspond to the fraction of ionic Cd present in the solution, which was subsequently concentrated in the eyes region. Another possibility would be that QDs could be more adsorbed over the zebrafish larvae peel than absorbed [65]. CdSe QDs are promising as drug carriers/delivers and fluorescent markers, with potential application in pulmonary drug delivery and lung imaging. LA-ICP-MS images of CdSe QDs in mouse lung that had inhaled these particles revealed distribution of Cd in the lungs, arising from QDs accumulated in the bronchiolar area. The LA-ICP-MS images obtained corroborated those of fluorescence spectroscopy imaging. Standards obtained by spiking slices of the lung tissue sample with Cd in solution allowed for quantifying the QDs in the lung [66].

Although the resolution of LA-ICP-MS is relatively low when compared to SXRF, EM, and SIMS, LA-ICP-MS enables quantitative bioimaging of elements, including free ions and NPs in single-cells. This can be possible by adjustment of the laser and ICP-MS instrumental parameters and calibration with standards similar to cells in order to avoid differences in the volume ablated and provide comparable mass load and responses. Quantification of NPs is based on the number of NPs in single-cell. In this context, LA-ICP-MS was employed for quantitative bioimaging of Ag and Au NPs in individual fibroblast cells. By improving the spatial resolution (low scan speed, higher pulse repetition rate, lower spot size and adequate dwell time), NPs localized in the cytosol were distinguished from those in the nucleus. The spatial resolution is better if ToF-ICP-MS or ICP-QMS with fast response employed. Drescher et al. [67] employed an SF-ICP-MS instrument for quantitative bioimage of Ag- and Au-NPs distribution in single 3 T3 cells incubated with the NPs. Nitrocellulose membrane doped with suspension of the NPs was used for calibration. The results obtained were useful to evaluate NPs and cell interactions.

Human $CD4^+$ T cells were labeled with the contrast agents Gd-DTPABMA (Dotarem) and Gd-DOTA (Omniscan) and analyzed by LA-ICP-MS. Single cells were plated onto slides and then identified by ablating 25 μm diameter areas in order to ensure sampling of the whole cell, without overlapping by neighboring cells. In mixed preparations of labeled and unlabeled cells, the enumerated labeled cells were close to the predicted ratio [68].

Human regulatory Au labeled macrophages (Mregs) were incubated with 50 nm AuNPs and injected in immunodeficient mice whose organs were afterward analyzed by LA-ICP-MS. The laser spot size was 8 μm in order to improve spatial resolution and make possible the Au-labeled Mregs detection. However, the laser beam diameter was 55 μm and ensured complete ablation of individual cells. The bioimages obtained revealed that more labeled Mregs were present in lung, liver, and spleen than in kidney, heart, brain, and intestine [69].

In single cell analysis using LA-ICP-MS the system can be operated to obtain images with resolution at subcellular level or single spot analysis of cells with a larger laser spot size. To evaluate both modes of analysis, adherent 3 T3 fibroblast cells stained with mDOTA-Ho or Ir-DNA-intercalator were analyzed. The signal-to-noise ratio observed for single spot mode was tenfold higher than that observed for the scanning mode. Matrix matched calibration based on standards spotted onto nitrocellulose membrane was used for Ir and Ho quantification in the cells. The limits of detection (LODs) of Ir and Ho were 12 fg and 30 fg, respectively. The amount of Ir and Ho found per single cell was 57 ± 35 fg and 1192 ± 707, respectively. The total concentration of Ho and Ir found by LA-ICP-MS was compared to that found by ICP-MS for digested single-cells [70]. In summary, calibration using matrix-matched standards is usually employed in LA-ICP-MS bioimaging metallomics. Internal standardization improves precision and accuracy but it is challenging to have an adequate IS. More difficulties are observed in quantitative single-cell bioimaging, mainly when NPs determination is intended.

The sample preparation procedure for bioimaging using LA-ICP-MS is also noteworthy; samples can be paraffin-embedded and cut into slices as in histological analyses or the native tissue can be cryocut. The slices thickness ranges from 1 to 50 μm. However, it was already demonstrated that cryogenic slicing is better than paraffin-embedded slicing because the former is less prone to analyte loss [71]. In the paraffin-embedded procedure, samples are sequentially treated with 4% paraformaldehyde and 30% sucrose, sliced with a microtome and fixed on glass. In cryogenic slicing, they are cut with a cryomicrotome at −18 to −16 °C and thaw-mounted onto adhesive microscope slides. Then, they are left drying and stored at room temperature [3, 15, 48, 72]. It was found that in murine brain treated with 4% paraformaldehyde and 30% sucrose followed by cryoprotection, several elements were lost; almost all K (99.78%) and Mg (78.01%) were leached from the brain tissue, while Ca and Sr either retained or absorbed additional metal ions from the sucrose solution. Leaching of less water-soluble transition metals was lower, being 26.61–31.62% for Fe, Cu, and Zn [71].

Bioimages can be plotted in 2D and 3D and created by repeatedly scanning of a specimen. For 2D images, a single slice of the sample is ablated line-by-line and the data (signal intensities) reconstructed by appropriate software. In 3D images, all slices of the sample are ablated line-by-line to construct the image [3, 4, 73] using a software written in-house or free software [74–82].

For imaging, the laser speed can be calculated by considering the relationship between the laser scan speed, the laser spot diameter, and the total scan cycle of the mass spectrometer [83]. Improved spatial resolution is achieved using smaller laser spot diameter and low laser scan rate, but at the expense of time and sensitivity. The measurement time for bioimaging could take a few minutes to several hours, depending on the size of the tissue area under analysis. Paul et al. [84] developed a method for 3D imaging of mouse brain by serial sectioning and subsequent volume reconstruction based on an image registration. Atlases depicting quantitative Fe, Cu, and Zn distribution in the mouse brain (cerebrum and brainstem) were constructed from the data obtained in LA-ICP-MS analysis of forty-six thin sections of the sample. In this way, visualization of metals distribution in the brain was improved and advances to elucidate their role in neurobiology were provided, but the analysis time for only one sample was 158 h [85]. In the case of hair, the hair sample can be directly analyzed; washing is the only treatment needed. Platinum was monitored in a single hair strand of a patient submitted to cisplatin treatment and Pt variation along the hair with reference to each cisplatin dose was clearly observed [25]. In a study about grizzly bears contamination [86], Hg concentration along the entire length of individual grizzly bear hair demonstrated a relationship between the Hg present in the hair and the Hg intake through the fish diet, as well as element elimination from the body during hibernation of the animal.

2.2 Secondary Ion Mass Spectrometry

The principle of SIMS is based on the sample bombardment by a focused primary ion beam under high vacuum to sputter neutral particles, elemental and molecular ions from the sample surface. The negative or positive ions generated are then analyzed by a mass spectrometer [87, 88]. For imaging, the primary ion beam scans the ROI on the sample surface. It is possible to collect the signal of the single ion, a few ions with specified *m/z*, or a whole mass spectrum ranging from a few to 1000 *m/z* or more, depending on the mass spectrometer employed. After scanning a specific area of the sample, images of the elements or compounds in the sample can be constructed, whose resolution is in general determined by the primary ion source focus. SIMS is very suitable for bioimaging due to its high resolution and capability of mapping metals, metalloids, and molecular compounds in biological tissues and single cells [89]. Similar to LA-ICP-MS, SIMS enables isotopic analysis. The spatial resolution of SIMS is at the nm scale for mass range from 1 to 2000 Da. However, the majority of ejected molecules are neutral and ionization is not efficient for all species, making the technique somewhat limited in metallomic applications involving transition metals because these elements are poorly ionized, less than low mass elements. Despite of that, SIMS is useful for light elements and molecular fragments imaging since it provides informative molecular context for metals.

SIMS spectrometers are in general equipped with ToF mass analyzers (ToF-SIMS) or double focusing sector field mass analyzers (NanoSIMS) [90]. Spatial resolution as low as 100 nm and 50 nm are achieved by ToF-SIMS and NanoSIMS, respectively. SIMS can be operated in dynamic and static modes; in dynamic mode the primary ion beam reaches the sample continuously and allows depth profiling and excessive fragmentation that leads to excellent elemental analysis; in static mode the primary ion beam is pulsed on the sample surface, which reduces sample erosion and provides surface-sensitive molecular information. In static mode, a low-energy and low-density ion beam is used (lower than 10^{12} ions cm^{-2} to minimize the interaction of the primary ions with the molecules on the top monolayer). As a consequence, only a sample layer of 1–2 nm of a biological sample is removed. In contrast, a high ion density and energy beam is used in dynamic SIMS, making it more destructive whereas the primary ions can deep 10 to 20 nm into the sample [90–93].

The primary ion beam can be of positive ions such as Ga^+, Cs^+, Bi^+, Au^+, O_2^+, Xe^+ or Ar^+, or negative ions like O^- and O_2^-. The primary ions are generated from duoplasmatron of Ar, O_2, Xe, and so on or liquid-metal gun (LMIG) of Au, Cs, Ga, or Bi. Negative ion beam is employed for metal determination given that it favors the formation of secondary positive ions like Al, Si, and rare earth elements [92–95]. When compared to duoplasmatron, LMIG yield more secondary ions and allows spot sizes down to 50 nm in diameter as well as working with high current density (1 A cm^{-2}). High-ion current density enables depth profiling and higher sample sputter rate, improving sensitivity. It is possible to sputter only the first and second atom or molecule layer from the sample surface when the current density is lower than 0.1 nA cm^{-2}. However, at this condition it is still possible mapping elements and/or molecules on a biological sample surface, but at the expense of sensitivity reduction [90, 92, 93].

Once SIMS works under high vacuum, only solid samples that are stable at this condition can be analyzed. Thus, due to the complexity of a biological tissue and the fragile nature of biological cells, they cannot be directly analyzed and special care has to be taken to maintain the structural and chemical integrity of biological tissues and cells.

Sample preparation procedures for biological tissues analysis using SIMS have been proposed [87]; typically, 300–500 nm (semi-thin) or 60–80 nm (thin) thick sections are cut and fixed on an appropriate support. Artifacts, structural damage, and element diffusion and loss can occur due to inadequate sample preparation. Inaccurate results are obtained in these cases and it is more critical in cell analysis, mainly for highly diffusible elements such as Na and K.

Native tissue mounted onto conventional glass slides, chemical fixation, freeze-drying, or freeze-drying with subsequent dehydration and infiltration with a resin are the most common procedures of sample preparation for SIMS imaging [96–101]. Freezing can preserve the chemical integrity of the biological sample and the respective hydration shell during sample fixation. Cells do not always freeze well and addition of freezing protectors such as sugars, polymers, and proteins are necessary when high-pressure freezing (HPF) is used. However, this treatment can change the secondary ion generation yield. For analysis of tissues like brain, samples of them are resin-embedded and cut into thin or semithin sections, loaded onto TEM grids, or silicon wafers, or glass slips, and finally gold-coated to avoid charge buildup on the sample surface [97, 102, 103]. Cryogenic fixation involves (a) frozen-hydration, (b) freeze-drying, and (c) freeze-fracturing. (a) Frozen-hydration comprises flash-freezing the sample in liquid nitrogen-cooled propane prior to freezing in liquid nitrogen. This procedure avoids physical damage of the sample by water crystallization and also preserves the chemical integrity of the cells and tissues. (b) In freeze-drying the sample is quickly frozen followed by slow warming under vacuum to remove residual water. Cell damage is often a major concern of freeze-drying as a consequence of water sublimation if the temperature of the sample is increased too quickly. Rearrangement of molecules in the cell and/or loss of some components may also occur during freeze-drying. Therefore, this sample preparation procedure is not appropriate when it is intended to map elements and other components in the cell. (c) In freeze-fracturing, a cell suspension is trapped between two shards of substrate in a sandwich form and then plunged into liquid propane. If the analysis is not carried out immediately, the sample can be stored in liquid nitrogen. However, if the sandwich gets separated at the moment of analysis and the ice within the sample is fractured, some cell components are deteriorated and/or lost. Freeze-fracturing is the procedure that better preserves the sample components in the cell [97, 104, 105].

Chemical fixation with glutaraldehyde, paraformalin/formaldehyde, or trehalose preserves the internal cellular structures by physically changing the cell. However, chemical fixation of the cell is not recommended for imaging of highly diffusible elements like Na, K, and Ca because redistribution of them in cell components are prone to occur. In addition, by chemical fixation other subcellular components may be not preserved in their native state [96]. Nevertheless, the procedure is suitable to determine elements bound to molecules, DNA-interacting drugs, and other not highly diffusible ions [97].

Due to its excellent spatial resolution, SIMS has been applied to a large variety of biological samples. In principle, any element and respective isotopes can be detected, enabling both qualitative and quantitative images. However, the secondary ion intensity depends on its ionization yield, which varies greatly from one element, to another and the chemical environment where the ion is localized. For that reason, direct quantification by SIMS is often hindered by high matrix effects. The most common approach to circumvent such effects is applying a relative sensitivity factors (RSFs). It is calculated by analyzing a reference matrix that is chemically similar to the sample having a known concentration of the element of interest. However, such approach is valid for samples with homogenous matrices and when the element of interest is a minor constituent. Nevertheless, biological tissues are generally heterogeneous and ionization yield of an element varies from one local to another, which induces variation of the RSFs. On the other hand, this approach would be feasible for quantification based on an average area, and has been described by Dérue et al. [106] in the determination of Na^+, K^+, Mg^{2+}, and Ca^{2+} in frozen-hydrated ionic solutions [99].

The potential of SIMS as a metal imaging technique in biomedical research has been demonstrated by mapping cisplatin [107], ^{12}C, ^{39}K, ^{23}Na, ^{40}Ca, and ^{10}B in human glioblastoma cells [108], in renal epithelial LLC-PK1 cells [109], and for imaging of Na, K, phosphocholine, choline, and cholesterol in mouse heart [75]. By determination using a ToF-SIMS instrument these species were detected in the analyzed heart tissue. NanoSIMS was used for elemental mapping of melanosomes of the choroid and retinal pigment epithelium (RPE), RPE lipofuscin and melanolipofuscin granules. Sulfur, Ca, Fe, Na, and Cu in each region of interest were mapped at resolution about 60 nm. Despite the good LOD of NanoSIMS, Zn (element that plays important role in physiology and also for characterization of diseases related to oxidative stress like age-related macular degeneration) was not detected in the analyzed samples [110].

MRI-relaxometry and SIMS were employed in a study dealing with distribution of Gd in blood flow within the microvascular network in ex vivo *Langendorff* isolated rat heart models perfused with the Omniscan contrast agent. The combined use of SIMS and MRI relaxometry allowed for qualitative mapping of Omniscan as well as quantification of the Gd concentration. The combination of chemical mapping and temporal determination of the Gd concentration in heart tissue provided new insights on the biomolecular mechanisms underlying the microcirculatory alterations in heart disease [111].

Arsenic and Hg species in metal-resistant oral bacteria were imaged through ToF-SIMS to elucidate how these bacteria take up and transform toxic elements inside the cells [112]. Strains of bacteria were isolated from the oral cavity of healthy volunteers and grown in the presence of Hg or As. In further steps, these bacteria were smeared on Al foil, air-dried, and then analyzed. Free Hg was found on the cell surface, methylmercury in the periplasmic space, and higher levels of methylated ^{200}Hg and ^{202}Hg in depth profiles of the cells. Arsenic in arsenate form was converted to arsenite and found inside the cells, close to the cell membrane. Mai et al. [113] applied ToF-SIMS for Cd, Cu, Cr, Hg, and Zn imaging in liver tumor cells doped with these elements. A protocol of sample preparation that combined rapid freezing, freeze-fracturing, and imprinting for transferring the cells to a silicon wafer was developed. It was observed that Cr and Cu diffused into the cell while doped Cd, Hg, and Zn were not detected in the cells. Multiple changes in brain structure and function are caused by Alzheimer disease (AD) [114]. Metal dyshomeostasis occurs due to anomalous binding of Fe, Cu, and Zn, or impaired regulation of redox-active metals that can induce formation of cytotoxic ROS and neuronal damage. In an attempt to understand the role of metals in AD, studies have been conducted using imaging techniques in order to find metal distribution in specific parts of the brain, especially in cellular structures [103]. SIMS images of the hippocampus of AD patients and the thalamus of transgenic mice revealed morphological and chemical modifications that took place in well-characterized pathological brain regions. It was also possible to image the pathological iron–ferritin–hemosiderin distribution in the brain AD patients' hippocampus, as well as Ca-Fe mineralization in thalamus amyloid deposits (AmD) in the transgenic mice brain. NanoSIMS was employed for Fe and Ca imaging with resolutions of 50–100 nm and 150–200 nm, respectively. Images of CN^-, P^-, and S^- distribution were acquired simultaneously by using Cs^+ LMIG as a primary ion beam source, while an O^- beam was employed for Ca and Fe detection imaging [102, 103]. NanoSIMS was also applied for metallodrugs imaging in cells as well as in organic ligands associated with the metal center in metallodrugs. However, the ligands could only be visualized through NanoSIMS images if they contained nonendogenously ubiquitous elements in their chemical structure such as Br or F or the ligands were isotopically labeled with ^{13}C or ^{15}N. For metal detection via NanoSIMS, semithin sections must be analyzed and a long acquisition time (>10 h) is necessary to obtain sufficient secondary ions [115]. NanoSIMS has provided information not only about internalization and subcellular localization of different Pt anticancer drugs, but also Pt and ligands exchange in platinum-amine antitumor drugs labeled with ^{15}N. In this way, clear differences in localization and time-course among cisplatin and a polynuclear Pt drug (TriplatinNC) could be observed [116]. NanoSIMS was employed in studies about Au [117], Pt [116, 118] and Ru [119] based anticancer drug imaging. In vitro inter- and intracellular distribution of an isotopically labeled Ru (II)–arene antimetastatic compound for human ovarian cancer cells was imaged at ultrahigh resolution. Images of ^{13}C, ^{15}N, and Ru indicated that the phosphine ligand remained coordinated with the Ru(II) ion but the arene detached. It was also demonstrated that Ru complexes were localized mainly on the membrane or at the interface of cells, which correlates with the antimetastatic effects [119].

NanoSIMS and energy-filtered transmission electron microscopy (EFTEM) were employed to image Au in situ in tumor cells of an individual submitted to treatment with Au(I) phosphine-based anticancer. NanoSIMS images of $^{12}C^{14}N^-$, $^{31}P^-$, $^{34}S^-$, and $^{197}Au^-$ showed cellular morphology and distribution of Au at subcellular level. By employing EFTEM, images of Au distribution and nuclear and mitochondrial morphology were obtained. The subcellular distribution of Au was associated with S-rich regions in the nucleus and cytoplasm, supporting evidence of the mechanism of action of Au(I) complexes based on inhibition of thiol-containing proteins [117].

Isotope selective NanoSIMS combined with confocal laser scanning microscopy (CLMS) was employed to evaluate ^{15}N-labeled ammine cisplatin complexes distribution in human colon cancer cells [118]. By doping the cells with ^{15}N-labeled cisplatin, it was possible to monitor the N/Pt stoichiometry of the compound, which suggested partial dissociation of Pt–N bonds, at least in nucleoli of cells treated with elevated amount of cisplatin. The colocalization of Pt with S and P-rich structures observed was consistent with the high affinity of Pt for S donors and binding to DNA. The cleavage of ligands and Pt colocalization with S-rich nucleoli corroborated the role of thiol-bearing molecules (affecting Pt–N bond stability) in cisplatin metabolism. The results obtained are relevant in studies about detoxification and/or mode of action of Pt drugs.

Direct detection of NPs in biological tissues is a current goal in nanotoxicology where SIMS is an important technique for that purpose [120]. Zinc oxide nanoparticles (ZnONPs) have found increasing use in sunscreen products and cosmetics. In view of the high resolution provided, ToF-SIMS and CLSM were used in a study concerning cytotoxicity of ZnONPs, being skin equivalent HaCaT cells used as a model system. The CLSM images revealed absorption and localization of ZnONPs in the cytoplasm and nuclei while the ToF-SIMS images demonstrated elevated levels of intracellular ZnO concentration and Zn concentration dependent of the Ca/K ratio, presumably caused by the dissolution of ZnONPs. The obtained images demonstrated partially resolved cytotoxicity relationship between intracellular ZnONPs, Ca–K ratio, phosphocholine fragments, and glutathione fragments. Changes of the ToF-SIMS spectra and images of ZnONPs treated HaCaT cells suggested possible mode of actions of ZnONP, involving cell membrane disruption, cytotoxic response, and ROS mediated apoptosis [121]. SIMS was employed for Au nanoparticle imaging in macrophage-like RAW 264.7 cells. The AuNPs and two medicine compounds, amiodarone and elacridar, were successfully imaged within the cells. To verify if SIMS could detect functionalized and nonfunctionalized NPs simultaneously, fluorophore-functionalized AuNPs were evaluated as a model system. The fluorescent characteristics of the functionalized NPs enabled their detection and localization within the cell. This unique capability of SIMS is useful in studies about targeted drug delivery through AuNPs [122].

As sample regions accessible to ToF-SIMS are limited to some hundred μm^2, the proper selection of regions containing NPs is mandatory for 3D imaging. Therefore, nondestructive techniques such as fluorescence microscopy and micro X-ray fluorescence spectrometry (μXRF) can be employed to screen and select ROI for further analysis by SIMS. A comparative study involving ToF-SIMS imaging and fluorescence microscopy imaging was conducted for SiO_2NPs detection in tissues; SiO_2NPs with average diameter of 25 nm, core-labeled with fluorescein isothiocyanate, were intratracheally instilled into rat lungs and afterward cryosections of the lungs were analyzed. The NPs were imaged in 3D by ToF-SIMS based on the SiO_3^- distribution and the NPs fluorescence pattern. The lateral distribution of protein (CN^-) and phosphate (PO_3^-) in the lung tissue could also be related to the SiO_3^- distribution. Therefore, ToF-SIMS is suitable for NPs imaging in biological tissue, besides providing additional information of the chemical environment of the NPs in the tissue [120]. In another study about 3D NPs imaging in biological tissues, analysis of ROI in cryosections from lungs of homogeneously laden containing CeO_2NPs (10–200 nm diameter) by ToF-SIMS also demonstrated the potential of this technique for NPs detection and imaging and shed light of other components of the sample tissue [123]. Using ToF-SIMS, AgNPs of 100 nm were detected in human mesenchymal stem cells as a model system. A combination of high lateral and high mass resolution was achieved by delayed extraction of the secondary ions; using delayed extraction mode for single cells, mass resolution up to 4000 at $m/z = 184.08$ and lateral resolution up to 360 nm were achieved. Cell compartments like the nucleus were visualized in 3D, whereas no realistic 3D reconstruction of intracellular AgNPs was possible due to the different sputter rates of inorganic and organic cell materials [124].

In summary, SIMS is a valuable technique to investigate the role of metals in neurodegenerative diseases, mechanisms of action of metal-based anticancer drugs, and cytotoxicity of metals and NPs at cellular levels. The capability of SIMS for elemental and NPs imaging in biological tissues and cells is remarkable, which may be one of the key tools for the identification of possible biomarkers when seeking diseases treatment, and also to detect absorption of NPs by living organisms.

2.3 X-Ray Based Techniques for Bioimaging

Energy-dispersive X-ray spectroscopy whose acronym is EDS, EDX, or EDXS, synchrotron X-ray fluorescence (SXRF and μXRF), and X-ray absorption spectroscopy (XAS) have been used to obtain high spatial resolution (μm to nm) images of elements in biological tissue, with good sensitivity (detection of μg g^{-1} to ng g^{-1}) and in a nondestructive way. The principle of these X-ray based techniques is that atoms of the sample are irradiated with X-ray photons from a X-ray source, including high energy synchrotron source. Core electrons of the irradiated atoms are excited and then X-ray photons are emitted (X-ray fluorescence) when these electrons decay to less energetic levels in the atom and element-specific X-ray fluorescence spectra are obtained. The X-ray fluorescence spectrum is element specific. Thus, scanning a sample with X-ray at a given energy can yield 2D or 3D images of elemental distribution. Sensitivity increases with the atomic number (Z) increase, making X-ray based techniques, in principle, particularly appropriate for heavier element determination. The spatial resolution achieved with conventional XRF is several orders of magnitude higher than that of μXRF. Conversely, higher sensitivity and resolution than conventional XRF spectrometry is achieved using SXRF spectrometry owing to the higher energy of X-ray photons from the synchrotron source whose intensity is one order of magnitude higher than that from X-ray tubes. Currently, μSXRF spectrometry enables elemental imaging with a few dozen nanometers spatial resolution. To achieve more detailed information, micro-SXRF spectrometry has been combined with other techniques, mainly micro-XAFS (X-ray absorption fine structure), micro-diffraction and micro-computed tomography (CT), atomic force microscopy (AFM), micro-Fourier transformed infrared microscopy (μ-FTIR), and so on.

SXRF microprobes are composed of several key elements: the magnet source, the monochromator, the focusing optics, the detection system, and the ancillary environment (vacuum, cryostat, etc.). None of the existing SXRF microprobe end stations are similar in terms of energy, resolution, limit of detection, and conditions of measurements; the radiation energy must be carefully selected depending on each case and purpose. Soft X-ray (energy ranging from 0.12 to 12 keV) is suitable for biological materials analysis as low-Z elements such as C, N, O, F, Fe, Zn, Mg, and others of fundamental importance for metabolism in biological systems can be detected. It is more difficult to focus photons with high energy than low energy, so hard X-ray (energy >12 keV) has been operated using a micro-focused beam for analysis of biological materials. Due to the high penetration of X-rays in the matter, 3D maps can be obtained by translating and rotating the sample. After mathematical reconstructions, elements distribution can be visualized in cells and tissues without any sectioning of the sample, and analysis can be conducted using a micro- or nano-focused X-ray beam [125–127].

X-ray-based techniques are nondestructive, but conservation of the cell structure and tissue during the sample preparation and measurements must be considered. Samples in their frozen hydrated state can be cryocut under helium or liquid nitrogen at −263 or −196 °C, respectively. As such, the sample preparation is minimal and does not involve chemical reagents and resin embedding, leading to detect metal species distribution and speciation changes, mainly of soluble species and low stable complexes [128]. To preserve the integrity of cell membranes and organelles as well as elemental and chemical species distribution, samples are frozen in liquid nitrogen-cooled isopentane or liquid ethane, and in frozen state they are transferred to microscope and then analyzed at low temperature [129]. Only when necessary a frozen tissue is cryocut into thin sections. Sample degradation due to X-ray exposure can be reduced by using fast detector and appropriate scan speed to diminish the time exposure during scanning [130, 131]. It is also possible to use X-ray methods to visualize cells without chemical fixation, dehydration, or staining of the specimen. As such, X-ray methods are more suited than optical and EM methods (excepting when sample stabilization with cryo-techniques is possible) for imaging of native-state specimens at spatial resolution of a few tens of nanometers, minimizing interventions that would change the metal oxidation state in the sample [132].

X-ray based techniques find large application in studies concerning distribution of biologically relevant elements and markers in tissues and cells. Some examples of a application are cited as follows. Images of the location and structural details of Au-labeled organelles were obtained using μXRF, which correlated well with the subcellular distribution image obtained using optical fluorescence microscopy [133]. Adherent mouse fibroblast cells were grown on silicon nitride windows (serving as biocompatible XRF support substrate) and labeled with organic fluorophore Au clusters in combination with primary antibodies specific for mitochondria or the Golgi complex. Scanning air dried cells by using scanning x-ray fluorescence microscopy (SXRFM) provided 2D maps with submicron resolution for Au and for most biologically relevant elements (K, Cl, Cu, Zn, Ca, P, S, and Mn). Additionally, bioimage of various elements in mitochondria was obtained by labeling the adenosine triphosphate with Au colloidal particles [134]. Nanobioimage of human cancer cells labeled with CdSe/ZnS QDs were obtained by means of nano-SXRF. These QDs were suitable for labeling and detection owing to the Se emission signal. Another example of application is AgNPs imaging in animal tissues using μXRF [135].

Benchtop μXRF instrument, operated under laboratory conditions, instead of more complex μSXRF instrument can alternatively be employed for elemental mapping [98]. However, μXRF spectrometry offers lower resolution and sensitivity than μ-SXRF spectrometry. Therefore, a benchtop μXRF instrument can be utilized for a coarse scan to identify ROI for further analysis using μSXRF. A benchtop μXRF instrument was employed for elemental mapping in teeth and bone tissue [136, 137], and to determine the ROI in liver biopsy specimen from a patient with Wilson disease (WD) [138]. The ROI in the liver biopsy was further analyzed by μSXRF with a beam size of 4 μm that led to resolution at the cellular level. Images of Cu, Fe, Zn, and Mn were obtained, whereas areas with high Cu, Fe, and Zn intensities showed an inverse correlation. In addition to μSXRF, X-ray absorption near edge structure spectroscopy (XANES) was employed to identify the oxidation states of Cu within the liver tissue, where both Cu(I) and Cu(II) were identified.

By means of μSXR and tomography approaches it is possible to construct 3D images of trace metals in tissues and cells [139], as done for Zn, Fe, and Cu in a zebrafish embryo. In order to preserve the native elemental composition, the zebrafish embryo was embedded in a methacrylate-based resin at cryo-temperature. Beam attenuation by the resin matrix was minimized by excising the specimen through fs laser sectioning before mounting it on the tomographic sample stage. Based on 60 tomographic projections acquired in 100 h of analysis, it was mathematically possible to construct the corresponding 3D image [140]. The constructed image showed Zn predominantly localized in the yolk and yolk extension and Fe in various regions of the brain and along the dorsal side of the embryo. The yolk syncytial layer exhibited higher concentration of Fe, likely due to enhanced expression of the Fe transporters ferroportin-1 and transferrin in that layer. The liver appeared as a Fe-rich anatomical feature of the embryo. As a major storage site of ferritin, the primary protein responsible for intracellular Fe storage and release, this organ was readily identified in the Fe distribution map. Higher levels of Fe in the retinal epithelium were associated with the presence of several Fe-dependent enzymes critical for cellular function. Higher levels of Zn were observed in the outer layers of the retina, probably due to metallothionein transcripts in the retina of zebrafish embryo. SXRFM and AFM were employed for Mg mapping in single cell and cell population. The Mg distribution inside the cells was similar except in one of the cell strains, where single cell analysis revealed two cells with an exceptionally high intracellular Mg content when compared to the other cells of the same strain. The high Mg content suggested that Mg subcellular localization correlated with oxygen differently in these cells when compared to the other sister cells of the same strain [141].

Images of Fe, Cu, and Zn concentrations in brain tissue obtained using μSXRF and LA-ICP-MS were compared; LA-ICP-MS is a more accessible technique and offers superior sensitivity while μSXRF offers far superior spatial resolution. However, both techniques demonstrated to be well suited for changes detection in regional biometal distribution when healthy and diseased brain tissues are analyzed [142]. Iron, Cu, and Zn, which are associated with neurodegenerative diseases, with aggregation of β-amyloid (Aβ) plaques in AD, which were examined in hippocampus of APP/PS1 mice. XRFM was used to assess how the anatomical location of Aβ plaques was influenced by the metal content in surrounding tissue. Individual plaques in four subregions of the hippocampus were identified and mapped. The study suggested that the in vivo binding of Zn to plaques was not simply due to increased protein deposition [143]. In another study, SXRF was applied to investigate Fe and Zn distribution within neuronal cell layers of hippocampal subregions in brain of mice and rats. Direct quantitative elemental mapping of total element pools, in situ within ex vivo tissue sections was possible—without any chemical fixation or addition of staining reagents to the tissue. Exact localization of Fe in pyramidal cell layers of the hippocampus and Fe gradient across neuron populations within the nondegenerating and pathology free rat hippocampus was demonstrated. Specific profiles of Fe and Zn within anatomical subregions of the hippocampus (cornu ammonis subregions) were determined [144].

Tissues and cells of individuals treaded with metallic compounds and NPs-based drugs in cancer therapy were analyzed using SXRF. High sensitive nanoimaging was obtained and intracellular distribution of noble metal NPs in cells was observed, making possible to evaluate the internalization of the drug in the cell and to investigate possible DNA damage and toxicity [145]. Synchrotron X-ray fluorescence nanoprobe, TEM and ICP-MS were employed in order to evaluate Os distribution in ovarian cancer cells of a patient treated with Os complex as anticancer drug. Data obtained for Os, Zn, Ca, and P suggested localization of Os in mitochondria and not in the nucleus, accompanied by mobilization of Ca from the endoplasmic reticulum, a signaling event for cell death [146].

SXRF and ICP-MS were used with the aim to examine Br distribution in bovine ovarian tissue, follicular fluid and aortic serum. XAS was additionally employed to identify chemical species of Br in tissues of a range of mammalian (bovine, ovine, porcine and murine) and fish, whole blood and serum of the mammalians. Bromine was found to be widely distributed across all tissues and fluids examined. In the bovine ovary in particular Br was more concentrated in the subendothelial regions of arterioles. Comparison of the near-edge region of the X-ray absorption spectra with a library of Br standards led to the conclusion that Br^- was the main species of Br present in all samples analyzed [147]. Bioimages of Fe and Zn in multiple sclerosis plaques were obtained by means of XANES and SXRF to know the implication of Fe and Zn on multiple sclerosis. It was found that distribution of Fe and Zn was heterogeneous in multiple sclerosis plaques, and with few remarkable exceptions they did not accumulate in chronic multiple sclerosis lesions [148]. Silver distribution in lung mice exposed to Ag, colocalization with Fe, Cu, and S, and speciation of elements were observed by combining micro-X-ray absorption near edge structure spectroscopy (μXANES) and micro-proton-induced X-ray emission (μPIXE) [135]. With spatial resolution down to 1 μm and good sensitivity, μPIXE can be applied to a large range of elements, including essential trace elements, toxic metals, and pharmacological compounds [149]. Quantitative image of Gd nanoparticles (GdNPs) in glioblastoma tumors grafted in mice was obtained by using μPIXE, whereas polysiloxane network surrounded by Gd chelates were used as model particle. The images obtained indicated heterogeneous diffusion of GdNPs, probably related to variations in tumor density, but all tumor regions contained Gd, suggesting the element diffusion to the healthy adjacent tissue [150].

In conclusion, X-ray techniques are essential in life sciences investigations. By employing these techniques, quantitative and qualitative 2D and 3D images of elements and nanoparticles in tissues and cells are obtained with high resolution and sensitivity. The information gained is important for understanding the effect, behavior, and role of metals and other elements in the organism.

2.4 Particle Beam Microscopy-Based Techniques for Bioimaging

Particle beam microscopy is based on physical interaction between high energetic electron and matter, enabling images with high spatial resolution (subnanometer, at individual atoms level) required for biological structural contexts, chemical mapping, including chemical elements localization in cells. When EM is coupled to EDXRF detectors or those used in electron-energy loss spectroscopy (EELS), elemental identification and speciation are possible. Coupling EM with PIXE provides increased sensitivity and fully quantitative elemental mapping, but at the expense of reducing spatial resolution. EM operates based on the same basic principles of light microscopy, but the resolution offered by EM is many orders of magnitude better and it is capable to reveal details of internal structures of cells. The technique can be operated in transmission, scanning transmission, and scanning modes. TEM and STEM are applied in analysis of biological tissue sections with thickness typically <100 nm, where transmitted electrons are measured. Through these modes of EM operation, the spatial resolution can be 0.05 nm. When EM is operated in SEM mode, backscattered electrons are measured and the spatial resolution achieved is about 0.4 nm. Thicker sample sections can be analyzed by means of SEM and when a focused ion beam is applied under cryo-environment, direct and fast 3D imaging of large native frozen tissue samples can be obtained. Elements constituents of the sample surface can be mapped when TEM and SEM are coupled to EDS

X-ray detectors and/or detectors for EELS. EDS is particularly sensitive to heavy elements up to Mo, whereas EELS is for elements ranging from carbon through transition metals. EELS permits to image chemical structures (through carbon, nitrogen, and oxygen) and provides information of coordinated chemical bonds.

The procedures of biological tissue preparation and cell analysis by EM are almost similar of those used for SIMS and/or X-ray spectrometry. However, for EM analysis the thickness of tissue section is in general lower and ranges from 50 to 200 nm. To maintain elements bound to macromolecules such as proteins, nucleic acids, carbohydrates, and lipids, the sample is usually treated with glutaraldehyde and osmium tetroxide solution, followed by dehydration with polar solvents and embedding in epoxy resins before sectioning. However, precipitation and cryo-fixation are recommended for specimens containing compounds that are soluble in the media [151]. For example, ammonium oxalate can be added to the specimen for Ca precipitation followed by immersion in liquid pentane, or liquid nitrogen or helium prior cryo-cutting [151, 152]. Thus, cryo-fixation is better than chemical fixation for sample integrity preservation. Cryo-fixation also reduces possible damages due to heating caused by electron incidence on the sample during analysis [151–153]. The following are examples of EM applications in the field of metallomic bioimages.

Frozen sections 100 nm thick from mouse cerebellar cortex were analyzed using STEM-EELS to evaluate Ca distribution in cells. The element was found in mitochondria and endoplasmic reticulum, which is the organelle responsible for Ca regulation in neurons and other cells [154]. Mouse brain was analyzed using STEM-EELS for Fe determination in ferritin in cells. The brain tissue was cut into 200 nm thick sections, chemically fixed and treated with osmium tetroxide. To avoid sample losses by the incident electron beam, the sample was maintained frozen at −160 °C through analysis. The number of Fe atoms in each ferritin particle ranged from 1200 to 3600 and was in agreement with the number of Fe atoms estimated previously (4500) [155]. Manganese in mitochondria of brain cells of healthy rats was mapped by means of STEM-EELS [156]. Samples were prepared by perfusion and slicing to 40 nm, followed by fixation with osmium tetroxide. The bioimage generated revealed Mn particles of about 0.2 nm diameter localized mainly in mitochondria.

Tridimensional images (3D) of element distribution in biological specimens can be obtained after computational treatment of the data collected by STEM-EELS [157–159].This approach was applied to obtain 3D images of Fe bound to melanin and ferritin in fungus cells and degenerating neurons in mice [154, 155].

Detection and mapping of metals in QDs and NPs bound to biomolecules permit molecule identification. STEM-EELS and

EFTEM were applied for identification of markers and nanomaterials used in human diagnostics and/or therapeutic applications. For instance, these techniques were employed to map luminescent QDs [160–162] as well as clusters attached to antibody fragments labeled with Au. In this case, it was possible localizing specific proteins in cells [163]. Mammalian tissue was analyzed using EDX where endogenous and labeled elements (Au, Cd-based nanoparticles) were mapped. The EDX image revealed that endocrine and exocrine vesicles exist in Islets of Langerhans in single cells [164]. Following other approach, a specific protein can be identified through fusion of metallothionein with the protein and treatment with Au solution for metal tagging. As a result, Au-thiolate clusters with 1 nm diameter [165] are produced, which are then detected by TEM and the image obtained shows the protein distribution in the tissue. Proteins in a cellular environment can also be accessed through cryo-electron tomography (cryo-ET) after labeling the proteins with AuNPs. Images (resolution about 2.2-nm) of Au-tagged peptides was achieved using cryo-ET [166].

The high energetic electron beam used in electron microscopy interacts with the electrons of the inner shell of atoms of the sample under analysis and generates X-ray photons with specific energy. The X-ray produced can be measured as energy dispersive (EDS) or wavelength dispersive (WDS). EDS is widely employed in electron microscopy and allows spatial resolution about 500 nm, which is, however, worse than that offered by STEM-EELS and EFTEM [167, 168]. The X-ray signal intensity is dependent of beam size, type of radiation source (a FEG is commonly used) and specimen thickness. The X-ray beam used in EM ranges from 1 to 100 nm for elemental mapping and quantification [169]. The EDS analysis can be divided into spot and elemental mapping. In spot analysis, the electron beam from the source is focused over a specific point of the sample and the X-ray photons generated are collected. In elemental mapping, the sample under analysis is scanned and X-ray photons generated across the specimen are collected [170]. The time of analysis depends on the sample size and instrumental parameters selected. Nevertheless, the acquisition time is usually long and it may take several hours to get good images [169].

Application of EDS for elemental image is more limited than STEM-EELS owing to the lower sensitivity of EDS. Consequently, most applications of EDS for elemental mapping in biological samples were dedicated to major elements such as Ca, K, Mg, Na, and Fe. For trace element mapping, EDS has been combined with TEM and STEM [171–174]. As such, more information about elemental composition and imaging of elements at cellular and subcellular levels are obtained by combining EDS with other techniques. Surface and internal structure of cells as well as metal species (mainly heavy elements) can be mapped with high sensitivity when

STEM is combined with EDS detectors [175]. In this context, EDS, TEM, and nanoSIMS were applied to investigate constituent elements and their distribution in unicellular green algae (Chlamydomonas reinhardtii). TEM coupled to EDS made possible morphology and size analysis while SIMS allowed imaging of Ca, P, K, Mg, Fe, and Zn in subcellular granules; resolution of TEM, EDS, and SIMS were 1 nm, 1 μm, and 50 nm, respectively [172]. EDS combined with EM techniques is also useful in studies of neurodegenerative diseases. TEM and EDS were used to examine the presence of Al in the brain of patients with Alzheimer disease. Senile plaques were identified by TEM and Al was detected by EDX in amyloid fibbers in the cores of senile plaques located in the hippocampus and the temporal lobe. Phosphorus and Ca were detected in the amyloid fibbers. Aluminum was not detected in the extracellular space in senile plaques or in the cytoplasm of nerve cells. It was concluded that Al would be involved in the aggregation of beta-amyloid peptides to form toxic fibrils and might facilitate iron-mediated oxidative reactions, which cause severe damage to brain tissues [176].

Particle beam microscopy-base techniques are the ones that allow for the highest spatial resolution; however, they are less sensitive than X-ray and mass spectrometric techniques. Spatial resolution of the order of nm can be achieved with EM, which can be applied as SEM, TEM, and STEM whereas elemental information is obtained when EM is combined with EDX. Therefore, particle beam microscopy allows obtaining multimodal applications in metallomic bioimaging.

2.5 Other Techniques for Bioimaging

Metals and other key elements are necessary in biological processes, directing cellular functions, cell signaling, (macro)molecular structure formation, and (co)enzyme activation. The elements concentration and localization in tissues and cell compartments must be strictly equilibrated to play their correct function in the body. Deregulation of them is often associated to diseases and researchers in this field are essential. Studies concerning localization, concentration, and binding of metals and other elements have been conducted using traditional imaging techniques. Besides that, new or alternative techniques such as hyperspectral imaging (HSI), magnetic resonance imaging (MRI), electron paramagnetic resonance imaging (EPRI), FTIR, and Raman spectroscopy (RS) imaging are being developed and/or applied in studies of metallomic imaging.

Hyperspectral imaging combines spectrophotometry, advanced optics, and algorithms to capture the signal correspondent to the 400–1000 nm interval of the electromagnetic spectrum by an enhanced dark-field microscope (EDFM). Hyperspectral mapping is particularly important for biological samples where NPs morphology is visually indistinct from surrounding tissue structures. EDFM-HSI in combination with RS, EDS, and SEM was used for identification and mapping metal oxide NPs (Ce oxide and Al oxide) in ex vivo histological porcine skin tissues. Results demonstrated that through the EDFM-HSI image it was possible locating and identifying NPs in the porcine tissue [177].

Molecular and cellular magnetic resonance imaging enables monitoring subcellular events. Molecular probes were developed based on a variety of MRI contrast mechanisms that cover a wide range of targets; from tracking therapeutic cells to monitoring enzyme activity and gene expression. A number of different sensors (probes) have been designed to investigate biologically relevant metals for MRI. In this context, redox-active first-row transition metals are central to biological homeostasis, and their marked electronic and magnetic changes upon oxidation/reduction have been used to develop MR sensors. Probes for K^+, Mg^{2+}, Cu^+, Cu^{2+}, Pb^{2+}, and Cd^{2+} are the most common [178]. Zinc is present in several enzymes and many secretory tissues whereas molecules release the element in response to external stimuli. MRI was used to study stimulated secretion of Zn ions from healthy prostate tissue from mice. Small malignant lesions were successfully detected by using a Gd-based Zn sensor, demonstrating that the MRI method is noninvasive and potentially useful to identify prostate cancer [179]. FTIR and RS are nondestructive as well and can provide specific molecular information. In general, the FTIR and RS spectra bring information about molecular conformation and aggregation of biomolecules, such as proteins and nucleic acids. These techniques allow for depiction of pathological changes in tissues and discrimination between viable and dead cells. FTIR and RS can be used for direct identification of molecules associated with metals and metallic NPs, allowing for 2D and 3D imaging of them in tissues and cells [88, 180–182].

3 Conclusion

The analytical techniques most used in metallomic bioimaging were cited and discussed, but there are other emerging techniques. Metallomics require understanding the role of metals (sodium, potassium, magnesium, calcium, iron, manganese, copper, zinc, molybdenum, lead, etc.) and other elements (selenium, arsenic, chloride, bromide, iodide, etc.) in tissues and cells. Understanding the role of them in biological systems can only be achieved through comprehensive analysis of tissues and cells using analytical techniques with high sensitivity (with detection at $\mu g\ g^{-1}$ to $ng\ g^{-1}$ level), selectivity and offering sufficient spatial resolution (μm to nm). Currently, a set of spectroscopic and nonspectroscopic techniques are available for investigations in the metallomic field and more detailed information is obtained when they are combined, which is dependent of the purpose in each case. Combination of various techniques is usually necessary when seeking to know the function and distribution of elements, metallic NPs, QDs, and organometallic compounds such as anticancer drugs in tissues, cells, and organelles. Metallomics bioimaging is complex, multidisciplinary and involves different areas of knowledge as chemistry, biology, medicine, physics, mathematics, and biochemistry.

References

1. Wang S, Brown R, Gray DJ (1994) Application of laser ablation-ICPMS to the spatially resolved micro-analysis of biological tissue. Appl Spectrosc 48:1321–1325
2. Kindness A, Sekaran CN, Feldmann J (2003) Two-dimensional mapping of copper and zinc in liver sections by laser ablation-inductively coupled plasma mass spectrometry. Clin Chem 49:1916–1923
3. Becker JS, Matusch A, Wu B (2014) Bioimaging mass spectrometry of trace elements-recent advance and applications of LA-ICP-MS: a review. Anal Chim Acta 835:1–18
4. Pozebon D, Scheffler GL, Dressler VL et al (2014) Review of the applications of laser ablation inductively coupled plasma mass spectrometry (LA-ICP-MS) to the analysis of biological samples. J Anal At Spectrom 29:2204–2228
5. Lear J, Hare DJ, Fryer F et al (2012) High-resolution elemental bioimaging of Ca, Mn, Fe, Co, Cu, and Zn employing LA-ICP-MS and hydrogen reaction gas. Anal Chem 84:6707–6714
6. Austin C, Hare D, Rozelle AL et al (2009) Elemental bio-imaging of calcium phosphate crystal deposits in knee samples from arthritic patients. Metallomics 1:142–147
7. Reifschneider O, Wehe CA, Diebold K et al (2013) Elemental bioimaging of haematoxylin and eosin-stained tissues by laser ablation ICP-MS. J Anal At Spectrom 28:989–993
8. Zoriy MV, Dehnhardt M, Matusch A et al (2008) Comparative imaging of P, S, Fe, Cu, Zn and C in thin sections of rat brain tumor as well as control tissues by laser ablation inductively coupled plasma mass spectrometry. Spectrochim Acta Part B 63:375–382
9. Becker JS, Zoriy M, Becker JS et al (2007) Elemental imaging mass spectrometry of thin sections of tissues and analysis of brain proteins in gels by laser ablation inductively coupled plasma mass spectrometry. Phys Status Solidi C 4:1775–1784
10. Fernandez B, Claverie F, Pecheyran C et al (2007) Direct analysis of solid samples by fs-LA-ICP-MS. TrAC–Trends Anal Chem 26:951–966
11. Van Malderen SJM, Managh AJ, Sharp BL et al (2016) Recent developments in the design of rapid response cells for laser ablation-inductively coupled plasma-mass spectrometry and their impact on bioimaging applications. J Anal At Spectrom 31:423–439
12. Giesen C, Wang HAO, Schapiro D (2014) Highly multiplexed imaging of tumor tissues with subcellular resolution by mass cytometry. Nat Methods 11:417–422
13. Wang HAO, Grolimund D, Giesen C et al (2013) Fast chemical imaging at high spatial resolution by laser ablation inductively coupled plasma mass spectrometry. Anal Chem 85:10107–10116
14. Feldmann J, Kindness A, Ek P (2002) Laser ablation of soft tissue using a cryogenically cooled ablation cell. J Anal At Spectrom 17:813–818
15. Zoriy MV, Kayser A, Izmer A et al (2005) Determination of uranium isotopic ratios in biological samples using laser ablation inductively coupled plasma double focusing sector field mass spectrometry with cooled ablation chamber. Int J Mass Spectrom 242:297–302
16. Konz I, Fernandez B, Fernandez ML et al (2014) Design and evaluation of a new Peltier-cooled laser ablation cell with on-sample temperature control. Anal Chim Acta 809:88–96
17. Konz I, Fernández B, Fernández ML et al (2014) Quantitative bioimaging of trace elements in the human lens by LA-ICP-MS. Anal Bioanal Chem 406:2343–2348
18. Hamilton JS, Gorishek EL, Mach PM et al (2016) Evaluation of a custom single Peltier-cooled ablation cell for elemental imaging of biological samples in laser ablation-inductively coupled plasma-mass spectrometry (LA-ICP-MS). J Anal At Spectrom 31:1030–1033
19. Bandura DR, Baranov VI, Ornatsky OI et al (2009) Mass cytometry: technique for real time single cell multitarget immunoassay based on inductively coupled plasma time-of-

flight mass spectrometry. Anal Chem 81:6813–6822

20. Chang Q, Ornatsky OI, Siddiqui I et al (2017) Imaging mass cytometry. Cytometry A 91A:160–169
21. Chang Q, Ornatsky OI, Siddiqui I et al (2016) Biodistribution of cisplatin revealed by imaging mass cytometry identifies extensive collagen binding in tumor and normal tissues. Sci Rep 6:36641
22. Ohata M, Tabersky D, Glaus R et al (2014) Comparison of 795 nm and 265 nm femtosecond and 193 nm nanosecond laser ablation inductively coupled plasma mass spectrometry for the quantitative multi-element analysis of glass materials. J Anal At Spectrom 29:1345–1353
23. Claverie F, Pecheyran C, Mounicou S et al (2009) Characterization of the aerosol produced by infrared femtosecond laser ablation of polyacrylamide gels for the sensitive inductively coupled plasma mass spectrometry detection of selenoproteins. Spectrochim Acta Part B 64:649–658
24. Sela H, Karpas Z, Cohen H et al (2011) Preparation of stable standards of biological tissues for laser ablation analysis. Int J Mass Spectrom 307:142–148
25. Pozebon D, Dressler VL, Matusch A et al (2008) Monitoring of platinum in a single hair by laser ablation inductively coupled plasma mass spectrometry (LA-ICP-MS) after cisplatin treatment for cancer. Int J Mass Spectrom 272:57–62
26. Luo R, Su X, Xu W et al (2017) Determination of arsenic and lead in single hair strands by laser ablation inductively coupled plasma mass spectrometry. Sci Rep 7:3426
27. Hare D, Reedy B, Grimm R et al (2009) Quantitative elemental bio-imaging of Mn, Fe, Cu and Zn in 6-hydroxydopamine induced Parkinsonism mouse models. Metallomics 1:53–58
28. Egger AE, Theiner S, Kornauth C et al (2014) Quantitative bioimaging by LA-ICP-MS: a methodological study on the distribution of Pt and Ru in viscera originating from cisplatin- and KP1339-treated mice. Metallomics 6:1616–1625
29. Lum TS, Ho CL, Tsoi YK et al (2016) Elemental bioimaging of platinum in mouse tissues by laser ablation-inductively coupled plasma-mass spectrometry for the study of localization behavior of structurally similar complexes. Int J Mass Spectrom 404:40–47
30. Togao M, Nakayama SMM, Ikenaka Y et al (2020) Bioimaging of Pb and STIM1 in mice liver, kidney and brain using laser ablation inductively coupled plasma mass spectrometry (LA-ICP-MS) and immunohistochemistry. Chemosphere 238:124581. https://doi.org/10.1016/j.chemosphere.2019.124581
31. Wu B, Chen Y, Becker JS (2013) Study of essential element accumulation in the leaves of a cu-tolerant plant Elsholtzia splendens after cu treatment by imaging laser ablation inductively coupled plasma mass spectrometry (LA-ICP-MS). Anal Chim Acta 633:165–172
32. Marshall J, Franks J et al (1991) Determination of trace elements in solid plastic materials by laser ablation-inductively coupled plasma mass spectrometry. J Anal At Spectrom 6:145–150
33. Todoli JL, Mermet JM (1998) Study of polymer ablation products obtained by ultraviolet laser ablation-inductively coupled plasma atomic emission spectrometry. Spectrochim Acta, Part B 53:1645–1656
34. Frick DA, Günther D (2012) Fundamental studies on the ablation behaviour of carbon in LA-ICP-MS with respect to the suitability as internal standard. J Anal At Spectrom 27:1294–1303
35. Frick DA, Giesen C, Hemmerle T et al (2015) An internal standardisation strategy for quantitative immunoassay tissue imaging using laser ablation inductively coupled plasma mass spectrometry. J Anal At Spectrom 30:254–259
36. Limbeck A, Galler P, Bonta M et al (2015) Recent advances in quantitative LA-ICP-MS analysis: challenges and solutions in the life sciences and environmental chemistry. Anal Bioanal Chem 407:6593–6617
37. Hoffmann E, Stephanowitz H, Ullrich E et al (2000) Investigation of mercury migration in human teeth using spatially resolved analysis by laser ablation-ICP-MS. J Anal At Spectrom 15:663–667
38. Kang D, Amarasiriwardena D, Goodman AH (2004) Application of laser ablation-inductively coupled plasma-mass spectrometry (LA-ICP-MS) to investigate trace metal spatial distributions in human tooth enamel and dentine growth layers and pulp. Anal Bioanal Chem 378:1608–1615
39. Arora M, Hare D, Austin C et al (2011) Spatial distribution of manganese in enamel and coronal dentine of human primary teeth. Sci Tot Environ 409:1315–1319
40. Farell J, Amarasiriwardena D, Goodman AH et al (2013) Bioimaging of trace metals in ancient Chilean mummies and contemporary Egyptian teeth by laser ablation-inductively

coupled plasma-mass spectrometry (LA-ICP-MS). Microchem J 106:340–346
41. Bellis DJ, Hetter KM, Jones J et al (2006) Calibration of laser ablation inductively coupled plasma mass spectrometry for quantitative measurements of lead in bone. J Anal At Spectrom 21:948–954
42. Castro W, Hoogewerff J, Latkoczy C (2010) Application of laser ablation (LA-ICP-SF-MS) for the elemental analysis of bone and teeth samples for discrimination purposes. Forensic Sci Int 195:17–27
43. Ranaldi MM, Gagnon MM (2009) Accumulation of cadmium in the otoliths and tissues of juvenile pink snapper (Pagrus auratus Forster) following dietary and waterborne exposure. Comp Biochem Phys C Toxicol Pharmacol 150:421–427
44. Austin C, Fryer F, Lear J et al (2011) Factors affecting internal standard selection for quantitative elemental bio-imaging of soft tissues by LA-ICP-MS. J Anal At Spectrom 26:1494–1501
45. Moraleja I, Esteban-Fernández D, Lázaro A et al (2016) Printing metal-spiked inks for LA-ICP-MS bioimaging internal standardization: comparison of the different nephrotoxic behavior of cisplatin, carboplatin, and oxaliplatin. Anal Bional Chem 408:2309–2318
46. Konz I, Fernandez B, Fernandez ML et al (2013) Gold internal standard correction for elemental imaging of soft tissue sections by LA-ICP-MS: element distribution in eye microstructures. Anal Bioanal Chem 405:3091–3096
47. O'Reilly J, Douglas D, Braybrook J et al (2014) A novel calibration strategy for the quantitative imaging of iron in biological tissues by LA-ICP-MS using matrix-matched standards and internal standardization. J Anal At Spectrom 29:1378–1384
48. Becker JS, Zoriy MV, Pickhardt C et al (2005) Imaging of copper, zinc, and other elements in thin section of human brain samples (Hippocampus) by laser ablation inductively coupled plasma mass spectrometry. Anal Chem 77:3208–3216
49. Becker JS, Zoriy MV, Dehnhardt M et al (2005) Copper, zinc, phosphorus and sulfur distribution in thin section of rat brain tissues measured by laser ablation inductively coupled plasma mass spectrometry: possibility for small-size tumor analysis. J Anal At Spectrom 20:912–917
50. Reifschneider O, Wentker KS, Strobel K et al (2015) Elemental bioimaging of thulium in mouse tissues by laser ablation-ICPMS as a complementary method to heteronuclear proton magnetic resonance imaging for cell tracking experiments. Anal Chem 87:4225–4230
51. Pugh JAT, Cox AG, McLeod CW et al (2011) A novel calibration strategy for analysis and imaging of biological thin sections by laser ablation inductively coupled plasma mass spectrometry. J Anal At Spectrom 26:1667–1673
52. Reifschneider O, Wehe CA, Raj I et al (2013) Quantitative bioimaging of platinum in polymer embedded mouse organs using laser ablation ICP-MS. Metallomics 5:1440–1447
53. Cronea B, Schlatta L, Nadar RA (2019) Quantitative imaging of platinum-based antitumor complexes in bone tissue samples using LA-ICP-MS. J Trace Elem Med Bio 54:98–102
54. Hare D, Austin C, Doble P et al (2011) Elemental bio-imaging of trace elements in teeth using laser ablation-inductively coupled plasma-mass spectrometry. J Dent 39:397–403
55. Sela H, Karpas Z, Zoriy M et al (2007) Biomonitoring of hair samples by laser ablation inductively coupled plasma mass spectrometry (LA-ICP-MS). Int J Mass Spectrom 261:199–207
56. Bonta M, Hegedus B, Limbeck A (2016) Application of dried-droplets deposited on pre-cut filter paper disks for quantitative LA-ICP-MS imaging of biologically relevant minor and trace elements in tissue samples. Anal Chim Acta 908:54–62
57. Bonta M, Lohninger H, Laszlo V et al (2014) Quantitative LA-ICP-MS imaging of platinum in chemotherapy treated human malignant pleural mesothelioma samples using printed patterns as standard. J Anal At Spectrom 29:2159–2167
58. Shariatgorji M, Nilsson A, Bonta M et al (2016) Direct imaging of elemental distributions in tissue sections by laser ablation mass spectrometry. Methods 104:86–92
59. Van Malderen SJM, Vergucht E, Rijcke M et al (2016) Quantitative determination and subcellular imaging of cu in single cells via laser ablation-ICP-mass spectrometry using high-density microarray gelatin standards. Anal Chem 88:5783–5789
60. Pozebon D, Dressler VL, Mesko MF et al (2010) Bioimaging of metals in thin mouse brain section by laser ablation inductively coupled plasma mass spectrometry: novel online quantification strategy using aqueous standards. J Anal At Spectrom 25:1739–1744

61. Douglas DN, O'Reilly J, O'Connor C et al (2016) Quantitation of the Fe spatial distribution in biological tissue by online double isotope dilution analysis with LA-ICP-MS: a strategy for estimating measurement uncertainty. J Anal At Spectrom 31:270–279
62. Jimenez MS, Luque-Alled JM, Gomez T et al (2016) Evaluation of agarose gel electrophoresis for characterization of silver nanoparticles in industrial products. Electrophoresis 37:1376–1383
63. Böhme S, Stärk H-J, Kühnel D et al (2015) Exploring LA-ICP-MS as a quantitative imaging technique to study nanoparticle uptake in Daphnia magna and zebrafish (*Danio rerio*) embryos. Anal Bioanal Chem 407:5477–5485
64. Scharlach C, Müller L, Wagner S et al (2016) LA-ICP-MS allows quantitative microscopy of europium-doped iron oxide nanoparticles and is a possible alternative to ambiguous prussian blue iron staining. J Biomed Nanotechnol 12:1001–1010
65. Zarco-Fernandez S, Coto-García AM, Munoz-Olivas R et al (2016) Bioconcentration of ionic cádmium and cadmium selenide quantum dots in zebrafish larvae. Chemosphere 148:328–335
66. Hsieh Y-K, Hsieh H-A, Hsieh H-F et al (2013) Using laser ablation inductively coupled plasma mass spectrometry to characterize the biointeractions of inhaled CdSe quantum dots in the mouse lungs. J Anal At Spectrom 28:1396–1401
67. Drescher D, Giesen C, Traub H et al (2012) Quantitative imaging of gold and silver nanoparticles in single eukaryotic cells by laser ablation ICP-MS. Anal Chem 84:9684–9688
68. Managh AJ, Edwards S, Bushell A et al (2013) Single cell tracking of gadolinium labeled $CD4^+$ T cells by laser ablation inductively coupled plasma mass spectrometry. Anal Chem 85:10627–10634
69. Managh AJ, Hutchinson RW, Riquelme P et al (2014) Laser ablation–inductively coupled plasma mass spectrometry: an emerging technology for detecting rare cells in tissue sections. J Immunol 193:2600–2608
70. Löhr K, Traub H, Wanka AJ (2018) Quantification of metals in single cells by LA-ICPMS: comparison of single spot analysis and imaging. J Anal At Spectrom 33:1579–1587
71. Hare DJ, George JL, Bray L et al (2014) The effect of paraformaldehyde fixation and sucrose cryoprotection on metal concentration in murine neurological tissue. J Anal At Spectrom 29:565–570
72. Kim P, Weiskirchen S, Uerlings R (2018) Quantification of liver iron overload disease with laser ablation inductively coupled plasma mass spectrometry. BMC Med Imaging 18:51–62
73. Pozebon D, Scheffler GL, Dressler VL (2017) Recent applications of laser ablation inductively coupled plasma mass spectrometry (LA-ICP-MS) for biological sample analysis: a follow-up review. J Anal At Spectrom 32:890–919
74. Zoriy MV, Dehnhardt M, Reifenberger G et al (2006) Imaging of Cu, Zn, Pb and U in human brain tumor resections by laser ablation inductively coupled plasma mass spectrometry. Int J Mass Spectrom 257:27–33
75. Becker JS, Breuer U, Hsieh H-F et al (2010) Bioimaging of metals and biomolecules in mouse heart by laser ablation inductively coupled plasma mass spectrometry and secondary ion mass spectrometry. Anal Chem 82:9528–9533
76. Pornwilard MM, Merle U, Weiskirchen R et al (2013) Bioimaging of copper deposition in Wilson's diseases mouse liver by laser ablation inductively coupled plasma mass spectrometry imaging (LA-ICP-MSI). Int J Mass Spectrom 354:281–287
77. Osterholt T, Salber D, Matusch A et al (2011) IMAGENA: image generation and analysis – an interactive software tool handling LA-ICP-MS data. Int J Mass Spectrom 307:232–239
78. Paul B, Paton C, Norris A et al (2012) CellSpace: a module for creating spatially registered laser ablation images within the Iolite freeware environment. J Anal At Spectrom 27:700–706
79. Pessoa GS, Capelo-Martínez JL, Fdez-Riverola F et al (2016) Laser ablation and inductively coupled plasma mass spectrometry focusing on bioimaging from elemental distribution using MatLab software: a practical guide. J Anal At Spectrom 31:832–840
80. Uerlings R, Matusch A, Weiskirchen R (2016) Reconstruction of laser ablation inductively coupled plasma mass spectrometry (LA-ICP-MS) spatial distribution images in Microsoft excel 2007. Int J Mass Spectrom 395:27–35
81. Hugo L-F, Pessoa GS, Arruda MAZ et al (2016) LA-iMageS: a software for elemental distribution bioimaging using LA-ICP-MS data. J Cheminformatics 8:65–74
82. Weiskirchen R, Weiskirchen S, Kim P (2019) Software solutions for evaluation and visualization of laser ablation inductively coupled plasma mass spectrometry imaging

(LA-ICP-MSI) data: a short overview. J Cheminformatics 11:16–36
83. Bonta M, Limbeck A, Quarles CD Jr et al (2015) A metric for evaluation of the image quality of chemical maps derived from LA-ICP-MS experiments. J Anal At Spectrom 30:1809–1815
84. Paul B, Hare DJ, Bishop DP et al (2015) Visualising mouse neuroanatomy and function by metal distribution using laser ablation-inductively coupled plasma-mass spectrometry imaging. Chem Sci 6:5383–5393
85. Hare DJ, Lee JK, Beavis AD et al (2012) Three-dimensional atlas of iron, copper, and zinc in the mouse cerebrum and brainstem. Anal Chem 84:3990–3997
86. Noël M, Spence J, Harris KA et al (2014) Grizzly bear hair reveals toxic exposure to mercury through salmon consumption. Environ Sci Technol 48:7560–7567
87. Wedlock LE, Berners-Price SJ (2011) Recent advances in mapping the sub-cellular distribution of metal-based anticancer drugs. Aust J Chem 64:692–704
88. Stewart TJ (2019) Across the spectrum: integrating multidimensional metal analytics for in situ metallomic imaging. Metallomics 11:29–49
89. Gyngard F, Steinhauser ML (2019) Biological explorations with nanoscale secondary ion mass spectrometry. J Anal At Spectrom 34:1534–1545
90. Pacholski ML, Winograd N (1999) Imaging with mass spectrometry. Chem Rev 99:2977–3005
91. Amstalden van Hove ER, Smith DF, Heeren RMA (2010) A concise review of mass spectrometry imaging. J Chromatogr A 1217:3946–3954
92. Boxer SG, Kraft ML, Weber PK (2009) Advances in imaging secondary ion mass spectrometry for biological samples. Annu Rev Biophys 38:53–74
93. Katz W, Newman JG (1987) Fundamentals of secondary ion mass spectrometry. MRS Bull 7:40–46
94. Waugh AR, Bayly AR, Anderson K (1984) The application of liquid metal ion sources to SIMS. Vacuum 34:103–106
95. Williams P (1985) Secondary ion mass spectrometry. Ann Rev Mater Sci 15:517–548
96. Dahl R, Staehelin LA (1989) High-pressure freezing for the preservation of biological structure: theory and practice. J Electron Micr Tech 13:165–174
97. Grovenor CRM, Smart KE, Kilburn M et al (2006) Specimen preparation and calibration for NanoSIMS analysis of biological materials. Appl Surf Sci 252:6917–6924
98. Qin ZY, Caruso JA, Lai B et al (2011) Trace metal imaging with high spatial resolution: applications in biomedicine. Metallomics 3:28–37
99. Cunha MML, Trepout S, Messaoudi C et al (2016) Overview of chemical imaging methods to address biological questions. Micron 84:23–26
100. Suhard D, Tessier C, Manens L et al (2018) Intracellular uranium distribution: comparison of cryogenic fixation versus chemical fixation methods for SIMS analysis. Microsc Res Tech 81:855–864
101. Agüi-Gonzalez P, Jähne S, Phan NTN (2019) SIMS imaging in neurobiology and cell biology. J Anal At Spectrom 34:1355–1368
102. Quintana C, Bellefqih S, Laval JY et al (2006) Study of the localization of iron, ferritin, and hemosiderin in Alzheimer's disease hippocampus by analytical microscopy at the subcellular level. J Struct Biol 153:42–54
103. Quintana C, Wu T, Delatour B, Dhenain M et al (2007) Morphological and chemical studies of pathological human and mice brain at the subcellular level: correlation between light, electron, and NanoSIMS microscopies. Microsc Res Tech 70:281–229
104. Winograd N, Bloom A (2015) Sample preparation for 3D SIMS chemical imaging cells. Methods Mol Biol 1203:9–19
105. Chandra S (2008) Challenges of biological sample preparation for SIMS imaging of elements and molecules at subcellular resolution. Appl Surf Sci 255:1273–1284
106. Dérue C, Gibouin D, Lefebvre F et al (2006) Relative sensitivity factors of inorganic cations in frozen-hydrated standards in secondary ion MS analysis. Anal Chem 78:2471–2477
107. Gulin AA, Pavlyukov MS, Gularyan SK et al (2015) Visualization of the spatial distribution of Pt^+ ions in cisplatin-treated glioblastoma cells by time-of-flight secondary ion mass spectrometry. Biochem Moscow Supp Ser A 9:202–209
108. Chandra S, Tjarks W, Lorey DR et al (2008) Quantitative subcellular imaging of boron compounds in individual mitotic and interphase human glioblastoma cells with imaging secondary ion mass spectrometry (SIMS). J Microsc 229:92–103
109. Chandra S (2019) Correlative microscopy of frozen freeze-dried cells and studies of intracellular calcium stores with imaging secondary ion mass spectrometry (SIMS). J Anal At Spectrom 34:1998–2003

110. Biesemeier A, Eibl O, Eswara S (2018) Transition metals and trace elements in the retinal pigment epithelium and choroid: correlative ultrastructural and chemical analysis by analytical electron microscopy and nano-secondary ion mass spectrometry. Metallomics 10:296–308
111. Bonechi C, Consumi M, Matteucci M (2019) Distribution of gadolinium in rat heart studied by fast field cycling relaxometry and imaging SIMS. Int J Mol Sci 20:1339–1352
112. Nygren H, Dahlén G, Malmberg P (2014) Analysis of As- and Hg-species in metal-resistant oral bacteria, by imaging ToF-SIMS. Basic Clin Pharmacol Toxicol 115:129–133
113. Mai F-D, Chen B-J, Wu L-C et al (2006) Imaging of single liver tumor cells intoxicated by heavy metals using ToF-SIMS. Appl Surf Sci 252:6809–6812
114. Braidy N, Poljak A, Marjo C et al (2014) Metal and complementary molecular bioimaging in Alzheimer's disease. Front Integr Neurosci 6:1–14
115. Wu K, Jia F, Luo Q (2017) Visualization of metallodrugs in single cells by secondary ion-mass spectrometry imaging. J Biol Inorg Chem 22:653–661
116. Wedlock LE, Kilburn MR, Liu R et al (2013) NanoSIMS multi-element imaging reveals internalisation and nucleolar targeting for a highly-charged polynuclear platinum compound. Chem Commun 49:6944–6946
117. Wedlock LE, Kilburn MR, Cliff JB et al (2011) Visualising gold inside tumour cells following treatment with an antitumour gold (I) complex. Metallomics 3:917–925
118. Legin AA, Schintlmeister A, Jakupec MA et al (2014) NanoSIMS combined with fluorescence microscopy as a tool for subcellular imaging of isotopically labeled platinum-based anticancer drugs. Chem Sci 5:3135–3143
119. Lee RFS, Escrig S, Croisier M et al (2015) NanoSIMS analysis of an isotopically labeled organometallic ruthenium(II) drug to probe its distribution and state in vitro. Chem Commun 51:16486–16489
120. Veith L, Vennemann A, Breitenstein D (2017) Carsten Engelhard, Martin Wiemann, Birgit Hagenhoff, detection of SiO_2 nanoparticles in lung tissue by ToF-SIMS imaging and fluorescence microscopy. Analyst 142:2631–2639
121. Lee P-L, Chen B-C, Gollavelli G (2014) Development and validation of TOF-SIMS and CLSM imaging method for cytotoxicity study of ZnO nanoparticles in HaCaT cells. J Hazard Mater 277:3–12
122. Bloom AN, Tian H, Winograd N (2016) C60-SIMS imaging of nanoparticles within mammalian cells. Biointerphases 11:02A306-1–02A306-7
123. Veith L, Dietrich D, Vennemann A (2018) Combination of micro X-ray fluorescence spectroscopy and time-of-flight secondary ion mass spectrometry imaging for the marker-free detection of CeO_2 nanoparticles in tissue sections. J Anal At Spectrom 33:491–501
124. Henss A, Otto S-K, Schaepe K (2018) High resolution imaging and 3D analysis of Ag nanoparticles in cells with ToF-SIMS and delayed extraction. Biointerphases 13:03B410
125. de Jonge MD, Vogt S (2010) Hard X-ray fluorescence tomography—an emerging tool for structural visualization. Curr Opin Struct Biol 20:606–614
126. Frascari I, Procopio A, Sargenti A (2018) Implementation of an iterative approach to optimize synchrotron X-ray fluorescence quantification of light elements in single cell. Spectrochim Acta Part B 149:132–142
127. Colvin RA, Jin Q, Lay B et al (2016) Visualizing metal content and intracellular distribution in primary hippocampal neurons with synchrotron X-ray fluorescence. PLoS One 11(7):e0159582
128. Matsuyama S, Shimura M, Fujii M et al (2010) Elemental mapping of frozen-hydrated cells with cryo-scanning X-ray fluorescence microscopy. X-Ray Spectrom 39:260–266
129. Sarret G, Smits EAHP, Michel HC et al (2013) Use of synchrotron-based techniques to elucidate metal uptake and metabolism in plants. Adv Agron 119:1–82
130. Ryan CG, Siddons DP, Moorhead G et al (2009) High-throughput X-ray fluorescence imaging using a massively parallel detector array, integrated scanning and real-time spectral deconvolution. 9th international conference on X-ray microscopy. J Phys Conf Ser 186:012013–012015
131. Lombi E, De Jonge MD, Donner E et al (2011) Trends in hard X-ray fluorescence mapping: environmental applications in the age of fast detectors. Anal Bioanal Chem 400:1637–1644
132. Collingwood JF, Adams F (2017) Chemical imaging analysis of the brain with X-ray methods. Spectrochim Acta Part B 130:101–118

133. McRae R, Lai B, Vogt S et al (2006) Correlative microXRF and optical immunofluorescence microscopy of adherent cells labeled with ultra small gold particles. J Struct Biol 155:22–29
134. Matsuyama S, Shimura M, Mimura H et al (2009) Trace element mapping of a single cell using a hard X-ray nanobeam focused by a Kirkpatrick-Baez mirror system. X-Ray Spectrom 38:89–94
135. Smulders S, Larue C, Sarret G et al (2015) Lung distribution, quantification, co-localization and speciation of silver nanoparticles after lung exposure in mice. Toxicol Lett 238:1–6
136. Dias AA, Carvalho M, Carvalho ML et al (2015) Quantitative evaluation of antemortem lead in human remains of the 18th century by triaxial geometry and bench top micro X-ray fluorescence spectrometry. J Anal At Spectrom 30:2488–2495
137. Blaske F, Reifschneider O, Gosheger G et al (2014) Elemental bioimaging of nanosilver-coated prostheses using X-ray fluorescence spectroscopy and laser ablation-inductively coupled plasma-mass spectrometry. Anal Chem 86:615–620
138. Hachmöller O, Buzanich AG, Aichler M et al (2016) Elemental bioimaging and speciation analysis for the investigation of Wilson's disease using μXRF and XANES. Metallomics 8:648–653
139. Bourassa D, Gleber S-C, Vogt S et al (2014) 3D imaging of transition metals in the zebrafish embryo by X-ray fluorescence microtomography. Metallomics 6:1648–1655
140. Bourassa D, Gleber S-C, Vogt S et al (2016) MicroXRF tomographic visualization of zinc and iron in the zebrafish embryo at the onset of the hatching period. Metallomics 8:1122–1130
141. Malucelli E, Procopio A, Fratini M (2018) Single cell versus large population analysis: cell variability in elemental intracellular concentration and distribution. Anal Bioanal Chem 410:337–348
142. Davies KM, Hare DJ, Bohic S (2015) Comparative study of metal quantification in neurological tissue using laser ablation-inductively coupled plasma-mass spectrometry imaging and X-ray fluorescence microscopy. Anal Chem 87:6639–6645
143. James SA, Churches QI, de Jonge MD et al (2017) Iron, copper, and zinc concentration in Aβ plaques in the APP/PS1 mouse model of Alzheimer's disease correlates with metal levels in the surrounding neuropil. ACS Chem Neurosci 8:629–637
144. Hackett MJ, Hollings A, Caine S (2019) Elemental characterisation of the pyramidal neuron layer within the rat and mouse hippocampus. Metallomics 11:151–165
145. Lewis DJ, Bruce C, Bohic S et al (2010) Intracellular synchrotron nanoimaging and DNA damage/genotoxicity screening of novel lanthanide-coated nanovectors. Nanomedicine 5:1547–1557
146. Sanchez-Cano C, Romero-Caneljn I, Yang Y (2017) Synchrotron X-ray fluorescence nanoprobe reveals target sites for organo-osmium complex in human ovarian cancer cells. Chem Eur J 23:2512–2516
147. Ceko MJ, Hummitzsch K, Hatzirodos N (2015) Distribution and speciation of bromine in mammalian tissue and fluids by X-ray fluorescence imaging and X-ray absorption spectroscopy. Metallomics 7:756–765
148. Popescu BF, Frischer JM, Webb SM (2017) Pathogenic implications of distinct patterns of iron and zinc in chronic MS lesions. Acta Neuropathol 134:45–64
149. Ortega R, Devès G, Carmona A (2009) Bio-metals imaging and speciation in cells using proton and synchrotron radiation X-ray microspectroscopy. J R Soc Interface 6:S649–S658
150. Carmona A, Roudeau S, L'Homel B (2017) Heterogeneous intratumoral distribution of gadolinium nanoparticles within U87 human glioblastoma xenografts unveiled by micro-PIXE imaging. Anal Biochem 523:50–57
151. Thompson RF, Walker M, Siebert A et al (2016) An introduction to sample preparation and imaging by cryo-electron microscopy for structural biology. Methods 100:3–15
152. Nagata T (2004) X-ray microanalysis of biological specimens by high voltage electron microscopy. Prog Histochem Cytochem 39:185–319
153. Mitsuoka K (2011) Obtaining high-resolution images of biological macromolecules by using a cryo-electron microscope with a liquid-helium cooled stage. Micron 42:100–106
154. Leapman RD, Hunt JA, Buchanan RA et al (1993) Measurement of low calcium concentrations in cryosectioned cells by parallel-EELS mapping. Ultramicroscopy 49:225–234
155. Zhang P, Land W, Lee S et al (2005) Electron tomography of degenerating neurons in mice with abnormal regulation of iron metabolism. J Struct Biol 150:144–153
156. Morello M, Canini A, Caiola MG et al (2002) Manganese detected by electron spectroscopy

imaging and electron energy loss spectroscopy in mitochondria of normal rat brain cells. J Trace Microprobe T 20:481–491

157. Aronova MA, Leapman RD (2012) Development of electron energy loss spectroscopy in the biological sciences. MRS Bull 37:53–62
158. Sousa AA, Leapman RD (2012) Development and application of STEM for the biological sciences. Ultramicroscopy 123:38–49
159. Collins SM, Midgley PA (2017) Progress and opportunities in EELS and EDS tomography. Ultramicroscopy 180:133–141
160. Nisman R, Dellaire G, Ren Y et al (2004) Application of quantum dots as probes for correlative fluorescence, conventional, and energy-filtered transmission electron microscopy. J Histochem Cytochem 52:13–18
161. Van Schooneveld MM, Gloter A, Stephan O et al (2010) Imaging and quantifying the morphology of an organic-inorganic nanoparticle at the sub-nanometre level. Nat Nanotechnol 5:538–544
162. Chen D, Monteiro-Riviere NA, Zhang LW (2017) Intracellular imaging of quantum dots, gold, and iron oxide nanoparticles with associated endocytic pathways. Wiley Interdiscip Rev Nanomed Nanobiotechnol 9:1–19
163. Sousa AA, Aronova MA, Kim YC et al (2007) On the feasibility of visualizing ultra small gold labels in biological specimens by STEM tomography. J Struct Biol 159:507–522
164. Scotuzzi M, Kuipers J, Wensveen DI et al (2017) Multi-color electron microscopy by element-guided identification of cells, organelles and molecules. Sci Rep 7:45970
165. Risco C, Sanmartın-Conesa E, Tzeng WP et al (2012) Specific, sensitive, high-resolution detection of protein molecules in eukaryotic cells using metal-tagging transmission electron microscopy. Structure 20:759–766
166. Dahan I, Sorrentino S, Boujemaa-Paterski R et al (2018) Tiopronin-protected gold nanoparticles as a potential marker for Cryo-EM and tomography. Structure 26:1408–1413
167. Ortega R (2005) Chemical elements distribution in cells. Nucl Instrum Meth B 231:218–223
168. Fiori CE, Leapman RD, Swyt CR (1988) Quantitative X-ray mapping of biological cryosections. Ultramicroscopy 24:237–250
169. Friel JJ, Lyman CE (2006) X-ray mapping in electron-beam instruments. Microsc Microanal 12:2–25
170. Wong JG, Wilkinson LE, Chen SW et al (1989) Quantitative elemental imaging in the analytical electron microscope with biological applications. Scanning 11:12–19
171. Fiori CE (1988) The new electron microscopy: imaging the chemistry of nature. Anal Chem 60:86R–90R
172. Penen F, Malherbe J, Isaure MP et al (2016) Chemical bioimaging for the subcellular localization of trace elements by high contrast TEM, TEM/X-EDS, and NanoSIMS. J Trace Elem Med Bio 37:62–68
173. Kowarski D (1984) Intelligent interface for a microprocessor controlled scanning transmission electron microscope with x-ray imaging. J Electron Micro Tech 1:175–184
174. Somlyo AP (1984) Compositional mapping in biology: X rays and electrons. J Ultra Mol Struct Res 88:135142
175. Wu S, Kim AM, Bleher R et al (2013) Imaging and elemental mapping of biological specimens with a dual-EDS dedicated scanning transmission electron microscope. Ultramicroscopy 128:24–31
176. Yumoto S, Kakimi S, Ohsaki A et al (2009) Demonstration of aluminum in amyloid fibers in the cores of senile plaques in the brains of patients with Alzheimer's disease. J Inorg Biochem 103:1579–1584
177. Del Pilar Sosa Peña M, Gottipati A, Tahiliani S (2016) Hyperspectral imaging of nanoparticles in biological samples: simultaneous visualization and elemental identification. Micr Res Tech 79:349–358
178. Allouche-Arnon H, Tirukoti ND, Bar-Shir A (2017) MRI-based sensors for in vivo imaging of metal ions in biology. Isr J Chem 57:843–853
179. Jordan MVC, Lo S-T, Chen S (2016) Zinc-sensitive MRI contrast agent detects differential release of Zn(II) ions from the healthy vs. malignant mouse prostate. Proc Natl Acad Sci 113:E5464–E5471
180. Balbekova A, Bonta M, Török S (2017) FTIR-spectroscopic and LA-ICP-MS imaging for combined hyperspectral image analysis of tumor models. Anal Methods 9:5464–5471
181. Krafft C, Schmitt M, Schie IW (2017) Label-free molecular imaging of biological cells and tissues by linear and nonlinear Raman spectroscopic approaches. Angew Chem Int Ed 56:4392–4430
182. Butler HJ, Ashton L, Bird B et al (2016) Using Raman spectroscopy to characterize biological materials. Nat Protoc 11:664–687

Index

A

Acute myeloid leukemia (AML) 89
Amidome 143–149, 151
Amines 49, 50, 143, 144, 174, 212, 221–227, 246, 247
Amino acids 2, 39, 42, 53, 54, 56, 80, 84, 144, 174, 200, 207, 222, 229, 230, 232, 236, 246

B

Bioimaging 270–272, 274–276, 278–281, 295–297
Biological specimens 155–169, 293
Biological tissues 270–275, 282, 284, 286, 287, 289, 293
Biomarkers 1, 2, 34, 61, 62, 67, 74, 89, 90, 97, 98, 114, 121, 122, 162, 169, 190, 191, 256, 273, 287
Blood 89–91, 114, 121, 122, 158, 166, 175–177, 183, 191, 196, 276, 284, 291
Brains 34, 36, 37, 63, 66, 67, 72–74, 90–93, 98, 122, 191, 193, 196, 200, 201, 203–207, 217, 229, 245–263, 269, 274–278, 280, 281, 283, 285, 290, 291, 293, 295

C

Capillary electrophoresis (CE) 39, 40, 113, 117, 118, 174, 179, 181, 183
Capillary electrophoresis-mass spectrometry (CE-MS) 114–119, 174, 175, 177–179, 181, 203
Carboxylic acids 50, 174, 221–227, 230, 232
Cells 49–59, 73, 74, 78, 80, 81, 89, 121, 122, 144–148, 150, 151, 160, 176, 190, 200, 236, 249, 269–273, 276, 277, 279, 280, 282–287, 289–297
Ceramides 69, 70, 72, 74, 247, 261
^{13}C-Glucose 78–80
Chemical isotope labeling (CIL) 49–59
Chiral chromatography 29, 101
Chirality 90, 91
Chloride adducts 69–76
Cholangiocarcinomas 89
Colorectal cancer (CRC) 163, 164
Comprehensive 27–31, 34, 38, 40, 41, 94, 118, 155, 156, 174, 185, 296

D

Derivatization 27, 49, 79, 81, 82, 87, 90–93, 95, 114, 156, 174, 180, 184, 185, 190, 193, 196, 200, 204, 214, 217, 221, 222, 226, 229, 230, 232, 236–238
Drosophila 145, 146

E

Enantioselective 39, 101, 102, 107, 110
Extractions 2, 3, 6, 50, 52, 53, 62–65, 70, 71, 76, 80, 81, 91, 92, 94–95, 99, 102, 104, 105, 107, 110, 122–125, 147–151, 159–161, 174–176, 181, 185, 192, 193, 196, 203, 204, 214, 222, 227, 230, 232, 234, 237, 239, 246, 249, 251, 252, 287

F

Fatty acid amides 143–151
Fatty acid hydroxyl fatty acids (FAHFA) 69–76
Fingerprinting 179–181, 189, 194–196, 203–206, 249–253
Flow injection analysis (FIA) 9–22, 71
Formate/acetate clusters 11, 21, 117, 191–194, 238, 240, 253
Fourier Transform Ion Cyclotron Resonance Mass Spectrometry (FTICR) 10

G

Gas chromatography (GC) 30, 40, 77–87, 89–99, 113, 174, 180, 190, 192, 195, 196, 200, 201, 203, 205, 206, 213, 214, 221, 222, 226, 227, 229, 230, 232, 234, 246, 248, 252–254
Gas chromatography-tandem mass spectrometry (GC-MS) 78, 79, 90–96, 99, 114, 174, 175, 177–181, 190, 195, 196, 200, 201, 203, 206, 207, 210, 211, 214, 222, 226, 229, 230, 246, 248, 249, 253, 254, 257, 262
GC-MS analysis 80, 82, 83, 193, 195, 205–207, 217, 218, 246, 252, 253, 257

Paul L Wood (ed.), *Metabolomics*, Neuromethods, vol. 159, https://doi.org/10.1007/978-1-0716-0864-7,

Gliomas 89, 98
Glycerophospholipids 69–76

H

Hair 274, 276, 277, 281
High mass accuracy GC-MS 2
High-resolution mass spectrometry (HRMS) 26, 36–38, 41–43
High throughput 9, 44, 66, 174, 189–196, 242, 246
Hippocampus 192, 193, 196, 201, 203, 205–207, 212, 246, 247, 249–253, 285, 291, 295
Hydroxyglutarate 89–99
2-hydroxyglutarate enantiomer 89–93, 95–99

I

Isocitrate dehydrogenase (IDH) 89–91, 98
Isotope reagent 50
Isotopic ration outlier analysis (IROA) 78–81, 83, 84

K

Ketamine 102, 110

L

LC x LC-MS 30
LC-MS analysis 49, 50, 239, 252, 257, 262
Lipidomics 6, 25–44, 69, 74, 122, 140
Liquid chromatography (LC) 25, 27–32, 35–44, 51, 53, 62, 64, 93, 102, 109, 113, 114, 117, 122, 139, 148, 174, 179, 190, 192, 236, 237, 239, 246, 252, 257, 259
Liquid chromatography in combination with mass spectrometry (LC-MS) 6, 29, 30, 49, 50, 52, 57, 76, 79, 90, 101, 114, 174, 175, 177–179, 182, 190, 192, 203, 227, 234, 237, 238, 246, 248, 249, 252, 253, 257–259, 261

M

Mass spectrometry (MS) 1, 3–6, 9, 10, 25, 26, 28, 29, 36–39, 41, 42, 56, 62, 64, 69–87, 89–99, 101, 103, 104, 107, 113, 114, 117, 125, 139, 156, 173, 174, 179–181, 189–196, 200, 214, 216, 222, 230, 232, 235–240, 242, 246, 248, 254, 255, 258, 259, 262, 270–296
MCF-7 breast cancer cells 52
Metabolic phenotyping 78
Metabolites identification 78, 79, 83, 84, 200
Metabolomics 1, 3, 7, 9–21, 25, 26, 29–32, 35–44, 50, 53, 61–67, 77, 79, 86, 87, 101–111, 113–119, 121, 155–160, 162–164, 167, 173–186, 189–196, 199–218, 221, 229, 245–263
Metallomics 270, 271, 280, 293, 295–297
Method development v
Molecular markers 156
MS[1] annotation 247, 258, 261
Multi-compartmental 191
Multiplatform metabolomics 203

N

N-acetyl amino acids 5
Nanoflow LC-MS 49–59
Negative ion electrospray mass spectrometry 118, 194
Negative ion mass spectrometry 2, 6, 11, 14, 15, 118, 194
Neural tissues 199–218
Neuroblastoma 144, 145
Neurodegeneration 121–140, 190, 235, 246, 249, 287, 291
Nontargeted metabolomics 200
N-phenylmaleimide 238
Nuclear magnetic resonance spectroscopy (NMR) 113, 155–169, 173, 174, 246

O

Optic nerve 200, 201, 204, 206, 207, 210
Oxylipins 249

P

Pancreatic cancers 114
Pentafluorobenzyl (PFB) 221–223, 226, 227
Polyamines 2, 229, 232
Positive ion mass spectrometry 2, 14, 73, 148, 281, 282
Postmortem interval (PMI) 245, 246, 249

R

Relative quantification 50

S

Saliva 114–117, 119, 190, 191
Serum 3, 4, 50, 52, 71, 91–94, 96–99, 114, 115, 146, 167, 175, 177, 190–193, 196, 222, 230, 276, 291
Social defeat model 62
Spectrum libraries 200
Supercritical fluid chromatography (SFC) 25, 26, 101–111
Supercritical fluid extraction (SFE) 102–105, 110, 111

T

Tandem mass spectrometry 26
Techniques for bioimaging.................................. 270–296
Tert-Butyldimethylsilyl (tBDMS) 229–234
Thiols 235–242
Two-dimensional liquid chromatography
(2D-LC)..............................26–34, 36, 38–41, 44

U

Ultra-high performance liquid chromatography
(UHPLC) 102, 179, 182, 190, 195, 196, 248
Untargeted metabolomics 49, 61, 62, 64–66, 173–175, 247, 249, 258
Urine................................40, 41, 50, 91, 92, 95, 96, 99, 101, 102, 104, 105, 109, 158, 162–164, 166, 167, 190–193

V

Validation....................1, 93–94, 98, 107, 111, 168, 200

Y

Yeast 80–82